T0176694

Green Mobile Networks

Green Mobile Networks

A Networking Perspective

Nirwan Ansari
New Jersey Institute of Technology, USA

Tao Han
University of North Carolina, Charlotte, USA

IEEE PRESS

WILEY

This edition first published 2017
© 2017 John Wiley & Sons Ltd

The right of Nirwan Ansari and Tao Han to be identified as the authors of this work has been asserted in accordance with law.

Registered Offices
John Wiley & Sons, Inc., 111 River Street, Hoboken, NJ 07030, USA
John Wiley & Sons Ltd, The Atrium, Southern Gate, Chichester, West Sussex, PO19 8SQ, UK

Editorial Office
The Atrium, Southern Gate, Chichester, West Sussex, PO19 8SQ, UK

For details of our global editorial offices, customer services, and more information about Wiley products visit us at www.wiley.com.

Wiley also publishes its books in a variety of electronic formats and by print-on-demand. Some content that appears in standard print versions of this book may not be available in other formats.

Library of Congress Cataloging-in-Publication data is available for this book.

ISBN 9781119125105 (hardback)

Cover Image: © Brandon Sillaro / EyeEm/Gettyimages
Cover Design: Wiley

Set in 10/12pt WarnockPro by Aptara Inc., New Delhi, India

Printed in the UK

Contents

Preface

Greening is not merely a trendy concept, but is becoming a necessity to bolster social, environmental, and economic sustainability. Naturally, green communications have received much attention recently. As mobile network infrastructures and mobile devices proliferate, an increasing number of users rely on cellular networks for their daily lives. As a result, mobile networks are among the major energy hogs of communication networks and their contribution to global energy consumption is increasing fast. Therefore, greening of cellular networks is crucial to reducing the carbon footprint of Information and Communications Technology (ICT). As a result, the field is attracting tremendous research efforts from both academia and industry.

This book is intended to provide a technical description of state-of-the-art developments in greening of mobile networks from a networking perspective. It discusses fundamental networking technologies that lead to energy-efficient mobile networks. These technologies include heterogeneous networking, multi-cell cooperation, mobile traffic offloading, traffic load balancing, renewable energy integrated mobile networking, device-to-device networking, and mobile content delivery optimization. The text is suitable for graduate courses in electrical and computer engineering and computer science. The authors have adopted some materials presented in this book for their graduate courses at New Jersey Institute of Technology[1] and University of North Carolina–Charlotte[2]. This book also includes many results and patented algorithms from our research, which makes this book a valuable reference for graduate students, practicing engineers, and research scientists in the field of green communications and networking.

The material is structured in a modular fashion with chapters being reasonably independent of each other. Individual chapters can be perused in an arbitrary order to the liking and interest of the reader, and they can also be incorporated as part of a larger, more comprehensive course. The first chapter provides an overview of existing networking technologies and solutions for greening mobile networks. The second to fourth chapters cover three major networking technologies in detail: multi-cell cooperation, green energy enabled mobile networking, and spectrum and energy harvesting. The fifth to ninth chapters present green mobile networking solutions including mobile traffic offloading, optimizing green energy enabled mobile networks, traffic load balancing, device-to-device networking, and content delivery optimization.

---

[1] ECE 639 Principles of Broadband Networks and ECE 788 Advanced Topics in Broadband Networks.
[2] ECGR 6120/8120 Wireless Communications and Networking.

Part I

Green Mobile Networking Technologies

1

Fundamental Green Networking Technologies

As cellular network infrastructures and mobile devices proliferate, an increasing number of users rely on cellular networks for their daily lives. Mobile networks are among the major energy guzzlers of information communications technology (ICT) infrastructure, and their contributions to global energy consumption are accelerating rapidly because of the dramatic surge in mobile data traffic [1, 2, 3, 4]. This growing energy consumption not only escalates the operators' operational expenditure (OPEX) but also leads to a significant rise of their carbon footprints. Therefore, greening of mobile networks is becoming a necessity to bolster social, environmental, and economic sustainability [5, 6, 7, 8]. In this chapter, we give an overview of the fundamental green networking technologies.

1.1 Energy Efficient Multi-cell Cooperation

The energy consumption of a cellular network is mainly drawn from base stations (BSs), which account for more than 50% of the energy consumption of the network. Thus, improving energy efficiency of BSs is crucial to green cellular networks. Taking advantage of multi-cell cooperation, energy efficiency of cellular networks can be improved from three perspectives. The first is to reduce the number of active BSs required to serve users in an area [9]. The solutions involve adapting the network layout according to traffic demands. The idea is to switch off BSs when their traffic loads are below a certain threshold for a certain period of time. When some BSs are switched off, radio coverage and service provisioning are taken care of by their neighboring cells.

The second aspect is to connect users with green BSs powered by renewable energy. Through multi-cell cooperation, off-grid BSs enlarge their service areas while on-grid BSs shrink their service areas. Zhou *et al.* [10] proposed a handover parameter tuning algorithm and a power control algorithm to guide mobile users to connect with BSs powered by renewable energy, thus reducing on-grid power expenses. Han and Ansari [11] proposed an energy aware cell size adaptation algorithm named ICE. This algorithm balances the energy consumption among BSs, enables more users to be served with green energy, and therefore reduces on-grid energy consumption. Envisioning future BSs to be powered by multiple types of energy sources, for example, the grid, solar energy, and wind energy, Han and Ansari [12] proposed optimizing the utilization of green energy for cellular networks by cell size optimization. The proposed algorithm achieves significant main grid energy savings by scheduling green energy consumption in the time

Green Mobile Networks: A Networking Perspective, First Edition. Nirwan Ansari and Tao Han.
© 2017 John Wiley & Sons Ltd. Published 2017 by John Wiley & Sons Ltd.

domain for individual BSs, and balancing green energy consumption among BSs for the cellular network.

The third aspect is to exploit coordinated multi-point (CoMP) transmissions to improve energy efficiency of cellular networks [13]. On the one hand, with the aid of multi-cell cooperation, energy efficiency of BSs on serving cell edge users is increased. On the other hand, the coverage area of BSs can be expanded by adopting multi-cell cooperation, thus further reducing the number of active BSs required to cover a certain area. In addition to discussing multi-cell cooperation solutions, we investigate the challenges for multi-cell cooperation in future cellular networks.

1.2 Heterogeneous Networking

The energy consumption of mobile networks scales with the provisioned traffic capacity. On deploying a mobile network, two types of BSs may be deployed. They are macro BSs (MBSs) and small cell BSs (SCBSs). As compared with SCBSs, MBSs provide a larger convergence area and consume more energy. SCBSs are deployed close to users, and thus consume less energy by leveraging such proximity. Owing to a small coverage area, in order to guarantee traffic capacity in an area, a very large number of SCBSs must be deployed. The total energy consumption of the large number of SCBSs may exceed that of the MBSs. Hence, in order to improve the energy efficiency of the network, a mixed deployment of both MBSs and SCBSs is desirable. In general, there are two SCBS deployment strategies: deployed at cell edges and at traffic hot spots.

- The users located at the edge of a macro cell usually experience bad radio channels due to excessive channel fading. In order to provide service to these users, MBSs could increase their transmit power, but this will result in a low energy efficiency. In a heterogeneous network deployment, SCBSs can be deployed at the edge of macro cells as shown in Figures 1.1–1.4. Depending on the traffic capacity demand, different SCBS deployment strategies can be adopted. For example, when the traffic capacity demand is relatively low, one SCBS may be deployed at the edge of a macro cell to serve the cell edge users as shown in Figure 1.1. As the traffic increases, additional SCBSs can

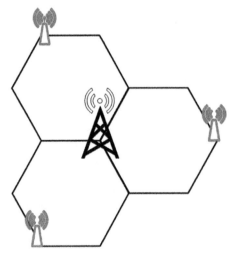

Figure 1.1 Scenario 1: One SCBS per macro site.

Figure 1.2 Scenario 2: Two SCBSs per macro site.

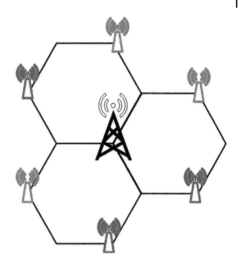

be deployed at the cell edge as shown in Figs. 1.2 and 1.3. When the traffic capacity demand is very high, additional SCBSs should be deployed. For example, five SCBSs are deployed for enhancing the energy efficiency of serving cell edge users in Figure 1.4. The number of SCBSs that are deployed to enhance the energy efficiency of serving users located at the edges of macro cells should be optimized based on traffic capacity demand at the cell edge.

- When the traffic capacity demand in mobile networks is inhomogeneous, deploying SCBSs at the edges of macro cells may not be optimal. Instead, SCBSs can be deployed in areas where there is high traffic capacity demand such as shopping areas, stadiums, and public parks. We define such areas as *hotspots*. Owing to proximity to the users, SCBSs can provide very high capacity at hotspots and serve the traffic demand with low energy consumption. In order to deploy SCBSs at traffic hotspots to enhance energy efficiency, the distribution of traffic capacity demand should be understood from network measurements. In addition, the traffic capacity demand should be localized so that a large portion of the traffic demand can be offloaded to

Figure 1.3 Scenario 3: Three SCBSs per macro site.

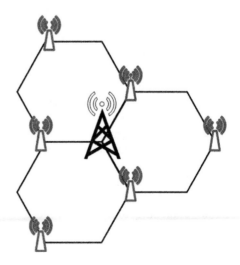

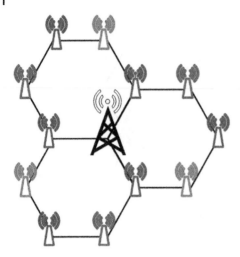

Figure 1.4 Scenario 4: Five SCBSs per macro site.

SCBSs. In the ideal case, MBSs are only serving users with high moving speed while all the other users are served by SCBSs. If the high traffic demand occurs indoors, the indoor deployment of SCBSs can significantly enhance the energy efficiency of mobile networks.

1.3 Mobile Traffic Offloading

Mobile traffic offloading, which is referred to as utilizing complementary network communications techniques to deliver mobile traffic, is a promising technique to alleviate congestion and reduce the energy consumption of mobile networks. Based on the network access mode, mobile traffic offloading schemes can be divided into two categories. The first category is infrastructure based mobile traffic offloading, which refers to deploying SCBSs, for example, pico BSs, femto BSs and WiFi hot spots, to offload mobile traffic from MBS [14, 15]. SCBSs usually consume much less power than MBSs. Therefore, offloading mobile traffic to SCBSs can significantly enhance the energy efficiency of mobile networks [6, 16]. However, the lack of cost-effective backhaul connections for SCBSs often impairs their performance in terms of offloading mobile traffic and enhancing the energy efficiency of mobile networks. The second category is ad-hoc based mobile traffic offloading, which refers to applying device-to-device (D2D) communications as an underlay to offload mobile traffic from MBSs. By leveraging Internet of Things (IoT) technologies, smart devices within proximity are able to connect with each other and form a communication network. Data traffic among the devices can be offloaded to the communication networks rather than delivering through MBSs. Moreover, in order to reduce CO_2 footprints, mobile traffic can be offloaded to BSs powered by green energy such as sustainable biofuels, solar, and wind energy [17, 12, 10, 18]. In this way, green energy utilization is maximized, and thus the consumption of on-grid energy is minimized. In this section, we briefly overview the related research on mobile traffic offloading and the solutions for user–BS associations in heterogeneous mobile networks.

1.3.1   Infrastructure Based Mobile Traffic Offloading

In infrastructure based mobile traffic offloading, the mobile traffic is offloaded to either pico/femto BSs or WiFi hot spots. Deploying pico/femto BSs improves the spectral and energy efficiency per unit area of cellular networks, and thus reduces the network congestion and energy consumption of cellular networks. Traffic offloading between pico/femto BSs and the MBS is achieved by adapting user–BS associations. Kim *et al.* [19] proposed a user–BS association to achieve flow level load balancing under spatially heterogeneous traffic distribution. Jo *et al.* [20] proposed cell biasing algorithms to balance traffic loads among pico/femto BSs and the MBS. The cell biasing algorithms perform user–BS association according to the biased measured pilot signal strength, and enable traffic to be offloaded from the MBS to pico/femto BSs.

WiFi hot spots are also effective in terms of offloading mobile traffic. Lee *et al.* [21] pointed out that a user is in WiFi coverage for 70% of the time on average, and if users can tolerate a two hour delay in data transfer, the network can offload about 70% of cellular traffic to WiFi networks. Balasubramanian *et al.* [22] proposed to offload the delay tolerant traffic such as email and file transfer to WiFi networks. When WiFi networks are not available or experiencing blackouts, data traffic is quickly switched back to 3G networks to avoid violating the applications' tolerance threshold. Han and Ansari [15] designed a content pushing system which pushes the content to mobile users through opportunistic WiFi connections. The system responds to a user's pending requests or predicted future requests, codes the requested content by using Fountain codes, predicts the user's routes, and prelocates the coded content to the WiFi access points along the user's route. When the user connects to these WiFi access points, the requested content is delivered to the user via the WiFi connections.

1.3.2   Ad-hoc Based Mobile Traffic Offloading

Ad-hoc based mobile traffic offloading relies on D2D communications to disseminate content. Instead of downloading content directly from BSs, User Equipment (UE) may retrieve content from their neighboring UEs. Han *et al.* [23] proposed a mechanism to select a subset of UEs based on either UEs' activities or mobilities, to deliver content to them through cellular networks, and to let these UEs further disseminate the content through D2D communications to the other users. Mashhadi *et al.* [24] proposed a proactive caching mechanism for UEs in order to offload the mobile traffic. When the local storage does not have the requested content, the proactive caching mechanism will set a target delay for this request, and explores opportunities to retrieve data from neighboring UEs. The proactive caching mechanism requests data from cellular networks when the target delay is violated. To encourage mobile users to participate in the traffic offloading, Zhou *et al.* [25] proposed an incentive framework that motivates users to leverage their delay tolerance for cellular data offloading.

1.3.3   User–BS Associations in Heterogeneous Mobile Networks

Heterogeneous networking is a promising network architecture which may significantly enhance the spectral and energy efficiency of mobile networks. One of the most important issues in heterogeneous cellular networks is to properly associate mobile users with the serving BSs, referred to as the "user–BS association problem." In heterogeneous cellular networks, the transmit power of SCBSs is significantly lower than that of MBSs. Thus, mobile users are more likely to be associated with the MBS based on the strength of their received pilot signals. As a result, SCBSs may be lightly loaded, and do not contribute much to traffic offloading. To address this issue, many user–BS association algorithms have been proposed [19, 20, 26]. Kim *et al.* [19] proposed a framework for user–BS association in cellular networks to achieve flow level load balancing under spatially heterogeneous traffic distribution. Jo *et al.* [20] proposed cell biasing algorithms to balance traffic loads among MBSs and SCBSs. The cell biasing algorithms perform user–BS association according to biased measured pilot signal strength, and enable traffic to be offloaded from MBSs to SCBSs. Corroy *et al.* [26] proposed a dynamic user–BS association algorithm to maximize the sum rate of the network and adopted cell biasing to balance the traffic load among BSs. Fooladivanda *et al.* [27] studied joint resource allocation and user–BS association in heterogeneous mobile networks. They investigated the problem under different channel allocation strategies, and the proposed solution achieved global proportional fairness among users. Madan *et al.* [28] studied user–BS association and interference coordination in heterogeneous mobile networks, and proposed heuristic algorithms to maximize the sum utility of average rates.

1.4 Device-to-Device Communications and Proximity Services

The surge in mobile data traffic brings about two major problems to current mobile networks. The first problem is that the significant data growth congests mobile networks and leads to long delays in content delivery [14]. The second problem is that a continuous surge of mobile traffic results in a dramatic increase in energy consumption in mobile networks from provisioning higher network capacity [5]. By leveraging IoT technologies, smart devices within proximity are able to connect with each other and form a communication network. Data traffic among the devices can be offloaded to the communication networks rather than delivering through MBSs. For example, by enabling D2D communications, some UE downloads content from MBSs while the other UEs may retrieve the content through D2D connections with their peers. In this way, D2D communications alleviates traffic congestion and reduces the energy consumption of mobile networks.

D2D communications may, however, suffer from several disadvantages which impair its performance in terms of offloading mobile traffic. First, UEs are battery powered devices, and therefore the additional energy consumption may prevent UEs from participating in D2D communications. Second, the transmission range for D2D communications among mobile devices is limited by its low transmission power. For example, if a mobile device experiences a shortage of battery power, the mobile device may restrict its power usage in D2D communications. This leads to a reduced transmission range for the mobile device's D2D communications. Especially when mobile devices operate at millimeter wavelengths, the transmission range is further limited by the channel propagation characteristics. Third, D2D communications may require complicated radio resource management schemes implemented in mobile devices to avoid the introduction of extensive interference to mobile networks. This will further increase mobile devices' power consumption in D2D communications. Fourth, sharing content through

D2D communications may reveal users' privacy. Hence, mobile devices may not be will-ing to participate in D2D communications in order to protect their users' privacy.

Much effort has been spent studying the related problems in device-to-device com-munications. Bletsas *et al.* [29] proposed a distributed relay node selection scheme which selects relay nodes based on the instantaneous channel condition at the nodes. Zhao *et al.* [30] studied the performance of a cooperative communications system where a source node communicates with a destination node with the help of multiple relay nodes, and showed that choosing one best relay node is sufficient to maintain full diver-sity order. Wang *et al.* [31] proposed a game theory based relay selection algorithm for multi-user cooperative communication networks. Sharma *et al.* [32] studied the relay assignment problem in cooperative ad hoc networks, and proposed a relay assignment algorithm to maximize the minimum data rate among all the source destination (SD) pairs. While these works investigated the relay assignment problem with different net-work settings and optimization objectives, they all assumed that a relay node is assigned to at most one source SD pair. Yang *et al.* [33] extended the relay assignment problem to consider a relay node being assigned to multiple SD pairs. The authors proposed maxi-mizing the summation of the data rates of all SD pairs, and proved that it is only neces-sary to assign one relay node to one SD pair in order to achieve the optimal solution.

1.5 Powering Mobile Networks With Renewable Energy

As green energy technologies advance, green energy such as sustainable biofuels, solar, and wind energy can be utilized to power SCBSs and MBSs [17]. Telecommunications companies such as Ericsson and Nokia Siemens have designed green energy powered BSs for cellular networks [34]. By adopting green energy powered secondary base sta-tions (SBSs), mobile service providers may reduce on-grid energy consumption and thus reduce their CO_2 emissions. For instance, Orange, a French mobile network operator (MNO), has already deployed more than two thousand solar powered BSs in Africa [35]. These BSs serving over 3 million people saved up to 25 million liters of fuel and reduced about 67 million kilograms of CO_2 emissions in 2011 [35].

The design and optimization of green energy powered mobile networks is challeng-ing. As shown in Figure 1.5, in addition to radio resource management, optimizing green energy enabled mobile networks involves the optimization of the utilization of green energy from both standalone power generators and green power farms. At the same time, smart grid techniques enable power trading among consumers via smart meters. As a result, power cooperation, which enables BSs to share their green power with each other, has been introduced to engineer green energy enabled mobile networks. The coupling of radio resource optimization and power utilization optimization intro-duces new research challenges on optimizing green energy enabled mobile networks. This chapter investigates the design and optimization issues involved in green energy enabled mobile networks.

1.6 Green Communications via Cognitive Radio Communications

The proliferation of wireless devices is driving the exponential growth of wireless data traffic over wireless and mobile networks. This leads to a continuous surge not only in

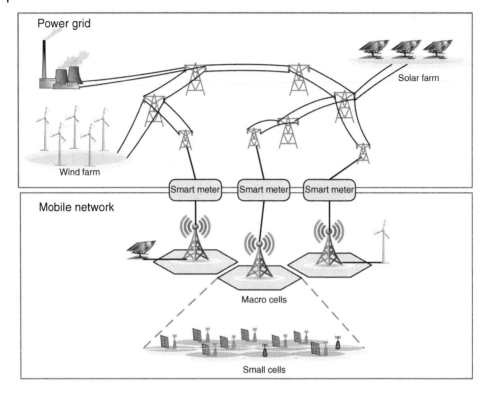

Figure 1.5 Green energy enabled mobile networks. *Source:* Han 2014 [17]. Reproduced with permission of IEEE.

network capacity demands but also in network energy consumption. Since a wireless system is spectrum limited, the ever-increasing capacity demands result in a spectrum crunch. Deploying additional network infrastructure such as BSs is an effective approach to alleviate spectrum shortage [36]. Thus, SCBSs will be widely deployed. SCBSs can provide high network capacity for wireless users by capitalizing on their close proximity to the users. However, an SCBS has a limited coverage area. Thus, the number of SCBSs will be an order of magnitude larger than that of MBSs for a large scale network deployment. As a result, the overall energy consumption of cellular networks will keep increasing [5]. Therefore, current wireless access networks are eventually constrained by spectrum scarcity and energy consumption. It is desirable to amalgamate spectrum harvesting and energy harvesting technologies to liberate wireless networks from these constraints.

Greening wireless access networks can capitalize on the broad paradigm of spectrum harvesting technologies. Spectrum harvesting technology, which is also known as cognitive radio (CR)[1], has been defined thus: "A cognitive radio transmitter will learn from the environment and adapt its internal states to statistical variations in the existing radio frequency (RF) stimuli by adjusting the transmission parameters (e.g., frequency band,

---

[1] In this book, we use the terms spectrum harvesting technology and cognitive radio technology interchangeably.

modulation mode, and transmission power) in [a] real time and online manner" [37]. With the capability to detect available spectrum and the reconfigurability to dynamically access parts of the spectrum over which less fading and interference is experienced, the intelligent CR communications system enhances spectrum agility and energy efficiency.

In addition to current CR networks powered by reliable on-grid or un-rechargeable energy sources, continuous advances in green energy have motivated researchers to investigate green powered cognitive radio (green CR) networks. The concept of the energy harvester has been proposed to capture and store ambient energy to generate electricity or other forms of energy that is renewable and more environmentally friendly than that derived from fossil fuels. If the green energy source is ample and stable in the sense of availability, a CR network can be powered to opportunistically exploit the underutilized spectrum by harnessing free energy, without requiring an energy supplement from the external power grid or batteries.

1.7 Green Communications via Optimizing Mobile Content Delivery

Owing to imminent fixed/mobile convergence, Internet applications are frequently accessed through mobile devices. Given limited bandwidth and unreliable wireless channels, content delivery in mobile networks usually experiences long delays. Inefficient content delivery significantly increases the energy consumption of mobile networks. Many solutions have been proposed to accelerate content delivery in mobile networks.

To understand the performance of mobile networks, many measurement studies have been presented. These studies unveil the obstacles that delay content delivery in mobile networks, and shed light on the research directions for enhancing the performance of mobile networks. Noticing the shortcomings of mobile networks, many solutions have been proposed to reduce content delivery latency and enhance subscribers' quality of experience (QoE) in mobile networks. Figure 1.6 shows a classification hierarchy of available content delivery acceleration solutions. We classify these solutions into three categories, namely, mobile system evolution, content and network optimization, and mobile data offloading. The techniques are further classified within each category. Mobile communications system evolution is one of the major solutions to enhance content delivery efficiency in mobile networks. On the one hand, to meet the increasing demands for mobile data services, the 3rd Generation Partnership Project (3GPP) has established evolution plans to enhance the performance of mobile communications systems. 3GPP LTE (Long Term Evolution) Advanced is a mobile communications standard for next generation mobile communications systems featuring high speed and low latency. LTE Advanced networks adopt multi-input-multi-output (MIMO) and orthogonal frequency-division multiple access (OFDMA) to enhance both the capacity and reliability of wireless links, introduce the Evolved Packet System (EPS) [38] to reduce the amount of protocol related processing, and integrate CR techniques to expand the available bandwidth in the system. On the other hand, mobile networks and content delivery networks are being integrated to provide end-to-end acceleration for mobile content delivery [39].

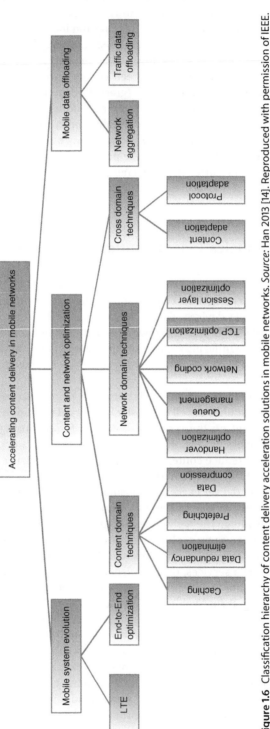

Figure 1.6 Classification hierarchy of content delivery acceleration solutions in mobile networks. *Source:* Han 2013 [14]. Reproduced with permission of IEEE.

The content and network optimization techniques are further classified into three categories based on their application domains. The first category pertains to the content domain techniques including caching, data redundancy elimination, prefetching, and data compression. These techniques aim to reduce traffic volume over mobile networks, thus reducing network congestion and accelerating content delivery. The second category refers to the techniques applied in the network domain. These techniques include handover optimization, queue management techniques, network coding, TCP optimization, and session layer optimization. The network domain techniques optimize the operation of mobile networks and communication protocols, and thus enhance network performance. The third category includes cross domain techniques such as content adaptation and protocol adaptation. Content adaptation adjusts the original content according to the mobile network conditions and the characteristics of mobile devices. Content adaptation can efficiently reduce the data volume over mobile networks and accelerate content delivery. Protocol adaptation optimizes communication protocols according to application behaviors. It reduces network chattiness, and thus reduces content delivery delay.

A significant data traffic increase may congest the mobile network, and lead to long delays in content delivery and low energy efficiency. Offloading data traffic from congested mobile networks is a promising method to reduce network congestion. Mobile data offloading techniques include two perspectives. The first is to directly offload mobile data to high speed networks, for example, WiFi. The second is network aggregation, which allows subscribers to simultaneously utilize their multiple radio interfaces, for example, 3G, WiFi, and Bluetooth, to retrieve content. Mobile data offloading techniques reduce the pressure on mobile networks in terms of data volume, thus enhancing network performance in terms of content delivery.

The rest of this book is organized as follows. Chapter 2 gives an overview of multi-cell cooperation solutions for improving the energy efficiency of cellular networks. Chapter 3 covers the research challenges and existing solutions for green energy enabled mobile networks. Chapter 4 discusses state-of-the-art research on spectrum and energy harvesting networks. Chapter 5 presents a novel energy spectrum trading (EST) scheme to enable MBSs to offload their mobile traffic to Internet service providers' (ISPs') wireless access points by leveraging CR techniques. Chapter 6 investigates a framework for optimizing green energy utilization for mobile networks with hybrid energy supplies. Chapter 7 presents an energy efficient traffic load balancing scheme in mobile networks. Chapter 8 covers device-to-device proximity services-based energy efficient communications schemes. Finally, Chapter 9 presents solutions for optimizing content delivery in mobile networks.

2

Multi-cell Cooperation Communications

It has been shown that, with the aid of multi-cell cooperation, the performance of a cellular network in terms of throughput and coverage can be enhanced significantly. However, the potential of multi-cell cooperation to improve the energy efficiency of cellular networks remains to be unlocked. In this chapter, we give an overview of multi-cell cooperation solutions for improving the energy efficiency of cellular networks. First, we introduce traffic intensity aware multi-cell cooperation, which adapts the network layout of cellular networks according to user traffic demands in order to reduce the number of active BSs. Then, we discuss energy aware multi-cell cooperation, which offloads traffic from on-grid BSs to off-grid BSs powered by renewable energy, and therefore reduces on-grid power consumption. In addition, we investigate how energy efficiency of cellular networks can be improved by exploiting CoMP transmissions. Finally, we discuss the characteristics of future cellular networks, and the challenges of achieving energy efficient multi-cell cooperation in future cellular networks.

2.1 Traffic Intensity Aware Multi-cell Cooperation

The traffic demand of cellular networks experiences fluctuations for two reasons. The first is the typical day-night behavior of users. Mobile users are usually more active in terms of cell phone usage during the day than during the night, and therefore traffic demand during the day is higher than at night. The second reason is the mobility of users. Users tend to move to their office districts during working hours and come back to their residential areas after work. This results in the need for a large capacity in both areas at peak usage hours but in reduced requirements during off-peak hours. However, cellular networks are usually dimensioned for peak hour traffic, and thus most BSs are working at low workload during off-peak hours. Owing to their high static power consumption,[1] BSs usually experience poor energy efficiency when they are operating at a low workload.

In addition, cellular networks are typically optimized for the purpose of providing coverage rather than for operating at full load. Therefore, even during peak hours, the utilization of BSs may be inefficient in terms of energy usage. Adapting the network layout of cellular networks according to traffic demands has been proposed to improve their

---

[1] The static power consumption of a BS refers to the power consumption of the BS when there are no active users in the coverage of the BS.

Green Mobile Networks: A Networking Perspective, First Edition. Nirwan Ansari and Tao Han.
© 2017 John Wiley & Sons Ltd. Published 2017 by John Wiley & Sons Ltd.

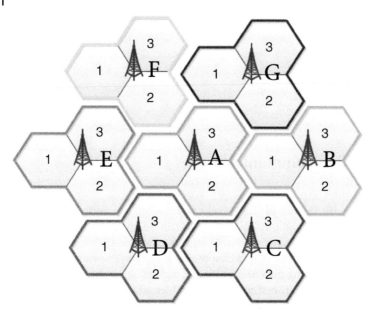

Figure 2.1 Original Network Layout. *Source:* Han 2013 [5]. Reproduced with permission of IEEE.

energy efficiency. The network layout adaptation is achieved by switching BSs on/off dynamically. Figs. 2.1, 2.2, and 2.3 show several scenarios of network layout adaptations. Figure 2.1 shows the original network layout, in which each BS has three sectors. For cell A, if most of the traffic demands on it are coming from sector three, and the traffic demands in sectors one and two are lower than a predefined threshold, cell A

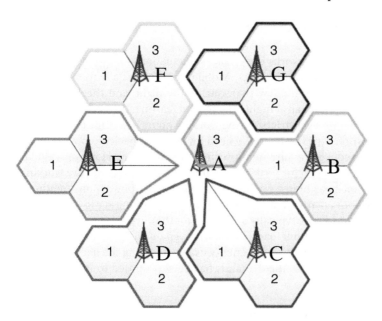

Figure 2.2 BS partially switched off. *Source:* Han 2013 [5]. Reproduced with permission of IEEE.

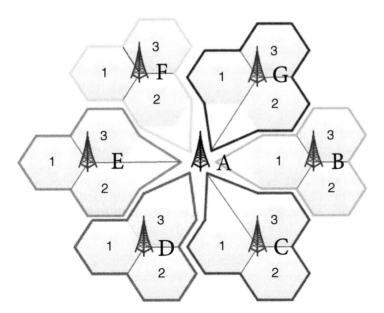

Figure 2.3 BS entirely switched off. *Source:* Han 2013 [5]. Reproduced with permission of IEEE.

could switch off sectors one and two to save energy, and the users in the sectors that are switched off will be served by the neighboring cells. In this case, the network layout after the adaptation is shown in Figure 2.2. If traffic demands from sector three of cell A also decrease below the threshold, the entire green cell will be switched off, and the network layout is adapted to the one shown in Figure 2.3. Under this scenario, the radio coverage and service provisioning in cell A are taken care of by its active neighboring cells. When a BS is switched off, the energy consumed by its radio transceivers, processing circuits, and air conditioners can be saved. Therefore, adapting the network layout of cellular networks according to traffic demands can reduce energy consumption significantly.

While network layout adaptation can potentially reduce the energy consumption of cellular networks, it must meet two service requirements: (1) the minimum coverage requirement, and (2) the minimum quality of service (QoS) requirements for all mobile users. Therefore, in carrying out network layout adaptation, multi-cell cooperation is needed to guarantee service requirements. Otherwise, it will result in a high call blocking probability and a severe QoS degradation. For example, two adjacent BSs may both experience low traffic demands. However, only one BS can be switched off to save energy, and the other BS should be active to sustain the service provisioning in both coverage areas. In this case, if both BSs are switched off according to their own traffic demands, their subscribers will lose connections. Therefore, cooperation among BSs is essential to enable traffic intensity aware network layout adaptation.

2.1.1 Cooperation to Estimate Traffic Demands

Network layout adaptation is based on the estimated traffic demands at individual cells. Traffic demand estimation at individual cells requires the cooperation of neighboring cells. To avoid frequently switching BSs on/off, BSs are switched off only when their traffic demands are less than a predefined threshold for a minimum period, T. Therefore, the estimated traffic demands should represent the traffic demands at individual cells for at least a time period of length T. Hence, traffic demands at a BS consist of three parts: traffic demands from users who are currently attached to the BS, traffic demands from users who will be handed over to other BSs, and traffic demands from users who will be handed over to the BS from its neighboring BSs. While the first two components can be measured and estimated by individual BSs, the estimation of the third component requires cooperation from neighboring cells. Two reasons contribute to the handover traffic from neighboring cells. The first is user mobility. A user's motion including the direction and velocity can be measured by various signal processing methods. Therefore, individual BSs can predict (1) how many users will hand over to other cells in the near future, and (2) to which cells these users are highly likely to hand over. Such information is important for their neighboring BSs to estimate their traffic demands. The second reason for handover traffic is the switching off of neighboring BSs. If one of the neighboring cells is switched off, the users under its coverage will hand over to its neighboring cells, thus increasing the traffic demands of the neighboring cells. Therefore, cooperation among radio cells is important for traffic demand estimation at individual cells.

2.1.2 Cooperation to Optimize Switching Off Strategy

With traffic demand estimation, cellular networks optimize the switching on/off strategies to maximize the energy saving while guaranteeing users' minimal service requirements. Currently, most of the existing switching on/off strategies are centralized algorithms, which assume that there is a central controller that collects the operation information of all BSs and optimizes network layout adaptations. Three methods have been proposed to determine which BSs to switch off. The first is randomly switching off BSs with low traffic loads. This method mainly applies to BSs in the night zone where few users are active. The method randomly switches off some BSs to save energy, and the remaining BSs provide coverage for the area. The second method is a greedy algorithm that enforces BSs with higher traffic loads to serve more users, and switches off BSs with no traffic load. The third method is based on the user–BS distance. The required transmission power of BSs for serving users depends on the distance between users and BSs. The longer the distance, the greater the transmission power required in order to meet users' minimal service requirements. Therefore, the user–BS distance can be an indicator for the energy efficiency of cellular networks: the shorter the average user–BS distance, the higher the energy efficiency. Hence, the algorithm tends to switch off BSs with the longest user–BS distance in order to improve energy efficiency.

Centralized algorithms, however, require the channel state information and traffic load information of every cell. Collecting this information centrally may impose tremendous communications overheads, and thus reduce the effectiveness of centralized algorithms in improving energy efficiency. Therefore, distributed algorithms are favored especially for heterogeneous networks, which consist of various types of cells such as macro cells, micro cells, pico cells, and femto cells. To enable distributed algorithms, individual cells may cooperatively form coalitions and share channel state information and traffic load information. Based on the shared information, individual BSs optimize their operation strategies to minimize the total energy consumption of the BS coalition.

2.2   Energy Aware Multi-cell Cooperation

With the worldwide penetration of distributed electricity generation at medium and low voltages, it is proven that renewable energy such as sustainable biofuels, solar, and wind energy can be effectively utilized to substantially reduce carbon emissions. Therefore, researchers have proposed to power BSs with renewable energy to decrease on-grid energy consumption and reduce carbon footprints. Nokia Networks has developed off-grid BSs [40] that rely on a combination of solar and wind power supported by fuel cell and deep cycle battery technologies. As compared to on-grid BSs, off-grid BSs achieve zero carbon emission and save a significant amount of on-grid energy. However, owing to the dynamic nature of renewable resources, the renewable energy generation is highly dependent on environmental factors such as temperature, light intensity, and wind velocity. In addition, because of the limited battery capacity, generated energy should be effectively utilized in order to avoid energy overflow. Therefore, optimizing the utilization of renewable energy in cellular networks is not a trivial problem. One intuitive solution to this problem is the energy aware cell breathing method. Here, "energy aware" pertains to two perspectives. The first is the awareness of energy sources such that users are guided to connect to off-grid BSs rather than on-grid BSs. This can be achieved by enlarging the cell sizes of off-grid BSs, and shrinking those of on-grid BSs [10]. The second perspective is the awareness of energy storage capacity such that off-grid BSs with higher energy storage capacities are forced to serve a larger area. In this way, energy overflow can be avoided, thus enabling the renewable energy generation system to harvest more energy from renewable sources [11]. The energy aware cell breathing technique is a step toward further traffic intensity aware network layout adaptations. In addition to the traffic load information and channel state information, energy aware cell breathing requires multiple cells to cooperatively share energy information including the estimated amounts of energy arrival and storage, and to coordinate to minimize the energy consumption of the network.

2.3   Energy Efficient CoMP Transmission

CoMP transmission is a key technology for future mobile communications systems. In CoMP transmission, multiple BSs cooperatively transmit data to mobile users to improve their receiving signal quality. BSs share the required information for cooperation via either high speed wired links or X2 interfaces (inter-eNode B interfaces). Based on the information, BSs determine joint processing and coordinated scheduling strategies. While CoMP transmission has been proved to be effective at improving data rates and spectral efficiency for cell edge users [41], its potential for improving energy efficiency of cellular networks has not been fully exploited. In this section, we discuss the potential of CoMP transmission for greening cellular networks.

2.3.1   Increasing Energy Efficiency for Cell Edge Communications

BSs usually have lower energy efficiency for serving the users at the cell edge than on serving the users within the cell. This is not only because cell edge users are further away from BSs and require more transmission power, but also because cell edge users experience more interference from neighboring cells. Multi-cell cooperation can improve energy efficiency for cell edge communication [13].

2.3.1.1 Joint Transmission

Joint transmission exploits the cooperation among neighboring BSs to transmit the same information to individual users located at the cell edge [42]. Figure 2.4(a) illustrates a joint transmission scheme. User 2 is associated with BS 2 while user 1 is associated with BS 1. Instead of serving their associated users independently, BS 1 and BS 2 cooperatively transmit information to their users. For example, if both BSs apply time division multiple access (TDMA), in the first time slot, both BS 1 and BS 2 transmit the same information to user 2. The user combines the signals from both BSs to decode the information. In the second time slot, the BSs cooperatively serve user 1. As a result of the joint transmission, cell edge users experience higher receiving signal strength and lower interference, and therefore less transmission power is required from both BSs. Thus, joint transmission improves the energy efficiency of cell edge communications.

2.3.1.2 Cooperative Beamforming

In cooperative beamforming, a BS forms radio beams to enhance the signal strength of its serving users while forming null steering toward users in its neighboring cells. As shown in Figure 2.4(b), BS 1 forms a radio beam toward its associated user, user 1, to increase the user's receiving signal strength. On the other hand, in order to reduce the interference to user 2 who is served by a neighboring BS, BS 1 forms null steering toward user 2. BS 2 executes the same operation as BS 1. Therefore, through cooperative beamforming, the signal to noise ratio (SNR) of both user 1 and user 2 are enhanced. Thus, less transmission power is required for BSs to serve cell edge users, and hence the energy efficiency of cell edge communications is increased.

2.3.1.3 Cooperative Relaying

By taking advantage of CR techniques, cooperation can be exploited between a primary cell and a secondary cell as shown in Figure 2.4(c). The secondary BS cooperatively relays the signal from the primary BS to the primary user. The primary user combines the signals from both the primary BS and the secondary BS to decode the information. To incentivize cooperative relaying, the primary BS grants the secondary BS some spectrum, which can be utilized for communications between the secondary BS and secondary users. By cooperating with secondary BSs, the transmission power required from primary BSs to serve cell edge users is reduced. Therefore, the cooperative relay scheme reduces the power consumption of primary BSs [43]. However, owing to time varying wireless channels, the scheduling delay involved in the relaying process may increase the outage probability. Fan *et al.* [44] showed that the outage probability caused by scheduling delay can be alleviated by optimizing the transmission power of the network nodes.

2.3.1.4 Distributed Space-Time Coding

Distributed space-time coding [45] exploits the spatial diversity of both the relay nodes and the mobile users to combat multipath fading, and can be applied to improve the spectral and energy efficiency of cell edge users. As illustrated in Figure 2.4(d), the distributed space-time coding scheme consists of two phases. In the first phase, the BS broadcasts a data packet to its destination and the potential relays. In the second phase, the potential relays that can decode the data packet cooperatively transmit the decoded

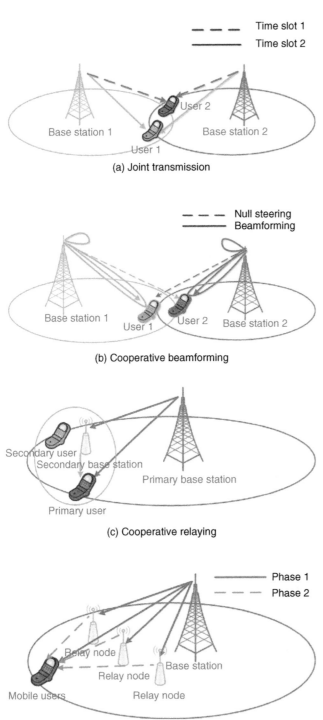

(a) Joint transmission

(b) Cooperative beamforming

(c) Cooperative relaying

(d) Distributed space-time coding

Figure 2.4 Energy utilization of the cellular network. *Source:* Han 2013 [5]. Reproduced with permission of IEEE.

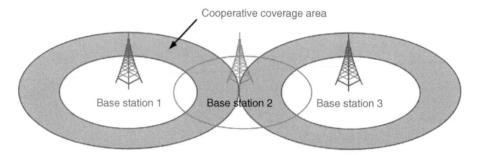

Figure 2.5 Cooperation to increase coverage area. *Source:* Han 2013 [5]. Reproduced with permission of IEEE.

data packet to the destination using a suitable space-time code. By exploiting spatial diversity, distributed space-time coded cooperative transmission improves the diversity gain of cellular networks. However, the multiplexing gain of the wireless system may be sacrificed [46]. Zou *et al.* [46] proposed an opportunistic distributed space-time coding (O-DSTC) scheme to increase the multiplexing gain. The spectral efficiency is increased by taking advantage of distributed space-time coded cooperative transmission. As a result, fewer transmissions are required in order to transmit a given number of data packets. Therefore, the energy efficiency of the cellular network is also enhanced.

2.3.2 Enabling More BSs Into Sleep Mode

As discussed in the previous subsection, CoMP transmission, for example, joint transmission, enhances the receiving SNR of cell edge users, implying that BSs are able to cover a larger area with the same transmission power. Therefore, under low traffic demands, adopting CoMP transmission can reduce the number of BSs required to cover an area, thus improving network energy efficiency [47]. As presented in Section 1, traffic intensity aware multi-cell cooperation enables BSs with low traffic demands to switch into sleep mode to save energy. The users in the off cells will be served by their active neighboring cells. However, owing to the limitation of BSs' maximal transmission power, a few BSs should be kept alive to provide coverage in the area. By applying CoMP transmission, the number of required active BSs can be further reduced. As illustrated in Figure 2.5, the inner circle of BS 1 is the original coverage area while the outer circle indicates the coverage area after cooperation with its neighboring BS, BS 3. The shaded area is the additional coverage area achieved by multi-cell cooperation. Under the conditions of lower traffic demands, if there is no cooperation among BSs, three BSs have to be kept alive to cover the whole area. However, if multi-cell cooperation is enabled, BS 1 and BS 3 can provide services to the users in the area by applying cooperative transmission, and thus BS 2 can be switched into sleep mode. Therefore, the total energy consumption of the network is reduced.

2.4 Summary and Future Research

This chapter has discussed how to reduce the energy consumption of cellular networks via multi-cell cooperation, and focuses on three multi-cell cooperation scenarios that enhance the energy efficiency of cellular networks. The first is traffic intensity aware

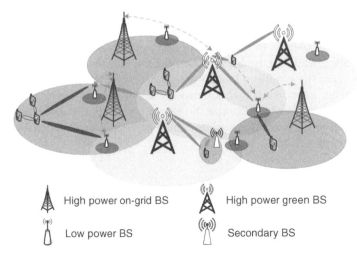

High power on-grid BS   High power green BS

Low power BS   Secondary BS

Figure 2.6 Future cellular networks with holistic cooperations. *Source:* Han 2013 [5]. Reproduced with permission of IEEE.

multi-cell cooperation, in which multiple cells cooperatively estimate traffic demands, and the network layout is adapted based on the estimated traffic demands. Through network layout adaptation, the number of active BSs can be reduced, thus reducing the energy consumption of the network. The second scenario is energy aware multi-cell cooperation, in which off-grid BSs powered by green energy are made to serve a large area to reduce the on-grid power consumption. The third scenario is energy efficient CoMP transmission, in which overall energy consumption is reduced by improving energy efficiency of BSs on serving cell edge users.

With advances in cellular network and communications techniques, future cellular networks will be heterogeneous in terms of both network deployments and communications techniques as shown in Figure 2.6.

Regarding network deployments, a heterogeneous network refers to deploying a mix of high power BSs and low power BSs in order to satisfy traffic demands of service areas. High power BSs, for example, macro and micro BSs, are deployed to provide coverage to a large area while low power BSs are deployed to provide high capacity within a small coverage area. In addition, based on energy supplies, BSs can be further divided into two types: on-grid BSs and off-grid BSs powered by green energy such as solar power and wind power. Therefore, future heterogeneous networks consist of four types of BSs: high power on-grid BSs, high power off-grid BSs, low power on-grid BSs, and low power off-grid BSs. In terms of technology diversity, heterogeneous networks consist of a variety of technologies such as MIMO, cooperative networking, and cognitive networking. By increasing network diversity, BSs have more cooperation opportunities, and thus can achieve additional energy savings. However, realizing optimal multi-cell cooperation is nontrivial.

2.4.1 Coalition Formation

The coalition formation problem is to determine the ideal coalition size and membership. Although the problem has been well studied in the field of economics, and some game theory methods have been applied to solve the problem in wireless networks, few of these models can be directly applied to the formation of coalitions of BSs for the purpose of reducing the energy consumption of cellular networks. This is because the original coalition formation problem is to form a coalition to maximize the benefit to each member in the coalition while the problem of forming the coalition of BSs is to maximize the total benefits of the coalition. In addition, owing to the diversity of BSs and radio access technologies, a successful coalition may consist of BSs and techniques that are complementary to each other to maximize the energy savings of the coalition. Hence, novel models or methods on forming coalition must be developed to facilitate energy efficient multi-cell cooperation.

2.4.2 Green Energy Utilization

For the sake of environmental friendliness, off-grid BSs powered by renewable energy are favored over on-grid BSs. In order to optimize the utilization of off-grid BSs, the fundamental design issue is how to utilize the harvested energy to sustain traffic demands of users in the network. The optimal utilization of green energy over a period of time depends on the characteristics of energy arrival and energy consumption both at the present time as well as in the future, and on the cooperation strategies of the neighboring or cooperating cells.

2.4.3 Incentive Mechanism

Note that in future heterogeneous networks, multi-cell cooperation may involve cells not owned by the same operator. For example, the coalition may consist of a BS from a different operator, a secondary BS operated via CR, and a relaying cell formed by mobile users. To incentivize cooperation among these cells, a simple but effective incentive mechanism should be developed.

2.5 Questions

2.1 How does traffic intensity aware multi-cell cooperation enhance energy efficiency of cellular networks?

2.2 How does energy aware multi-cell cooperation work?

2.3 How does energy efficient CoMP transmission enhance the performance of cellular networks?

3

Powering Mobile Networks with Green Energy

This chapter provides a timely overview of the research challenges and existing solutions for green energy enabled mobile networks and lays the foundations for powering mobile networks with green energy. First, we introduce green power generation and prediction models and mobile network energy consumption models. Then, we present proposals on how to design and optimize green energy powered BSs including provisioning the green energy system and optimizing resource management in BSs. Finally, a discussion is provided on how to optimize green energy enabled mobile networks under different network power supply scenarios.

3.1 Green Energy Models: Generation and Consumption

Green power generation, which is highly dependent on the power generators' geolocations and weather conditions, is rather dynamic. Meanwhile, the energy consumption of mobile networks is also highly dynamic. Thus, understanding the characteristics of green power generation and the dynamics of energy consumption of mobile networks are essential for designing and optimizing green energy enabled mobile networks.

3.1.1 Green Power Generation

Since green power generation is highly dynamic, a green energy powered system should be designed and optimized to incorporate the dynamics of power generation. For system implementations, the availability of green energy is predicted based on statistical data by using various prediction models [48]. For example, the availability of solar energy is predicted based on the statistical data that provide the solar energy expected under a clear sky condition, and the cloud coverage estimation that forecasts the percentage of the sky covered by the cloud. Define E, E_c, and β as the amount of predicted solar energy, the amount of solar energy under a clear sky condition, and the cloud coverage estimation, respectively. Then, $E = E_c(1 - \beta)$. For theoretical analysis, the energy harvest process is modeled as a stochastic process. The first order Markov stochastic process is an analytically simple and practically accurate model for solar energy generation [49].

3.1.2 Mobile Network Energy Consumption

Base stations, which consist of multiple components such as antennas, power amplifiers, radio frequency transceivers, baseband processing units, power supply units, and

Green Mobile Networks: A Networking Perspective, First Edition. Nirwan Ansari and Tao Han.
© 2017 John Wiley & Sons Ltd. Published 2017 by John Wiley & Sons Ltd.

cooling units, account for the major energy consumption of a mobile network. In general, a BS's power consumption can be modeled as the sum of its static power consumption and its dynamic power consumption. The static power consumption is the power consumption of a BS without any traffic load. The dynamic power consumption refers to the additional power consumption incurred by the traffic load in the BS, which can be well approximated by a linear function of the traffic load or the output radio frequency power [3]. The BS power consumption model can be adjusted to model the power consumption of either MBSs or SCBSs by incorporating and tweaking the BS's static power consumption and the linear coefficient that reflects the relationship between the BS's dynamic power consumption and its traffic load.

3.2 Green Energy Powered Mobile Base Stations

In order to maximize green power utilization, green energy powered BSs should be properly designed and optimized to cope with the dynamics of green power and mobile data traffic. Figure 3.1 shows a simplified diagram of a green energy powered BS. In order to utilize green energy, five energy related components may be integrated into a BS. These components are the green power generator, for example, a solar panel, the charge controller which regulates the output voltage of the green power generator, the DAC inverter, the battery, and the smart meter, which facilitates the power transmission between BSs and the power grid.

3.2.1 Green Energy Provisioning

By integrating green energy into mobile networks, mobile service providers may save on-grid power consumption and thus reduce their CO_2 emissions. However, equipping a BS with a green energy system incurs additional capital expenditure (CAPEX) which is determined by the size of the green power generator, the battery capacity,

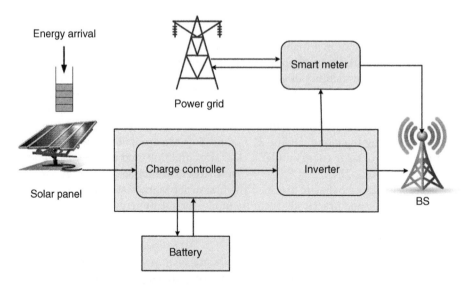

Figure 3.1 A green energy powered BS. *Source:* Han 2014 [17]. Reproduced with permission of IEEE.

and other installation expenses. It is desirable to minimize the CAPEX on provisioning green energy for BSs. Here, green energy provisioning refers to determining the maximum required capacity of the green power generator and the battery. Although the green energy provisioning problem for off-grid loads are well studied [50], the existing solutions do not directly apply to provisioning green energy for mobile networks. The process of green energy provisioning involves three basic models: the load model, the battery model, and the green power generation model. The existing solutions usually take the statistical load information as an input and evaluate the loss of load probability (LOLP) and the loss of energy probability (LOEP). LOLP is defined as the probability of not being able to satisfy the energy demands while LOEP measures the amount of energy loss due to the inability to charge the battery beyond its maximum capacity. Based on the evaluation results, these methods adjust the green power generator size and battery capacity until the system performance (in terms of LOLP and LOEP) is satisfied. These methods relying only on statistical load information do not optimize energy utilization, and may result in over-provisioning.

In a mobile network, a BS's power consumption can be adapted according to the availability of green energy. On the one hand, a BS's transmission strategies can be optimized to reduce energy demands without degrading the QoS of the network [49]. On the other hand, owing to seamless deployment of BSs, a mobile user may be covered by multiple BSs. The traffic load of a BS can be reduced by offloading its associated users to neighboring BSs. In this way, the BS's power consumption is adaptive. Therefore, by optimizing the BS's transmission strategies and mobile network layout, the power consumption of the BSs can be optimized to minimize the size of the green energy system.

3.2.2 Base Station Resource Management

Since renewable energy is highly dynamic, a green power unit may not always be able to guarantee sufficient power supplies to BSs even though the green power system is well provisioned. Therefore, the BS's resource management, including energy management and radio resource management, should be designed to optimize the BS's performance with constrained green energy. A BS's optimal resource management depends on five dynamic processes: the energy arrival dynamics, the battery dynamics, the power grid dynamics,[1] the traffic dynamics, and the wireless channel dynamics. Owing to the complex coupling of the energy allocation and radio resource allocation, it is challenging to achieve optimal resource management.

3.2.2.1 Packet Scheduling Optimization

While it is mathematically intractable to obtain optimal packet scheduling with consideration of all of these dynamic processes, several contributions have provided some insights into solving this difficult problem. Considering a single user communications system with an energy harvest transmitter, Yang and Ulukus [51] proposed optimizing the packet transmission policy to minimize the packet transmission completion time. For this problem, the most challenging aspect is the causality constraint: a packet cannot be delivered before it has arrived and the energy cannot be consumed before it is harvested. This constraint introduces a trade-off between the energy harvest time and

---

[1] The term power grid dynamics refers to the dynamics of power generation and power demand in the power grid.

the packet transmission time in determining the transmission rate and power. In general, considering a network with a single transmitter, using more transmission power may enable a higher transmission rate and thus reduce the packet transmission time. However, since the causality constraint is introduced by green energy, in order to adopt a higher transmission rate, the transmitter has to wait until enough green energy has been harvested.

For a multi-user system, the multi-user diversity in terms of channel conditions can be exploited to enhance green energy utilization. Within a given completion time, transmitting a packet toward a user with better channel conditions usually requires less transmission power. Therefore, scheduling different users in a given time slot may require different amounts of green energy. By exploiting multi-user diversity, a packet scheduling algorithm may be able to shape the BS's energy demands to match the green power generation [52]. If a BS's energy demands perfectly match the green power generation, no additional energy sources are required to sustain the traffic demands in the BS.

3.2.2.2 Energy Allocation Optimization

Since mobile traffic shows temporal dynamics, a BS's energy demands change over time. Green power generation varies along the time dimension. Thus, in order to optimize their performance, BSs should determine how much energy is to be utilized at the current time and how much energy is to be reserved for the future. If more energy is utilized at the current time, the BS may provision a larger capacity. However, the BS may suffer from service outages due to energy shortages in future periods. In order to satisfy the network's outage constraint, Farbod and Todd [53] proposed reducing BS's power consumption at certain periods by reducing their instantaneous capacity, in order to alleviate service outages. The proposed on/off proportional capacity deficit algorithm satisfies the outage constraint with the minimum capacity deficit.

3.3 Green Energy Powered Mobile Networks

Green energy enabled mobile networks may consist of BSs with different power supply configurations, as classified in Figure 3.2. Based on whether a BS is connected to the power grid, the BS is classified as an on-grid BS or an off-grid BS. The power supplies for an on-grid BS can be grid power and green power from either a standalone green power generator or from a green power farm, for example, a solar/wind farm. An

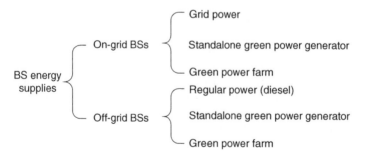

Figure 3.2 A BS's power supplies. *Source:* Han 2014 [17]. Reproduced with permission of IEEE.

off-grid BS may be powered by regular power (non-renewable energy such as diesel) and green power from either individual generator or green power farm. Considering BSs' various power sources, optimizing green energy enabled mobile networks is challenging.

This section discusses network optimization for off-grid green mobile networks, on-grid green mobile networks, and mobile networks consisting of both green BSs and grid powered BSs.

3.3.1 Off-Grid Green Mobile Networks

Off-grid green mobile networks can either be powered by standalone green power generators or green power farms.

3.3.1.1 Powered by Standalone Green Power Generator

When a BS is powered by a standalone green power generator, the green power generated in the BS is not shared with other BSs. Under this scenario, the fundamental design issue is to utilize the harvested energy to sustain traffic demands of users in the network. The optimal utilization of green energy over a period of time depends on the characteristics of the energy arrival and power consumption at the current time slot as well as in future time slots. Optimizing green energy utilization involves two aspects. The first aspect is to optimize the energy allocation in multiple time slots by determining how much energy should be used at the current time slot, and how much energy is reserved for future time slots for individual BSs. To solve the energy allocation problem, parameters such as current energy arrival and consumption and estimations of future energy arrival and consumption should be considered. The second aspect is to maximize the utilization of the allocated green energy in individual time slots. The BSs' power consumption depends on the intensity of the mobile traffic, which exhibits spatial diversity.[2] Thus, the power consumption in BSs may be different. In order to sustain the traffic demand of all users, green energy utilization should be optimized by balancing the power consumption among BSs according to the availability of green energy. The power consumption of BSs is balanced by balancing the traffic loads among the BSs [11].

3.3.1.2 Powered by Green Power Farm

When BSs in the mobile network are powered by green power farms, the total amount of green energy is budgeted by the capacity of the green power farms. Given a total power constraint, the network performance can be improved by optimizing spatial and temporal power sharing [54]. Since mobile traffic exhibits spatial diversities, the mobile user distributions in different BSs are usually different. Thus, the mobile users in different BSs may experience different signal interference noise ratios (SINRs). In order to save energy, the BSs whose mobile users have better SINR may reduce their transmit power. Since the achievable data rates for users with high SINR logarithmically decreases as the SINR decreases, reducing the BSs' transmit power will not significantly impair the users' achievable data rates. In addition, reducing the BSs' transmit power may increase the SINR of their neighboring BSs' cell edge users. As a result, neighboring BSs may also

---

[2] The spatial diversity of mobile traffic implies that BSs at different locations experience different traffic demands.

reduce their transmit power to save energy. Therefore, the budgeted green energy should be properly shared across the BSs. On the other hand, since mobile users' channel conditions vary over time, scheduling a user in different time slots results in different power consumption in a BS. For example, if a user experiences severe channel conditions, in order to save energy, the BS may not schedule the user until its channel condition is satisfactory. Thus, for individual BSs, the power consumption in different time slots should be optimized via user scheduling.

3.3.2 On-Grid Green Mobile Networks

On-grid green mobile networks refer to mobile networks whose BSs are not only connected to the power grid but also equipped with standalone green power generators. Green power is utilized to reduce the on-grid power consumption while grid power serves as a backup power source for when power demand exceeds the green power capacity.

For on-grid green mobile networks, the energy storage (the battery) is not necessarily located in individual BSs' green energy systems because the BSs are connected to the power grid for energy backup. Thus, the optimal green energy utilization strategy is different depending on whether BSs have energy storage. On the one hand, when the BSs have energy storage, green energy can be stored and utilized to shift the peak power demands, thus reducing OPEX as well as alleviating CO_2 emissions. For example, as the smart grid advances, the electricity price is highly correlated with demand. In peak power demand hours, the electricity price is usually higher than that in off-peak power demand hours. In this case, in order to reduce OPEX, green energy is utilized when the electricity price is higher than a threshold, and is stored in the batteries when the electricity price is low.

On the other hand, when the BSs do not have energy storage, green energy should be utilized when it is generated. In this case, in order to maximize green energy utilization, traffic loads are directed to the BSs with larger green power generation capacity. In other words, the BSs with larger green energy capacity serve more traffic loads while the BSs with smaller green energy capacity serve less traffic load.

3.3.2.1 Green Energy Aware User Association

Green energy aware user association means directing traffic loads to BSs with higher green energy capacity. However, such user association strategies may result in traffic congestion in BSs with larger green energy capacity. Therefore, a proper traffic offloading scheme is desirable not only to optimize the utilization of green energy but also to avoid excessive traffic congestion in BSs. Han and Ansari [55] proposed a green energy aware and latency aware (GALA) user association scheme that minimizes the sum of the weighted traffic delivery latency of BSs in a heterogeneous mobile network. The weight of a BS reflects the green power capacity of the BS: a large weight indicates a small green power capacity in a BS.

3.3.2.2 Green Energy Aware Base Station Sleeping

Traffic demands in mobile networks exhibit highly dynamic temporal behavior, thus requiring a large capacity in peak usage hours but reduced requirements during off-peak hours [5]. Mobile networks are usually dimensioned for peak hour traffic, and thus most

BSs operate at low workload during off-peak hours. Owing to the large static power consumption, these BSs have poor energy efficiency. In this case, optimally switching some of these BSs into sleep mode will enhance the network energy efficiency while maintaining sufficient network capacity. When BSs in the mobile network are powered by green power generated from their standalone power generators, the optimal BS sleeping strategies should take green power into account. If a BS's green power cannot be shared with other BSs or be stored, the BS should remain alive as long as it has sufficient green power to sustain its operation even though the BS has low traffic loads. However, if the BS's green power can be shared among other BSs via a smart grid, the power transmission efficiency of the power grid also determines whether the BS should sleep or not. When the power transmission efficiency of the power grid is high, the green power can be losslessly transferred among BSs. Thus, the optimal BS sleeping strategies are similar to those applied in mobile networks without green power supplies. When the power transmission loss is considerable, the trade-off between the power transmission loss and BSs' static power consumption should be examined in determining the optimal BS sleeping strategies.

3.3.2.3 Green Energy Aware CoMP

CoMP transmission is a promising technique that enhances the network efficiency and overall user QoS for next generation mobile networks. By applying CoMP, multiple BSs either jointly transmit data to mobile users or jointly schedule their data transmissions. There are two basic design issues in optimizing CoMP. The first is the cluster formation (CF) problem that determines which transmit points[3] should be clustered to perform CoMP transmission. The other issue is the resource allocation (RA) problem which optimizes the spectrum and power allocation among multiple transmit points within a cluster. In on-grid green mobile networks, different BSs may have different amounts of green power. Thus, the available green power for the transmit points is different. If the power consumption of a transmit point is larger than the available amount of green power, the transmit point consumes on-grid power. Given the green power in each BS, optimizing CoMP to minimize the on-grid power consumption involves both the CF and RA problem. Considering the green power constraint in individual BSs, the CF and RA problems are highly coupled. On the one hand, if the cluster formation is given, the optimal power allocation that minimizes the on-grid power consumption while satisfying the users' QoS requirements can be derived. On the other hand, if the power allocation is given, the optimal cluster formation can be obtained. Thus, the optimal power allocation depends on the cluster formation, and vice versa. Owing to the coupling of the CF and RA problems, it is challenging to solve the CoMP transmission problem.

3.3.3 Mixture of Green Base Stations and Grid Powered Base Stations

A green BS is defined as the BS having green power supplies from either standalone green power generators or green power farms. A grid powered BS is defined as a BS without any green power supply. When a mobile network consists of both green BSs and grid powered BSs, the mobile network optimization is to minimize the on-grid power

---

[3] Each BS may be equipped with multiple transmit antennas and each transmit antenna can be recognized as a transmit point.

consumption and maximize the utilization of green energy. Therefore, the mobile network guides more data traffic to the green BSs. Two approaches have been proposed to encourage mobile users to access the BSs powered by green energy. The first approach is to adjust the handover parameters to prioritize the green BSs [10]. This approach adjusts the handover parameters of BSs to enable mobile users to hand over more easily to BSs powered by green energy than to those powered by grid energy. The other approach [10] is to increase the transmit power of the BSs powered by green energy, thus enlarging the coverage area of these BSs. As a result, more mobile traffic will be offloaded to the BSs powered by green energy.

3.4 Summary

In this chapter, we have discussed the technology of using green energy to power mobile BSs and networks. Green energy generation and consumption models have been introduced. The design of green energy powered BSs has been presented. We have also investigated the optimization of green energy powered mobile networks under different energy supply scenarios.

3.5 Questions

3.1 How does one model the energy consumption of a BS?

3.2 How does one model the energy generation of a green power source such as solar power?

3.3 Does packet scheduling impact the performance of green energy powered BS? How does one optimize the packet scheduling to improve the green energy utilization?

3.4 How does one optimize off-grid green mobile networks?

3.5 How does one optimize on-grid green mobile networks?

4

Spectrum and Energy Harvesting Wireless Networks

It is not a trivial matter to design and optimize spectrum and energy harvesting networks owing to the opportunistic arrival of idle spectrum and free energy in reality. This chapter discusses state-of-the-art research on this topic. Although green powered wireless networks are not currently deployed on a large scale owing to the higher cost per watt than grid energy, powering wireless networks by green energy is imminent and is becoming a sustainable and economically attractive solution. The aim of this chapter is to provide some insights for future research in green powered dynamic spectrum access-based CR networks, which are capable of liberating wireless access networks from spectral and energy constraints.

4.1 Spectrum Harvesting Techniques

Spectrum harvesting techniques explore the underutilized licensed spectrum to boost the capacity of wireless and mobile networks and enable dynamic access, allocation, and aggregation of licensed, unlicensed, and TV white space frequency channels. Spectrum harvesting techniques consist of the following three major functions:

- *Spectrum Sensing and Analysis [56]:* One of the essential functions of spectrum sensing is to identify the spectrum opportunities, for example, spectrum holes. The identified spectrum holes can be utilized by secondary transceivers without interfering with the primary users.[1] The other function of spectrum sensing is monitoring the primary users' activities on using the spectrum holes. When primary users resume their transmissions over the spectrum holes, the secondary users should evacuate from the spectrum holes to avoid interference. Spectrum analysis uses the information obtained from spectrum sensing to schedule and make a decision to access the spectrum by secondary users.
- *Spectrum Management and Handover [57]:* This function manages spectrum sensing and access strategies. Spectrum sensing strategies specify when and which spectrum to sense. Based on the spectrum sensing results, the spectrum access strategies evaluate the spectrum holes and decide whether they are accessible. This function also enables secondary users to vacate the channel when the primary user is detected to

---

[1] Primary users refers to the users who own the license to the spectrum.

Green Mobile Networks: A Networking Perspective, First Edition. Nirwan Ansari and Tao Han.
© 2017 John Wiley & Sons Ltd. Published 2017 by John Wiley & Sons Ltd.

reclaim it [58], and therefore addresses to some extent the challenges imposed by the fluctuating nature of the available spectrum, as well as the diverse QoS requirements of various applications [57].

- *Spectrum Allocation and Sharing [57]:* This function coordinates spectrum allocation among users to mitigate inference and collisions among users. Moreover, this function aggregates available frequency channels according to users' capacity demands.

We will investigate spectrum harvesting networking technologies from two perspectives: (1) achieving power aware functionality in spectrum harvesting communications systems, and (2) designing energy efficient wireless access systems via spectrum harvesting.

4.1.1   Energy Efficiency in Spectrum Harvesting Networks

In traditional spectrum licensing schemes, explicit rules are set for use of the spectrum. As a result, severe underutilization of the static licensed spectrum presents great challenges, and opportunities exist for the resource constrained wireless network to handle the growing popularity of new wireless devices and applications. Dynamic spectrum access techniques have been proposed to solve this spectrum inefficiency problem, by allowing unlicensed users to access the radio spectrum under certain restrictions.

Spectrum harvesting technologies have emerged as a key enabler for dynamic spectrum access. In a spectrum harvesting-based dynamic spectrum access (DSA) network, the primary system owns the spectrum rights, while unlicensed users can dynamically share the licensed spectrum in an opportunistic manner. Although CR enables spectrum sharing with smart operation and agile spectrum access [59], a major limitation of a practical CR network is the increased power consumption introduced by the cognitive capability and reconfigurability. Energy aware CR systems have been investigated from three general perspectives: (1) Energy minimization minimizes the power consumption for a given performance requirement; (2) Performance maximization maximizes the performance for a given limited power budget; and (3) Utility maximization takes the power consumption cost into account during sensing/transmission, with utility functions expressed in the form of either the difference or the ratio between the performance reward and cost of power as shown in Eqs. (4.1) and (4.2).

$$\eta = F(P) - \mu P, \tag{4.1}$$

$$\eta = \frac{F(P)}{\mu P}, \tag{4.2}$$

where specific performance criteria of utility η, such as *bit/Joule*, can be found in [60], F is the performance function with respect to power consumption P, and μ is a price parameter. Based on the various approaches and criteria mentioned above, we will introduce the functionalities of energy efficient spectrum harvesting networks.

4.1.1.1   Energy Efficient Spectrum Sensing and Analysis

The primary goal of spectrum harvesting is to access the spectrum opportunistically. This makes spectrum sensing crucial because it enables users to detect the activities of a primary user in a frequency band.

To design energy efficient spectrum sensing methods, we should first understand the performance indicators used to measure detection performance: the detection probability p_d, that is, the probability of declaring the presence of a primary user when they are in fact occupying the spectrum, and the false alarm probability p_f, that is, the probability of declaring the presence of primary users when they are actually idle. The higher the p_d, the better the primary users are protected. The lower the p_f, the more spectrum opportunities are exploited by secondary users [61]. Note that the complementary probability of p_d is called miss-detection probability, and sometimes referred to as collision probability when the secondary user starts transmission immediately once the sensing result indicates the channel is idle [62].

To improve the quality of spectrum sensing under various conditions, such as whether prior information about the probability of the primary user's presence on the target channel, defined in Eq. (4.3), is available or not, and the target channel is narrow or wide, a collection of spectrum sensing methods have been proposed [63].

$$\pi_0 = Prob(\mathcal{H}_0),$$
$$\pi_1 = Prob(\mathcal{H}_1), \tag{4.3}$$

where $\mathcal{H}_1$ and $\mathcal{H}_0$ are the hypotheses that primary user is active and idle, respectively. π_0 and π_1 are the probabilities of the primary user's presence or not on the target channel under the corresponding prior information, respectively.

In addition to sensing performance, the energy efficiency of the spectrum sensing algorithms depends on the power consumption, as shown in Eq. (4.4), which involves the scheduling of a spectrum sensing activity in the time and spatial domains by the media access control (MAC) layer. The sensing power P_s and the power consumed for reporting the sensing result P_r are

$$P_s = F_s(T_s) = F_s(N_s),$$
$$P_r = F_r(N_c, d), \tag{4.4}$$

where F_s and F_r are the sensing and reporting power functions, respectively. The sensing power P_s is increasing with the sensing time T_s, which is further proportional to N_s, the number of collected samples. P_r increases with the transmission distance D and the number of nodes participated in the sensing N_c.

Besides the energy consumed to obtain the sensing result, the major energy wastage in spectrum harvesting networks is due to the miss-detection of primary users, which can lead to collisions and retransmissions. As a result, the power consumed for data transmission, P_T, is related to the detection probability p_d and SNR, according to a decreasing function F_T defined as follows.

$$P_T = F_T(\gamma, p_d). \tag{4.5}$$

The research challenges in spectrum sensing and analysis include:

1) Sensing Duration and Frequency Problem: To balance the energy-inefficiency of continuous spectrum sensing (high P_S) and the unreliability issue of periodic spectrum sensing (low p_d and high p_f), Gan *et al.* [64] optimized the sampling rate (N_S/T_S) and sensing time T_S to minimize the total sensing power across multiple potential channels, within the constraints of detection performance.

2) Sensing Architecture Design Problem: With multiple secondary users sensing and sharing information, power consumed by sensing P_S and reporting P_R can both be decreased if every CR can randomly turn off its sensing device; this energy saving approach is referred to as sleeping or on-off sensing. The probability of switching off is called the sleeping rate. The second approach to reduce P_R is censoring, where a sensing result is sent only if it is deemed informative. The censor rate refers to the probability of the sensing results being located in the censor region, which is defined as a signal interval for the locally collected energy [65]. Another method of reducing P_R is clustering: instead of sending local sensing results directly to the fusion center (FC), they are sent to the assigned cluster heads (CHs), which then make local cluster decisions and send them to the FC. In this way the network energy consumption is reduced by means of distance reduction. Extensive centralized detection performance optimization and overhead minimization algorithms can be found in [66, 67], where three approaches are implemented in a complementary way. Besides centralized spatial diversity scheduling, each secondary user can decide independently whether to sleep or share results. However, the broadcasting nature of the wireless environment provides the opportunity for each secondary user to be a free rider, who overhears the sensing result and does not contribute to the sensing process. Various games have been modeled in [68, 69] to study the above mentioned incentive issue of cooperative spectrum sensing where each selfish secondary user aims to maximize its own utility.

3) Multi-channel Scheduling Problem: With the introduction of scheduling in the frequency domain, the spectrum opportunity can be further exploited from a new degree of freedom.

- Cooperative Sensing Scheduling: When periodic spectrum sensing is adopted for multiple channels, the sensing needs to address the cooperative sensing scheduling (CSS) problem [70]: how to assign secondary nodes to different channels? Except for the energy-performance trade-off (greater N_C leads to a better sensing performance but more energy will be consumed in both sensing and reporting), a performance-opportunity trade-off exists in CSS, that is, assigning more secondary users to one channel will yield a more reliable sensing outcome, but the spectrum opportunity is not fully exploited since less potential channels are sensed [71].

- Sequential Channel Scheduling: As far as the sensing process is concerned, the only task for periodic spectrum sensing is to find idle spectrum. After getting the sensing result, the secondary user can either transmit when the sensed channel is free, or wait for the channel which is currently occupied by the primary user. In sequential channel sensing, the task is to find an idle channel with good quality. So, the secondary user can switch to a different channel and perform sensing again even when the current sensed channel turns out to be idle. A channel-specific access decision has to be made based on the analysis of both the sensing result and possible channel quality estimation afterwards if the channel is unoccupied [72].

Since channel switching and sensing require additional time and energy, the secondary user needs to efficiently locate an idle channel with satisfactory channel quality. More specifically, a channel sensing order scheme and a channel selection protocol (stopping rule) have to be designed in order to optimize CR networks.

Any performance metrics for sequential sensing scheduling will be related to transmission performance, such as throughput and delay, since they are the criteria to judge whether the channel is good enough for the secondary user. Transmission performance related design, which is the function of the spectrum analysis and access process, will be further discussed later in the following subsections.

4.1.1.2 Spectrum Management and Handoff

Spectrum analysis and spectrum access are two major components of the spectrum management mechanism [58]. First, the sensing strategy specifies whether to sense and where in the spectrum to sense. Then, spectrum analysis will estimate the characteristics of the spectrum holes that are detected through spectrum sensing. Third, the access strategy will make a decision on whether to access or not, by judging whether this is the best available channel. While the focus of the previous subsection (except for sequential channel scheduling) was finding a spectrum opportunity in an energy efficient way, this section will further elaborate on locating a good quality spectrum opportunity. The channel characteristics (mostly channel capacity) determine the data transmission power level appropriate to the user requirements. As a result, energy efficiency plays an important role in the design of the spectrum management mechanism.

1) Spectrum Access: The sequential decision process of sensing and access strategies gives rise to a new option for the secondary user, that is, keeping idle on the current channel without sensing or transmission. The reason is that when energy consumption is considered, instead of sensing a channel that may turn out to be in use by the primary user or obtaining an idle channel with poor fading conditions, it is better for the secondary user to do nothing at all. The partially observable Markov decision process (POMDP) framework, which permits uncertain information in modeling the operation mode selection problem, is suitable for CR networks where primary users' traffic is not fully observable to the secondary network. By jointly considering three conflicting objectives—gaining immediate access, gaining spectrum occupancy information, and conserving energy (for future sensing, transmission, and idling) on a single-channel—the POMDP framework [73] indicates that the optimal sensing decision and access strategy is one that uses a threshold structure in terms of channel free probability and channel fading condition. That is to say, a secondary user with an un-rechargeable battery and limited power should sense the channel when the conditional probability that the channel is idle in the current slot is above a certain threshold, and it will access the channel if the fading condition of this idle channel is better than a certain threshold.

2) Spectrum Mobility: Spectrum availability in CR networks varies over time and space. So, except for the traditional mobility which refers to mobile users traversing across cells or current channel conditions becoming worse [74], CR networks incur another dimension of mobility, spectrum mobility, in which the secondary user has to move from one spectrum hole to another to avoid interference in case of the reappearance of the primary user. Cognitive radio networks need to perform mobility management adaptively depending on the heterogeneous spectrum availability, which is a result of the highly unpredictable mixed multimedia primary traffic in today's wireless networks. In general, spectrum handoff, in which the secondary user can switch to another vacant channel and continue data transmission when the current

channel is sensed as busy, can be categorized into two major types. One is reactive-sensing spectrum handoff, in which the target channel is sensed or selected only after the spectrum handoff request is made. The other is proactive-sensing spectrum handoff, where the target channel for spectrum handoff is predetermined [75]. Reactive spectrum handoff pays the cost of sensing time to guarantee the accuracy of the selected target channel. By pre-determining the target channel, proactive spectrum handoff avoids the sensing time but loses certain accuracy. Extensive performance analysis of spectrum handoff can be found in [75, 76].

As shown in Eq. (4.6), tuning on the radio frequency of the cognitive device will result in additional power consumption.

$$P_{HO} = F_{HO}(f_t - f_c), \tag{4.6}$$

where F_{HO} is an increasing function and the handing off power P_{HO} is related to the frequency difference between the target channel f_t and the current channel f_c.

4.1.1.3 Spectrum Sharing and Allocation

According to the coordination behavior between primary users and secondary users, existing solutions for spectrum sharing can be classified into three major categories [77]: (1) Spatial spectrum sharing, for example, spectrum underlay, always allows secondary users to access the spectrum subject to the interference temperature (IT) constraint, which is the threshold of interference in impairing the primary user receiver [78]; (2) Temporal spectrum sharing, for example, spectrum overlay, allows secondary users to utilize the spectrum only when it is idle; (3) Hybrid spectrum sharing, in which secondary users initially sense for the status (active/idle) of a frequency band (as in the temporal spectrum sharing) and adapt their transmit power based on the decision made by spectrum sensing, to avoid causing heavy interference (as in spatial spectrum sharing) [79].

Spectrum sharing can be implemented in two ways: cooperative and non-cooperative, depending on the secondary users' behavior within the secondary network [80, 81].

Regardless of the spectrum sharing model, spectrum allocation mechanisms have a great impact on the energy consumption and performance of each individual secondary user and the whole secondary network. As a result, energy efficient designs, which take into account the diversity of secondary users' power budgets, the channel conditions, and the QoS requirements, are of great interest to the spectrum sharing and allocation algorithms.

1) Spatial Spectrum Sharing: Since IT is subject to space-time-frequency variation, spectrum sharing, along with power control, bit rate, and antenna beam allocation, which are all implemented by the dynamic resource allocation (DRA), have become essential techniques for energy efficient underlay CR networks [82]. Moreover, to access the same spectrum being used by primary users, secondary users normally provide incentives to compensate for the additional interference induced to primary users. The incentive can be in the form of spectrum trading, in which radio resources such as spectrum in a CR environment can be sold and bought [83]. So, efficient spectrum sharing and allocation can generate more revenue for the spectrum owner and also enhance the satisfaction of primary users. Illanko *et al.* [84] considered the centralized cooperative DRA problem, where the downlink power allocation of

OFDM CR networks is modeled as a concave fractional program. Near optimal solutions, which maximize *bit/Joule* energy efficiency while meeting secondary users' rate demands and primary users' IT-constraint, are given in closed form. In addition to centralized DRA, which requires a centralized entity having full knowledge of the network, distributed DRA is more suitable for scenarios with no such infrastructure [85]. We have previously studied spectrum trading between the primary and secondary network from the primary users' viewpoint [86]. Primary users' traffic is offloaded to the secondary network, which in exchange will gain spectrum for its own traffic transmission. In order to minimize the primary energy consumption subject to power budget and rate requirements, an auction-based energy spectrum trading scheme is proposed to approximate the behavior of the primary and secondary network for distributed spectrum allocation.

2) Temporal Spectrum Sharing: In temporal spectrum sharing, the issue is to explore transmission opportunities. After a spectral hole is identified, spectrum exploration addresses the issue of how efficiently secondary users can access and utilize the spectrum without interfering with the primary users. Because of the spectrum sensing process, the IT-constraint is less stringent as compared to the spatial spectrum sharing model. However, interference still exists due to frequency reuse in today's wireless systems and power leakage into the adjacent frequency band being occupied by the primary user [87]. Although spectrum sensing is up to secondary users, varying spectrum conditions and diverse QoS requirements still challenge the energy aware DSA. Hasan *et al.* [87] considered centralized power allocation over multiple OFDM subcarriers subject to an IT-constraint and limited transmission power budget. Suboptimal schemes have been proposed to maximize the corresponding convex utility function which incorporates the achievable data rate of the secondary user and the expected rate loss due to sensing error or spectrum reoccupation by the primary user. Gao *et al.* [88], on the other hand, proposed a framework of distributed DRA for energy constrained OFDMA based CR networks.

4.1.2 Enhancing Energy Efficiency Through Spectrum Harvesting

In this subsection, we discuss the possibilities of adopting spectrum harvesting technologies into the wireless access network to improve energy efficiency. Given the complexity of the topic and the diversity of existing technical approaches [89], we focus on the major applications of cooperative and CR communications, and heterogeneous CR for the emerging Long Term Evolution-Advanced (LTE Advanced) networks.

4.1.2.1 Green Relaying and Cooperative Cognitive Radio

The major challenge of CR networks is to guarantee the QoS requirements while not causing unacceptable performance degradation for primary users. In some cases, reliable end-to-end transmission within the secondary network requires a large amount of power, leading to harmful interference to primary users [90]. The key enabling technology for boosting the overall performance while saving energy is relay technology [80]. For relay based CR networks, the path loss is less due to the shorter transmission range, and the interference caused to the primary network is potentially reduced due to the lower transmission power. Furthermore, when no common available spectrum between a pair of secondary users exists, the relay node can help establish end-to-end communications through the dual-hop channel [90] (two sets of available spectra: one for the source-relay link and one for the relay-destination link). Unlike pure relay systems, the cooperative node acts as both an information source and a relay. Network coding based two-way relay schemes with decoding (decode-and-forward) and without decoding (amplify-and-forward, denoise-and-forward, or compress-and-forward) have been introduced to implement cooperative communications [80]. The inherent cooperative diversity can save energy by combining the signals received from different spatial paths and consecutive time slots [91].

In the context of CR networks, two basic cooperative transmission scenarios exist: (1) cooperative transmission between secondary users that aims to increase the secondary throughput for a given spectral hole; (2) cooperative transmission between primary and secondary users that aims to increase the spectrum opportunities of secondary users. Given the fact that under the specification of relay protocols, relay nodes will work differently from regular cognitive devices as illustrated in Figure 4.1, spectrum sensing and resource allocation designs need to take into account the features of relaying cognitive radio and cooperative CR systems.

1) Relay enhanced Cooperative Spectrum Sensing: With secondary relay enhanced CR networks, besides the fact that cooperation can be used to enhance the transmission

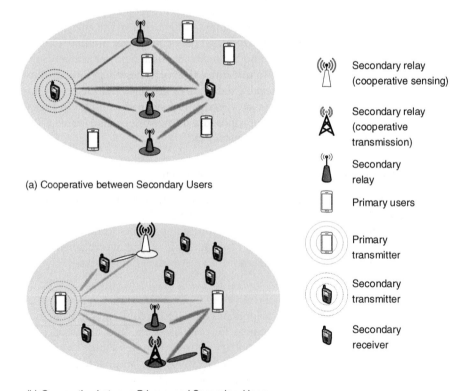

(a) Cooperative between Secondary Users

(b) Cooperative between Primary and Secondary Users

Figure 4.1 Relaying and cooperation in CR systems. *Source:* Huang 2015 [100]. Reproduced with permission of IEEE.

performance, the capability of sensing can be improved when relays perform spectrum sensing cooperatively with secondary users.

Huang *et al.* [92] investigated the trade-off between cooperative sensing performance and energy consumption for the scenario shown in Figure 4.1(b), where the relays operate with the amplify-and-forward (AF) protocol, and one of the relays will forward the signals of the primary user to the secondary user in order to help to determine the presence and absence of the primary user. Given p_d and p_f, the energy consumption for spectrum sensing is minimized over the number of samples (proportional to sensing time) and the amplification gain.

2) Resource Allocation for Cooperative Transmission between secondary users: With the introduction of the secondary relay, relay selection, power allocation and subcarrier matching become challenging issues in resource allocation.

Ge and Wang [90] investigated the power allocation issue for the scenario shown in Figure 4.1(a), where multiple primary users co-exist in the three node AF-relay enhanced CR network. The optimization problem is formulated to maximize the overall *bits/Joule* of the OFDM based CR relaying system under the consideration of many practical limitations, such as transmission power budget of the CR source node (SN) and the relay node (RN), minimal capacity requirement, and interference threshold of the primary users, where the energy consumption includes a constant circuit energy. Similarly, Shaat and Bader [93] considered the resource allocation problem in the decode-and-forward (DF) relayed OFDM based CR system. The power allocation is optimized jointly with the subcarrier matching under individual power constraints in the source and relay. The sum rate is maximized while the interference temperature is kept below a pre-specified threshold. Similarly, Zhao *et al.* [94] investigated power and channel allocation for a three node DF relay enhanced CR network, where cooperative relay channels are divided into three categories: direct, dual-hop, and relay channels, providing three types of parallel end-to-end transmission.

Chen *et al.* [95] studied the distributed relay selection scheme to achieve the goal of maximizing the received reward contributed by system spectrum efficiency and energy consumption. The problem is formulated as a multi-armed restless bandit problem, which has been widely used for stochastic control. The secondary relay node in Figure 4.1(a) is equivalent to the arm, in which the state of an arm is characterized by the time varying channel condition of all related links, spectrum usage and residual energy states. Each arm chooses to be passive or active for relaying the data, and performs the corresponding adaptive modulation and coding (AMC) strategy if the arm is selected.

Similarly, Luo *et al.* [96] jointly considered the relay selection scheme and optimal power allocation scheme to achieve a good trade-off between the achievable data rate and network lifetime. Furthermore, Chen *et al.* [97] adopted the multi-armed restless bandit model to improve multimedia transmissions over underlay CR relay networks, while ensuring primary users had a minimum rate for a certain percentage of time. The cross-layer design approach is used to minimize multimedia distortion, increase spectral efficiency, and prolong network lifetime, by jointly considering optimal power allocation, relay selection, adaptive modulation and coding, and intra-refreshing rate.

3) Resource Allocation for Cooperative Transmission between primary users and secondary users: With the introduction of cooperative relay in Figure 4.1(b), a primary user engages appropriate secondary users to relay its transmission so as to improve primary transmission performance, for example, enhancing achievable throughput/reliability and/or saving energy. In return, primary users yield a portion of spectrum access opportunities to relaying secondary users for secondary transmissions.

The strategy of cooperation can be (a) TDMA based three-phase cooperation, that is, primary users broadcast in the first phase, secondary users relay in the second phase, and secondary users transmit in the third phase; (b) FDMA-based two-phase cooperation, in which the primary user divides its spectrum into two orthogonal sub-bands, and broadcasts on the first sub-band in the first phase. Secondary users relay on the same sub-band in the second phase, and continuously transmit in both phases on the second sub-band.

The work in [98] is concerned with enhancement of the spectrum-energy efficiency of a cooperative CR network shown in Figure 4.1(b). The relay selection and parameter optimization are formulated as a Stackelberg game, which is suitable for problems with a sequential structure of decision making [99]. Since primary users have higher priority over secondary users, primary users are leaders when cooperating with secondary users, that is, primary users decide whether to cooperate and how to cooperate in the first stage. In the second stage, secondary users aim to maximize their own utility by playing the non-cooperative power control game.

4.1.2.2 Green Cognitive Small Cells

Although relay and cooperative networks can improve the signal quality of cell edge indoor users to a certain degree, the rapid growth of mobile service usage accelerates the need for novel cellular architectures to meet such demands [101]. Instead of relying on careful deployment of conventional BSs, the LTE Advanced or beyond standards have proposed heterogeneous networks (HetNets) to resolve the capacity demand issue [102].

A HetNet consists of a macro cell network overlayed by small cells. The macro-tier guarantees the coverage, while the overlay network provides means to satisfy local capacity demand. The small cells in this two-tier architecture can be micro cells, pico cells, or femto cells, where the distinction between them can be found in the size of the cells and their capability for auto-configuration and auto-optimization. The reduced cell size can lead to higher spatial frequency reuse and lower power consumption. However, cross-tier interference will occur if small cells and the macro cell overlap in their assigned spectrum; intra-tier interference among multiple small cells will also impair user performance in both cell tiers. Efficient and agile spectrum access enabled by CR can avoid interference and overcome the coexistence issues in multi-tier networks [103].

As illustrated in Figure 4.2, both macro cells and small cells can be equipped with cognitive functionality. Depending on whether the priorities of secondary users are provisioned for, the system could be open access or closed access [102]. When cognitive cells are configured as closed access, only registered secondary users can communicate with their cognitive BSs and primary users can only access their BSs even in a shared sub-channel. When cognitive cells are configured as open access, they can be accessed by both its users and all co-channel primary users, and there is no notion of priority for the transmissions of MBSs [103]. Improving overall energy efficiency, which is severely

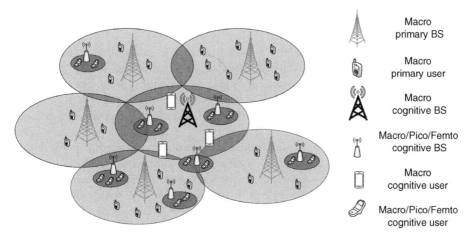

Macro
primary BS

Macro
primary user

Macro
cognitive BS

Macro/Pico/Femto
cognitive BS

Macro
cognitive user

Macro/Pico/Femto
cognitive user

Figure 4.2 Heterogeneous CR system. *Source:* Huang 2015 [100]. Reproduced with permission of IEEE.

affected by the installation of the additional small cell tier, and how to utilize the cognitive capabilities to save energy and mitigate interference are still research challenges.

1) Open access architecture: The major function of secondary small cells is to offload traffic from the macro cells. Since the traffic demand fluctuates significantly over space, time, and frequency, the small cells do not always have users to serve. By utilizing spectrum sensing results to determine whether there is traffic to offload, the small cell can decide whether or not to enter sleep mode. For HetNets with open access configuration, Wildemeersch *et al.* [104] evaluated a distributed sleep mode strategy to minimize the energy consumption of the small cells. Optimal sensing duration and sensing probability are derived under various detection performance constraints and traffic offloading requirements.

2) Closed access architecture: The transmission of the primary BS has higher priority, and the secondary network has its own users to serve. As compared with the traditional spatial/temporal spectrum sharing scenario, the differences in this case are the confined cell size and transmission power of the cognitive small cells. Similar to the spatial spectrum sharing scenario in [81], the secondary cells can grant access to the primary users. In return for the extra power consumption, the secondary cells are guaranteed a certain revenue as the incentive mechanism. Yang *et al.* [105] studied joint power allocation of the macro primary BS and secondary small cells to guarantee energy saving and the throughput of the primary network. In particular, the utility of the primary users is maximized and the pricing function is incorporated to reflect the virtual money to be paid to the small cells. Xie *et al.* [106] fully explored the cognitive capability by considering a wireless network architecture with macro and femto cognitive cells. The *bits/Hz per Joule* energy efficient resource allocation problem is modeled as a three-stage Stackelberg game, where the primary network leads the game by offering the selling price in the first stage. The cognitive MBS will decide whether to buy the spectrum size in the second stage, and allocate the procured spectrum among femto cells and macro secondary users. In Stage III, the femto cell performs power allocation for the femto cell secondary users. A gradient based iteration algorithm is proposed to obtain the Stackelberg equilibrium.

4.2 Energy Harvesting Techniques

While all the above techniques optimize and adapt energy usage to achieve energy efficient CR networks, energy related inhibition has not been precluded. Battery powered nodes can only operate for a finite duration as long as the battery lasts. For main grid powered nodes, the primary rationale still pertains to cost efficiency and reduced carbon footprints. As a result, CR has to be optimized to provision green communications by using green energy.

Energy harvesting has been applied to address the problem of tapping energy from readily available ambient sources that are free for users, including wind, solar, biomass, hydro, geothermal, tides, and even radio frequency signals [107]. Depending on the application scenarios, secondary users can be sensor nodes in a sensor network, or mobile devices and even BSs in a cellular network; all these devices can be equipped with energy harvesting capability [108, 17].

The energy arrival rate is a significant metric to evaluate energy harvesting capability. To guarantee a certain level of stability in energy provisioning, energy sources are normally implemented in a complementary manner. As a result, passively powered devices (energy harvesting generators that do not require any internal power source [107] and sometimes use different green energy sources) and hybrid powered devices (energy harvesting generators that have a backup non-renewable energy source in case the power provided by energy harvesting is insufficient [12, 109]) are attractive as inexhaustible replacements for traditional wireless electronic devices in CR networks.

4.2.1 Green Energy Harvesting Models

Depending on whether there is a storage capability for the power output of the harvesting system, the generic system model is classified as: (1) the harvest-use architecture, which mandates that the instantaneous energy harvesting rate should always be no less than the energy consumption rate [108]; or (2) the harvest-store-use architecture with a storage component (e.g., rechargeable batteries) to hoard the harvested energy for future use. As illustrated in Figure 4.3, the energy harvesting process and energy consuming process (i.e., sensing, transmission, reception, etc.) can be scheduled simultaneously, or in a time switching way [110]. The unified model of the energy harvesting process of these architectures is given in Eq. (4.7).

$$E_h = \alpha \eta E, \tag{4.7}$$

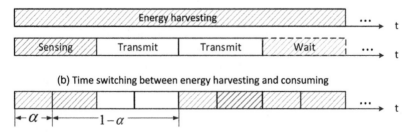

Figure 4.3 The harvest-store-use mechanism of a cognitive device. *Source:* Huang 2015 [100]. Reproduced with permission of IEEE.

where E_h is the harvested energy, and E represents the green energy source, which is normally modeled as a Markov process [111]. $0 < \alpha \leq 1$ is the time switching ratio that is used for energy harvesting ($\alpha = 1$ for simultaneous energy harvesting and consumption), and $0 < \eta < 1$ is the energy conversion efficiency.

For a green CR network, RF harvesting is an energy form of particular potential because green CR can transmit data on the idle spectrum while harvesting energy from the busy spectrum. Another advantage of RF energy is the simultaneous transfer of wireless information and power [110]. So, except for the separated energy harvester and information receiver similar to Figure 4.3(a), the co-located energy harvester and information receiver can adopt two practical architectures: power splitting and time switching. As depicted in Figure 4.4(a), when the RF signal reaches the receiver, part of it is used for power extraction and the rest for simultaneous data detection. Note that Figure 4.4(b), which requires only one set of antenna for both energy and data, is a special case of Figure 4.3(b). The unified model of the RF energy harvesting process of these architectures is given in Eq. (4.8):

$$E_h = \alpha\eta|h|^2E, \tag{4.8}$$

where h is the channel condition between the RF energy harvester and the RF energy source, $|h|^2E$ represents the energy of the received RF signals, and $0 < \alpha \leq 1$ is the time switching ratio or power splitting ratio that is used for energy harvesting.

For the architectures shown in Figure 4.3(a) and Figure 4.4(a), the existing literature assumes that energy harvested in the current time slot can only be used in subsequent time slots, owing to the energy half-duplex constraint [112]. So, before performing the cognitive functionality, the available residual energy is observable in all of the architectures. However, as compared with CR systems powered by traditional on-grid energy, EH CR is different in the sense of the dynamic nature of energy supply, that is, opportunistic energy harvesting makes the energy arrival rate no longer constant. Accordingly, the power budget is dynamic, represented by the energy causality constraint (EC-constraint), also referred to as the energy neutrality constraint. The EC-constraint requires that the total consumed energy should not exceed the total harvested energy, which may further be limited by finite battery capacity [113].

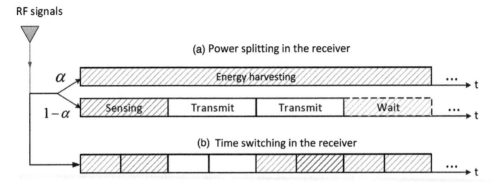

Figure 4.4 The RF energy harvesting mechanism. *Source:* Huang 2015 [100]. Reproduced with permission of IEEE.

4.2.2 Green Energy Utilization and Optimization

Conventional energy efficient techniques cannot be applied directly to the case where the transmitter or the receiver is subject to intermittent and random harvested energy. For passively powered devices, it is more reasonable to adopt a performance maximization approach: optimize the system function according to harnessed green energy. Although minimizing current energy consumption may bring future reward in performance, the energy minimization approach alone cannot guarantee system performance, which is the overall design objective. For hybrid powered devices, the three approaches can still be adopted because on-grid non-green power remains part of the energy source. However, energy minimization and utility maximization schemes should factor in the fact that green energy consumption is free for a given deployed EH system.

1) Transmission Policy with Energy Harvester: For the architecture with separated energy harvester and information transmitter, as illustrated in Figure 4.3(a), maximizing the utilization of green energy means tailoring the power management scheme for the random harvested source. Ho and Zhang [49] considered a point to point wireless data transmission system with an energy harvesting transmitter. They assumed that two types of side information (SI) about the harvested energy and time varying channel conditions are available: causal SI (of the past and present slots) or full SI (of the past, present and future slots). Optimal energy allocation is developed to maximize the throughput over a finite horizon. Similarly, optimal energy management schemes are provided to maximize the throughput by a deadline and minimize the transmission completion time of the communications session [114]. Also, a throughput optimal policy is proposed for the constrained setting where only online knowledge of the energy arrival process and the channel fade level is available [115].

   Energy management policies can be extended beyond this in various directions. For example, Huang *et al.* [116] explored joint source and relay power allocation over time to maximize the throughput of a three node DF relay system, in which both the source and relay nodes transmit with power drawn from energy harvesting sources. We [109] studied the power allocation scheme for the system with hybrid energy supplies. By utilizing green power thoughtfully, on-grid power consumption is minimized while the QoS of users in terms of SINR is guaranteed.

   Although energy cannot be consumed before it is harvested, traffic cannot be delivered before it has arrived. Yang and Ulukus [51] proposed a packet transmission policy to minimize the packet transmission completion time, in which the transmitter has to wait until enough green energy is harvested in order to adopt a higher transmission rate. For a multi-user system, the multi-user diversity in terms of channel conditions can be exploited to enhance green energy utilization. Therefore, scheduling different users at a given time slot may require different amounts of green energy. By exploiting multi-user diversity, a packet scheduling algorithm may shape a BS's energy demands to match the green power generation profile [52].

2) Reception Policy with Energy Harvester: To address opportunistic wireless energy harvesting at the receiver side, the goal is to design the optimal splitting and switching rule for the architectures shown in Figure 4.4.

   Since co-channel interference is useful for a wireless RF energy harvester, although it limits the quality of data reception, the time switching rule needs to incorporate the time varying channel condition. To minimize the outage probability or the ergodic capacity for wireless information transfer while maximizing the average harvested energy, joint optimization of transmit power control of the transmitter (which is plugged into a traditional on-grid power source) and scheduling of information and energy transfer in the receiver is considered in [117].

The design of the harvesting rule in the receiver is of particular importance to relay networks because the amount of harvested energy determines the forwarding ability of the relay, while the information received in the relay acts like the data source in the forwarding phase. According to numerical analysis [118], various system parameters, such as power splitting ratio, energy harvesting time, source transmission rate, noise power, source to relay distance, and energy harvesting efficiency, affect the performance of the three node AF relay channel, where the relay node harvests energy from the received RF signal and uses that harvested energy to forward the source information to the destination.

3) Challenges on Green Energy Utilization and Optimization: Currently, research on green energy utilization and optimization mainly relies on having knowledge of the instantaneous energy arrival rate, either assuming that it is already available (offline optimization), or assuming that causal information or statistical knowledge can be obtained (online optimization) [111].

For offline optimization, the closed form optimal transmission/reception can normally be obtained, such as variants of the water-filling algorithm [49, 115]. Since the assumption of having complete information of the energy arrival rate (past, present, and future) is normally unrealistic due to the intermittent nature of green energy sources, various learning algorithms have been studied to estimate the statistical parameters of the energy arrival processes [119]. When the energy harvesting profile is modeled as a Markov process, the online energy management scheme is cast under the framework of Markov decision processes (MDPs). However, such decision policies are complex, and thus a simpler alternative is required to analytically balance energy consumption and harvesting [111].

4.2.3 Cognitive Functionalities in Energy Harvesting

Envisioning green energy as an important energy resource, the design and optimization of the CR network is challenging because the network performance and the user QoS are highly dependent on the dynamics of the available spectrum and green energy. To alleviate the spectral and energy constraints on the network, the inherent relationships between system performance and the availability of power and spectrum are investigated, as illustrated in Figure 4.5.

1) Harvesting strategy: In the time switching architecture, a harvesting strategy is needed in the time domain, prior to identifying the sensing strategy and access strategy, to specify whether to harvest and how long to harvest; this is because it affects the EC-constraint, which will influence the subsequent sensing and possible transmission. Yin *et al.* [120] studied the harvesting duration and number of channels to be sensed next. The mixed integer non-linear programming (MINLP) problem is formulated to maximize the expected achievable throughput of a secondary user over all of the idle channels in one time slot.

Meanwhile, when considering the architecture of the receiver in Figure 4.4, another research challenge is how to split the power between RF harvesting and data

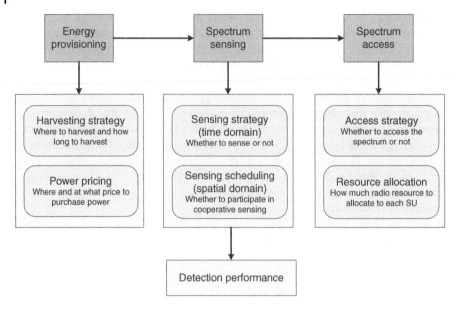

Figure 4.5 Cognitive functionalities in energy harvesting. *Source:* Huang 2015 [100]. Reproduced with permission of IEEE.

detection. This is of particular importance for relaying and cooperative CR networks because the detected data needs to be transmitted to the destination by utilizing the harvested energy.

2) Sensing strategy: The non-constant EC-constraint will affect the sensing strategy because the instantaneous low residual energy may not be able to fully support a complete sensing process, or not be able to complete the data transmission successfully, even through the channel will be indicated as idle if the secondary user indeed decides to sense. So, should the secondary user (1) remain idle until the accumulated energy reaches a threshold that is greater than the power needed for a single sensing and transmission; (2) perform sensing and possible transmission when the residual energy is sufficient for a single sensing and transmission; (3) perform sensing and gain spectrum occupancy information when the accumulated energy is above a pre-defined sensing power threshold; or (4) keep sensing as long as there is energy left? Although the work in [73] has already considered the sensing decision policy, the corresponding zero energy arrival rate of the un-rechargeable cognitive device makes the third and forth policies trivial because when the residual energy is less than the power required for a successful sensing and transmission, any sensing strategy afterwards will bring zero payoff. For the cognitive device equipped with an energy harvester, however, the fourth policy may be preferable because spectrum sensing not only gains spectrum occupancy information, but also leaves room in the battery for the newly arrived free energy.

3) Detection performance: The detection probability and the false alarm probability in Eq. (4.3) should factor in the sensing strategy, owing to the fact that p_d will increase if the cognitive device stays idle without sensing when the channel is busy. On the other hand, when the channel is actually idle, p_f will increase. Modeling the energy arrival process as an independent and identically distributed (i.i.d.) sequence of

random variables with fixed average rate, Park *et al.* [121] adopted the sensing decision policy where the secondary user with sufficient energy to carry out spectrum sensing and transmission will sense the channel. The optimal single spectrum detection threshold λ is derived to maximize the expected total throughput of the secondary user under energy causality and collision constraints.

4) Sensing scheduling: In the spatial domain, the diversity of residual energy among cooperative devices makes the scheduling problem more complex. For instance, to maximize the life time of sensors which are dedicated to performing cooperative spectrum sensing using harvested energy, an online algorithm is required for the scheduling of each sensor's active time, and the scheme proposed in [122] is insufficient because the energy arrival rate is not zero anymore.

When spectrum sensing is shared in a distributed non-cooperative way, the existing literature makes the implicit assumption that no explicit relationship exists between the sensing/reporting power and available transmission power, that is, the secondary user can always transmit with maximum power. This is reasonable since a steady positive energy arrival rate is guaranteed by the on-grid power source. So, the expected throughput will be a function of sensing policies (whether to contribute to sensing or not), sensing results, and channel condition. In the EH CR environment, however, each secondary user's utility function should take the available residual energy into account, which will further affect their sensing strategy. In the frequency domain, the secondary system or every secondary user still has to take the residual energy into consideration in scheduling the sensing process because the subsequent transmission will be affected by it.

5) Access strategy: With a varying energy arrival rate, the user may be forced to remain idle on the unoccupied spectrum if the remaining energy is insufficient to perform a successful data transmission, and so the access strategy has to incorporate the dynamic EC-constraint. Park *et al.* [123] investigated how energy harvesting capability affects the expected total throughput of a secondary user. The optimal sensing and access policy determining which channel to sense and possibly access is formulated as a partially observable Markov decision process, because the spectrum occupancy state is not fully observable to the secondary user. With a given sensing duration and fixed power required for data transmission in one time slot, a sub-optimal myopic policy, which maximizes the expected immediate throughput of the secondary user, is proposed. Furthermore, Park *et al.* [124] considered a scenario where idle secondary users can harvest RF energy from active primary users. The optimal sensing decision policy and access policy are also formulated as POMDP, and a sub-optimal myopic policy is proposed as well.

6) Resource Allocation: The harvesting strategy will have an impact on the sensing decision, which further affects the available power for subsequent data transmission. To achieve better performance of secondary networks, the dynamic resource allocation mechanism should be able to reveal these internal relationships. Adopting a similar sensing decision policy as [121], Chung *et al.* [125] studied the power control scheme which maximizes the average throughput during a slot, subject to the collision constraint imposed by the primary user. Gao *et al.* [126] jointly considered the sensing strategy and power allocation strategy to maximize the throughput of the secondary user over multiple consecutive time slots. A sub-optimal online algorithm is designed based on the dynamic spectrum state, harvested energy, and the channel fading level.

7) Power Pricing: Although current wireless access networks mainly rely on the power grid, which is a large interconnected infrastructure for delivering electricity from power plants to end users, the majority of research endeavors on green CR has been focused on locally generated green power, whether by the cognitive device itself or through trade with other green BSs. In fact, faced with the emerging challenges of rising energy demand, aging infrastructure, and increasing greenhouse gas emission [127], the traditional power grid is becoming smart and can integrate with renewable green energy resources such as wind and solar power [128]. The work in [129] represents cognitive heterogeneous mobile networks in the smart grid environment. A three-level Stackelberg game is developed to model the problems of electricity price decision, energy efficient power allocation, and interference management. Then, a homogeneous Bertrand game with asymmetric costs is used to model price decisions made by the electricity retailers [99]. A backward induction method is used to analyze the proposed Stackelberg game. Simulations are conducted to evaluate the reduction in the operational expenditure and CO_2 emissions in cognitive heterogeneous mobile networks.

4.3 FreeNet: Spectrum and Energy Harvesting Wireless Networks

By leveraging spectrum and energy harvesting technologies, FreeNet [130] typically comprises various spectrum harvesting BSs powered by renewable energy from the energy harvesters, as shown in Figure 4.6. These spectrum harvesting BSs are mainly powered by renewable energy while the power grid is used as the backup. On the one hand, by taking advantage of spectrum harvesting technologies, FreeNet liberates the wireless network from the spectrum constraint, in which wireless nodes sense and utilize the available spectrum for data communications. In this way, FreeNet enhances the spectrum agility and energy efficiency of the network because wireless nodes are able to sense and utilize the spectrum over which they experience less fading and

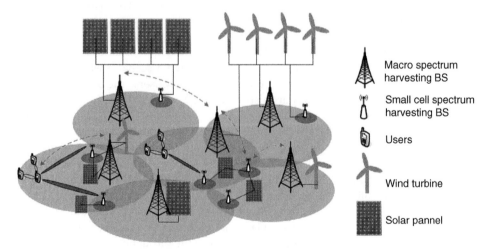

Macro spectrum harvesting BS

Small cell spectrum harvesting BS

Users

Wind turbine

Solar pannel

Figure 4.6 FreeNet: spectrum and energy harvesting wireless network. *Source:* Ansari 2016 [130]. Reproduced with permission of IEEE.

interference, improves network ubiquity because wireless nodes are able to access the network as long as they can detect the available spectrum, and easily incorporates new technologies as long as rules on sensing and utilizing the spectrum are followed. On the other hand, envisioning renewable energy as an important energy resource for future wireless access networks, FreeNet explores the usage of renewable energy that is sustainable. Hence, FreeNet reduces grid power consumption, and thus enhances network energy efficiency as well as reducing carbon footprints. Moreover, FreeNet, by making use of renewable energy, can be featured as a drop-and-play network that improves network ubiquity: FreeNet is able to provide data communications in areas without a power grid such as remote rural areas or areas affected by natural disasters.

To provision high capacity wireless networks, FreeNet exploits the heterogeneity of wireless access networks and integrates both the infrastructure based and mesh-based access networks to provide network access to users. Moreover, FreeNet adopts advanced communications technologies such as bandwidth aggregation and smart antennas to enhance the spectrum and energy efficiency of the network. Based on the availability of the spectrum, renewable energy, and traffic demands, FreeNet adapts its operation strategies in terms of network topology, operation bandwidth, transmission techniques, and communication protocols. For infrastructure based access networks, the network capacity is adapted by changing the transmission techniques and coverage areas of BSs. For the mesh-based access network, the network topology is adapted by optimizing routing algorithms.

FreeNet aims to build a self-organized and self-optimized wireless network in which the renewable energy powered network nodes can dynamically utilize the available spectrum for their data communications. The major concern of energy harvesting CR networks is to design spectrum sensing and access strategies for secondary network nodes that are powered by harvested energy [131]. In comparison, FreeNet, though it has to address the same issue, is more advanced: it enables a higher order of freedom in terms of spectrum and energy usage and realizes a self-organized and self-optimized wireless network.

4.3.1 FreeNet Application Scenarios

In FreeNet, wireless nodes are able to dynamically access the available spectrum to enhance the spectrum efficiency of the network. In addition, wireless nodes are able to utilize renewable energy as power supplies. In this way, FreeNet liberates the network from the constraints of the spectrum and energy. FreeNet will be deployed for three major purposes: (1) alleviating network congestion in urban areas, (2) provisioning broadband services in rural areas, and (3) enhancing emergency communications capability in unexpected critical situations caused by natural disasters, such as hurricanes and earthquakes, or by terrorist attacks, such as the 9/11 attack.

4.3.1.1 Alleviating Network Congestion in Urban Areas

With the rapid development of radio access techniques and mobile devices, an increasing number of Internet applications are carried over mobile and wireless networks. Urban areas are characterized by high population density, and thus demand high capacity wireless networks. Tremendous research efforts have been made to enhance the capacity and QoS of wireless access networks [14]. FreeNet, which dynamically exploits the underutilized spectrum to enhance spectrum efficiency of wireless networks, can potentially alleviate network congestion in urban areas. Moreover, without the strict constraint of power supplies, the deployed locations of wireless access points can be optimized based on traffic demands and spectrum availability instead of being restricted by the power outlets [132].

In urban areas, FreeNet adopts a heterogeneous networking architecture consisting of both MBSs and SCBSs. The BSs are able to harvest spectrum and utilize renewable energy. Based on the traffic demands and their geological distribution, traffic loads are balanced among the BSs according to the renewable energy status and the availability of spectrum. In addition, mesh-based networks and device-to-device (D2D) communications networks may be dynamically formed to further enhance the capacity and efficiency of the networks [133].

4.3.1.2 Provisioning Broadband Services in Rural Areas

Fiber or xDSL access may not be available in rural areas. To provide connectivity to communities in mountainous, undulating and remote terrains, WindFi BSs are designed to provide basic network connectivity in rural areas [134]. WindFi BSs operate entirely on renewable energy, and wireless access options for users are realized via point to point (P2P) hopping backhaul links and point to multi-point (P2MP) customer last mile access. However, the traffic demand from rural areas is also increasing. WindFi, with its static frequency planning, may not be able to efficiently provision high network capacity. Since FreeNet is a drop-and-play network, multiple infrastructure based energy harvesting BSs can be deployed to provide broadband services in remote rural areas. These BSs may dynamically utilize the available spectrum to provide users the network access as well as to work as the backhaul network. In addition, multiple mesh access networks can be formed in FreeNet to provide users with alternative network accesses. FreeNet, by optimizing the utilization of the available spectrum and renewable energy, will significantly increase the quality of the broadband services in remote rural areas.

4.3.1.3 Enhancing Emergency Communications Capability in Critical Situations

Emergency communications refers to providing users with network accesses in unexpected critical situations [135]. Usually, the destruction or extremely limited availability of the communications infrastructure of the region in distress exacerbates the challenges of rescue and recovery operations. Voice services, used to communicate among responders and with headquarters for command and control, or used by those affected by the emergency situation, may be severely restricted and unreliable even when available. Furthermore, it is often impossible to share and use all the relief resources through advanced information technologies—such as accessing remote databases, web sites, and web based applications—and exchange data with agency headquarters and other field command centers. Without the constraints of the radio frequency and energy supplies, FreeNet is a viable solution for provisioning emergency communications capability in these critical situations. Since the major goal of emergency communications is to disseminate surveillance information, FreeNet is able to provide users with basic communications connectivity and to extend the lifetime and coverage area of the access network.

4.3.2 Dynamic Network Architecture Optimization

FreeNet consists of both infrastructure based wireless access networks and mesh-based wireless access networks. To optimize the utilization of FreeNet, the network architecture should be adapted according to the spectrum opportunities, the availability of renewable energy, and the traffic load distribution.

In order to provision high network capacity, mobile and wireless access networks have adopted several advanced techniques such as bandwidth aggregation, the smart antenna system, and small cell networks. The bandwidth aggregation technique is able to extend bandwidth and thus enhance network capacity. However, as the bandwidth increases, the transmit power should also be increased to keep the transmit power per unit bandwidth constant. The smart antenna system enhances the capacity of wireless access networks by increasing the SINR. The MIMO system is one of the smart antenna systems that will be widely adopted in future wireless access networks. MIMO applies multiple antennas at both the transmitter and receiver to mitigate channel fading, and thus increases the access network capacity. Smart antenna systems consume extra power since multiple transmit antennas are employed. Each additional antenna consumes extra DC power, which is constant irrespective of the transmission rate in the circuit. This additional DC power consumption will affect the energy consumption of smart antenna systems. Small cell networks deploy low power BSs with close proximity to users. Because of the proximity, SCBSs experience much less fading than MBSs do. Therefore, given the same bandwidth and transmission power, an SCBS can provide a higher network capacity. However, owing to the low transmission power, the coverage area of an SCBS is limited. As a result, a large number of SCBSs have to be deployed in order to cover the service area. The densely deployed SCBSs may interfere with each other, and subsequently reduce the achievable network capacity of individual BSs.

FreeNet adapts its capacity by using the above methods according to the dynamics of the availability of the spectrum and renewable energy, and traffic demands. For infrastructure based access networks, BSs with higher renewable energy may adopt advanced transmission techniques and aggregate more bandwidth to serve more traffic demands. Since FreeNet also integrates mesh-based wireless access networks, the routing strategies are optimized to maximize the utilization of available spectrum and renewable energy while reducing grid power consumption. Badawy *et al.* [136] studied energy aware routing in renewable energy powered mesh networks. By exploiting spectrum harvesting techniques, the routing strategies in FreeNet should also be aware of the available spectrum.

The availability of spectrum and renewable energy is highly dynamic. Since the optimal network architecture for FreeNet depends on the availability of spectrum and renewable energy, the optimal network architecture changes dynamically. However, adapting network architecture requires wireless nodes to negotiate and exchange operation parameters such as cell sizes and routing information; this introduces additional overhead to the access networks. Thus, adapting the network architecture frequently may result in excessive overheads and reduce the network efficiency. Therefore, the trade-off between obtaining the optimal network architecture and the overhead incurred during the network architecture adaptations should be determined. To obtain optimal control of the network architecture adaptation, the inherent relationships between the availability of spectrum, the generation of renewable energy, and the data traffic demand should be investigated by using advanced probability theory, queuing theory, and machine learning techniques.

4.3.2.1 Optimal Network Resource Management

Since FreeNet adapts its network architecture according to the dynamics of the available spectrum and renewable energy, network resource allocation and management should be optimized under each network architecture to enhance the spectrum and energy efficiency of the network. FreeNet adopts spectrum and energy harvesting techniques. Thus, both spectrum and renewable energy are considered as network resources. It is desired to optimize the joint allocation of the spectrum and harvested energy to enhance the spectrum and energy efficiency of the network. However, it is challenging to harness spectrum dynamics and renewable energy dynamics simultaneously. Thus, we translate the joint spectrum and renewable energy allocation problem into two coupled sub-problems. The first is to optimize spectrum usage with the given renewable energy dynamics, while the other is to optimize the renewable energy usage with the given spectrum dynamics.

4.3.2.2 Spectrum Sensing and Sharing

FreeNet relies on spectrum harvesting techniques to dynamically utilize the spectrum. Thus, spectrum sensing and sharing is essential to FreeNet. Unlike traditional CR networks, FreeNet is powered by renewable energy. Therefore, the spectrum sensing and sharing scheme for FreeNet should be optimized according to the dynamics of renewable energy. The spectrum sensing problem consists of two sub-problems: the first is to form coalitions among wireless nodes to enhance the probability of sensing the available spectrum; the second is to decide which part of the spectrum each coalition should sense. The spectrum sensing problem can be formulated as:
   Given:

- The topology of a wireless access network
- The dynamics of renewable energy

   Obtain: The optimal spectrum sensing strategies.
   Objective: To maximize the total amount of the sensed spectrum with minimal grid power consumption.

   There are three major spectrum sharing strategies: spatial spectrum sharing, temporal spectrum sharing, and hybrid spectrum sharing [100]. These solutions do not consider the renewable energy status in optimizing the spectrum sharing strategies, which may lead to inefficient spectrum allocation. For example, a wireless access point with a very limited amount of green energy may be allocated a large amount of spectrum that cannot be fully utilized because of the energy shortage. Thus, in FreeNet, the spectrum sharing strategies should be aware of the renewable energy status. Assuming that the nodes always have data to transmit, the spectrum sharing problem can be formulated as:
   Given:

- The available spectrum
- Interference relationships among different wireless nodes
- The dynamics of renewable energy

Obtain: The spectrum allocation among the wireless nodes.
Objective: To maximize the network capacity.
Subject to: The data rate requirements of individual nodes being satisfied.

4.3.2.3   Renewable Energy Sharing
Renewable energy as an important network resource for FreeNet should be optimized to minimize the grid power consumption of the network. Two scenarios of renewable energy sharing in FreeNet should be investigated:

1) Multiple wireless nodes share the same renewable energy generator: In this scenario, renewable energy is shared among wireless nodes attached to the renewable energy generator. The objective of renewable energy sharing is to maximize the network capacity with minimal grid power consumption [17].
2) Individual wireless nodes exclusively use their own renewable energy generators: In this case, renewable energy cannot be shared among wireless access points directly. However, the power consumption among wireless access points can be balanced by adapting the coverage area of individual wireless access points [109], in which the wireless access points with higher amounts of renewable energy increase their coverage areas to absorb more traffic. In this way, the probability that the network consumes grid power is reduced, and the utilization of renewable energy is improved. In addition to cell size adaptation, wireless access points with higher amounts of renewable energy may activate advanced transmission techniques such as MIMO and transmission beamforming to enhance spectrum efficiency. Therefore, these wireless access points can satisfy their users' traffic demands with less spectrum. The saved spectrum can be allocated to wireless access points that are short of renewable energy. Given the users' traffic demands, increasing the bandwidth allocation toward a wireless access point reduces that access point's energy consumption, thus reducing the probability of the access point consuming grid power.

4.3.2.4   Renewable Energy Sharing
Traffic scheduling reshapes the traffic injected into the network, and hence affects users' data rates, spectrum usage, and the energy consumption of wireless access networks. Thus, context aware traffic scheduling is desirable in FreeNet to optimize the utilization of spectrum and renewable energy while satisfying the users' QoS requirements.

4.3.2.5   Grid Power Consumption vs. Delay
The power consumption of wireless access points is closely related to their transmission rates. A higher transmission rate usually consumes more energy [137]. Figure 4.7 shows an example of the trade-off between grid power consumption and traffic delay. Assume the energy consumption is proportional to the traffic transmission rate. Figure 4.7(a) shows the energy consumption of transmitting the incoming traffic at its original traffic arrival rate. Figure 4.7(b) shows the grid power consumption and the renewable energy consumption when the traffic is transmitted at its arrival rate. When the traffic rate is higher than a threshold, wireless access points require more energy than the available renewable energy, and thus draw energy from the power grid. If the arrival traffic is shaped as shown in Figure 4.7(c), the consumption of grid power is avoided at the cost of

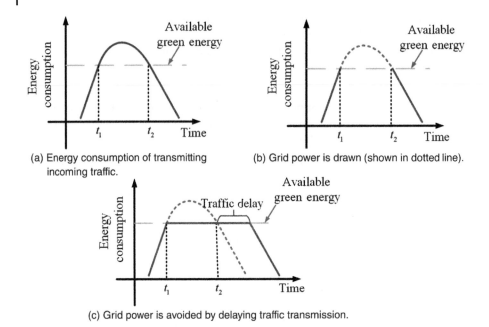

(a) Energy consumption of transmitting incoming traffic.

(b) Grid power is drawn (shown in dotted line).

(c) Grid power is avoided by delaying traffic transmission.

Figure 4.7 Grid power consumption vs. input traffic. *Source:* Ansari 2016 [130]. Reproduced with permission of IEEE.

additional packet delay. To guarantee QoS requirements of a user's session, each incoming packet can only be delayed for a limited duration. If the packet delay exceeds a threshold, wireless access points have to utilize grid power to transmit the packet. Therefore, a trade-off between grid power consumption and packet delay should be determined to minimize the grid power consumption while satisfying users' QoS requirements. In addition, reshaping the outgoing traffic may create spectrum holes: a wireless access point may not fully utilize its allocated spectrum. In this case, traffic scheduling may be designed to enhance spectrum efficiency. For example, when a wireless access point seeks additional bandwidth, other wireless nodes may delay their traffic to create spectrum holes for the wireless access point.

4.3.2.6 Grid Power Consumption vs. Delay of Diversified Applications

In the above discussion, we assume that all user sessions have the same delay requirement and that traffic of all sessions exhibits the same characteristics. In reality, wireless access networks carry diversified applications which impose different delay requirements and exhibit different traffic characteristics. For example, the legacy telephone service has strict delay requirements and traffic arrival is constant bit rate (CBR), while the web-surfing service has less stringent delay requirements and the traffic is bursty. For the first kind of traffic, delaying the traffic may not be able to achieve a low average rate, but may exacerbate the delay performance instead; delaying the traffic of the second class, however, may enable smoothing out the traffic and produce a lower average rate, thus avoiding consuming grid power. It is challenging to design a scheduling algorithm to minimize grid power consumption when considering these diversified requirements and traffic characteristics.

4.3.2.7   Grid Power Consumption vs. Loss of Diversified Applications

In the above analysis, network delay performance is traded for reducing grid power consumption. Loss performance can also be traded for energy saving. From the end users' perspective, their QoE may not be degraded when the packet loss ratio is below some upper bound. Taking the voice application, for example, the maximum allowable packet loss ratio can be as high as 3% (this is attributed to human self error-correction capability). Therefore, deliberately dropping some traffic may not degrade the user service but can help to avoid switching the network to grid power supplies. By including the packet loss ratio, the traffic scheduling problem is further complicated, and the challenge of solving it is intensified.

4.3.3   Communication Protocol Suite Design

We have presented optimal operation strategies for adapting the architecture of access networks, managing network resources, and scheduling traffic for FreeNet. To realize FreeNet, a communication protocol suite is required to coordinate wireless access points. Designing the communication protocol suite to enable FreeNet with minimum overhead is challenging. The protocol suite is designed in three different planes: the control plane, energy plane, and user plane.

4.3.3.1   Control Plane Protocol

The control plane protocol enables dynamic network architecture adaptation, spectrum sensing, and spectrum sharing. FreeNet adapts its network architecture according to the dynamics of the available spectrum, renewable energy, and traffic demands. Therefore, under different conditions, FreeNet may adopt different network operation strategies in terms of network topology, routing algorithms, and transmission techniques. The control plane protocol should be designed to enable network architecture adaptation with limited protocol overheads. In addition, FreeNet, by adopting spectrum harvesting techniques, dynamically senses and accesses the available spectrum to enhance the spectrum efficiency of the wireless access network. The control plane protocol should enable wireless access points to form coalitions to sense the spectrum cooperatively, and allow wireless access points to optimally share the available bandwidth in their coalitions.

4.3.3.2   Power Plane Protocol

The power plane protocol facilitates the sharing of renewable energy information and renewable energy. Since FreeNet is powered by renewable energy, this is therefore an important network resource which affects the operation of FreeNet. For this reason, renewable energy information such as energy generation rates, renewable energy consumption, and renewable energy storage should be shared among the wireless access points in order to optimize the network architecture and network resource allocation. Thus, a power plane protocol is required to share renewable energy information and control renewable energy sharing and distribution among wireless access points, in order to enhance the utilization of renewable energy.

4.3.3.3   User Plane Protocol

The user plane protocol is responsible for data traffic scheduling, packet routing, and user handover among various types of wireless access points. Traffic scheduling is essential to yield the optimal trade-off between grid power consumption, packet delay, and packet loss. In order to optimize traffic scheduling, wireless access points should be aware of the renewable energy and traffic characteristics. The user plane protocol is proposed to enable wireless access points to obtain such information for traffic scheduling. In addition, FreeNet consists of both infrastructure based wireless access networks and mesh-based wireless access networks. The user plane protocol should enable packet routing via heterogeneous networking, and allow handover between the two networks with minimal packet delay and packet loss.

4.4 Summary

This chapter has introduced FreeNet, which liberates the spectrum and energy constraints of the wireless networks by exploiting spectrum and energy harvesting technologies. We have discussed the deployment scenarios of FreeNet including alleviating network congestion in urban areas, provisioning broadband services in rural areas, and upgrading emergency communication capability in critical situations. Since the radio spectrum and renewable energy are highly dynamic, designing FreeNet is challenging. We have also briefly analyzed the related research issues on network architecture optimization, network resource management, and context aware traffic scheduling. These discussions shed light on designing and optimizing future spectrum and energy efficient wireless networks.

4.5 Questions

4.1 What are the advantages of spectrum and energy harvesting wireless networks?

4.2 How do energy efficient spectrum sensing and analysis technologies enhance the energy efficiency of mobile networks?

4.3 Why are spectrum management and handoff important in CR networks?

4.4 How does one enhance energy efficiency via spectrum harvesting technologies?

4.5 Please briefly describe energy harvesting technologies and discuss how they may be integrated into mobile networks.

Part II

Green Mobile Networking Solutions

5

Energy and Spectrum Efficient Mobile Traffic Offloading

With strong revenue growth in wireless data markets, ISPs such as Comcast and Optimum are densely deploying WiFi hot spots to provide WiFi connectivity to their customers in urban and suburban areas [138]. Therefore, it is desirable to utilize the hotspots deployed by ISPs to offload mobile data traffic. However, since carrying mobile traffic introduces additional operational costs to ISPs' networks, without proper incentives, the ISPs are not willing to open their networks to mobile network subscribers. In this chapter, we propose a novel EST scheme to enable MBSs to offload their mobile traffic to ISPs' wireless access points by leveraging CR techniques. The EST scheme exploits the merits of both mobile networks and ISP networks. One of the advantages of mobile networks is that the networks are operating on licensed spectra, which are not accessed by unlicensed users. Therefore, by proper spectrum management, mobile networks are able to provide their subscribers a variety of services with different QoS levels. However, compared with the hotspots deployed by ISPs, mobile network BSs are usually sparsely deployed. Such deployments are not efficient in terms of energy and spectral utilization. One of the merits of ISPs' hotspots is that they are densely deployed, and are able to provide high speed data rates to their subscribers. However, operating on an unlicensed spectrum, the QoS of data services cannot be guaranteed.

The EST scheme enables mobile networks to offload data traffic to ISPs' networks to improve energy and spectral efficiency, and allows ISPs access to the licensed spectrum to provide data services with different QoS levels. However, in the EST scheme, achieving optimal mobile traffic offloading in terms of minimizing the energy consumption of the MBSs is NP-hard. We thus propose a heuristic algorithm to approximate the optimal solution with low computational complexity. We also design an auction-based decentralized energy spectrum trading scheme that exploits the cooperation between primary base stations (PBSs) and SBSs to enhance both the energy and spectrum efficiency of cellular networks. In the cooperation, by leveraging CR, PBSs share the licensed spectrum with SBSs, and the SBSs provide data services to the primary users under its coverage utilizing shared bandwidth. The cooperation between PSBs and SBSs can significantly improve the energy and spectral efficiency of cellular networks.

The proposed scheme is illustrated in Figure 5.1, where the primary BS (PBS) is defined as the MBS owned by the mobile network operator while the secondary BSs (SBSs) are the hotspots owned by ISPs. We assume both the PBS and SBSs are able to dynamically access the spectrum by leveraging CR techniques. There are two types of users: primary users (PUs) and secondary users (SUs). PUs are subscribers of the mobile networks while SUs are subscribers of ISPs. Different SUs may subscribe to different

Green Mobile Networks: A Networking Perspective, First Edition. Nirwan Ansari and Tao Han.
© 2017 John Wiley & Sons Ltd. Published 2017 by John Wiley & Sons Ltd.

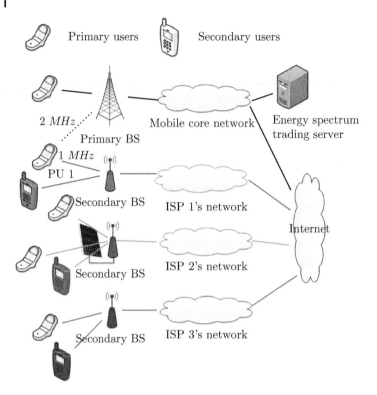

Figure 5.1 Illustration of energy spectrum trading scheme. *Source:* Han 2014 [139]. Reproduced with permission of IEEE.

ISPs. The energy spectrum trading server manages the spectrum sharing and mobile data offloading between the mobile networks and ISPs' networks.

The PBS has exclusive access to the licensed band. However, owing to wireless channel fading between the PBS and PUs, providing high data rates to the PUs, especially to those located at the cell edge, is both bandwidth and power intensive. As compared with the PBS, the SBSs which are closer to the PUs may experience less wireless channel fading and have higher spectral and energy efficiency in providing data services to the PUs. In the EST scheme, the PBS shares a certain amount of licensed bandwidth with, while SBSs provide data services to PUs within their coverage area using the allocated bandwidth. Since SBSs are close to PUs, the SBSs can satisfy PUs' QoS requirements by utilizing only a portion of the allocated bandwidth. The residual bandwidth can be utilized to fulfill SUs' data rate requirements. For example, in Figure 5.1, if PU 1 is associated with the PBS, the PBS should allocate 2 *MHz* bandwidth to the PU to satisfy its minimum data rate requirement. If associated with the SBS, PU 1 may only require 1 *MHz* to ensure its minimum data rate. If the PBS offloads PU 1 to the SBS and grants SBS 2 *MHz* bandwidth, then the SBS spends 1 *MHz* bandwidth to serve PU 1, and the other 1*MHz* bandwidth can be utilized to enhance the QoS of its SUs. Therefore, the EST scheme enables the PBS to reduce its power consumption by offloading some of the PUs to SBSs, and allows the SBSs to enhance their QoSs to SUs by utilizing the licensed bandwidth. Since SBSs usually have a low transmit power, the power consumption and the spectrum usage of the mobile network in providing data services to PUs is reduced.

Thus, the EST scheme enhances both the energy efficiency and the spectral efficiency of the mobile network.

The EST between the PBS and SBSs can be either event driven or traffic driven. For event driven EST, the PBS triggers an EST process when a cell edge user initiates data service requests. For traffic driven EST, the PBS monitors its traffic intensity from cell edge users. When the traffic intensity is beyond a threshold, an EST process is triggered. In this chapter, assuming the PBS experiences heavy traffic load from the cell edge users, we design algorithms to optimize the EST between the PBS and SBSs to minimize the PBS's energy consumption when an EST process is triggered. However, minimizing the energy consumption of the PBS in the EST scheme is not trivial. On the one hand, in order to minimize the power consumption, the PBS has to maximize the number of users offloading to SBSs. Meanwhile, since the total amount of licensed spectrum is limited, the PBS aims to minimize the amount of bandwidth allocated to SBSs because the less bandwidth is allocated to SBSs, the more bandwidth is reserved for the PUs associated with the PBS, and therefore the less power is consumed by the PBS. On the other hand, the PBS has to give the SBS sufficient incentive, in terms of the amount of licensed spectrum made available to them, to incentivize SBSs to provide data services to the PUs. Therefore, solving the power consumption minimization (PCM) problem requires finding user–BS associations and bandwidth allocations that minimize the power consumption of the PBSs while satisfying PUs' minimum data rates and SBSs' bandwidth requirements. In fact, the PCM problem is an NP-hard problem. To solve the PCM problem, we propose a heuristic algorithm to approximate the optimal solution achieved by the brute-force search as well as an auction-based decentralized energy spectrum trading scheme.

The heuristic algorithm is a centralized algorithm which first finds the PUs whose user–BS associations are not determined, and then iteratively associates the PU whose power-bandwidth ratio is the largest with SBSs. If the power consumption of the PBS is reduced, the PU is associated with SBSs; otherwise, the PU is associated with the PBS. The heuristic algorithm achieves at least 50% power consumption savings compared to brute-force search when the PBS experiences heavy traffic load from cell edge users.

In the auction-based decentralized solution, at the beginning of each time slot, the SBSs bid for the spectrum usage. Then, the PBSs decide the winning bids according to the proposed selection algorithm. After each round of auction, the SBSs adjust their bids by adapting the data requirements according to the auction result of the previous time slot, in order to enhance their probabilities of winning the next auction. Meanwhile, the PBSs also adjust the amount of bandwidth that is provided for auction in order to maximize their energy savings; they do this by adapting the power consumption weight of individual PUs. In the auction-based EST scheme, truth telling is proven to be a dominant strategy for SBSs to enable the auction game to converge to the Nash equilibrium.

5.1 Centralized Energy Spectrum Trading Algorithm

Consider an area consisting of one PBS and several SBSs from various ISPs as shown in Figure 5.1. The PUs are randomly distributed in the area. Denote the set of PUs and SBSs as $\mathcal{U}$ and S, respectively. The PBS provides a data service to the PUs within its coverage area via the licensed spectrum. SBSs are randomly deployed in the area. We assume that SBSs are able to dynamically access the licensed spectrum by utilizing CR techniques.

5.1.1 System Model and Problem Formulation

5.1.1.1 Communications Model

In the EST scheme, the PBS aims to offload data traffic to SBSs to reduce its energy consumption, and is willing to grant a portion of the licensed spectrum to incentivize SBSs to allow PUs to access their networks. Meanwhile, SBSs aim to dynamically utilize the licensed spectrum to enhance the QoS of data services for their subscribers. Thus, SBSs are willing to allow PUs to access their networks in exchange for access of the licensed spectrum.

We assume the total amount of licensed spectrum is W, which can be split into orthogonal channels via OFDMA protocol, with variable amounts of bandwidth to avoid interference. Each channel is allocated to an individual PU as needed. For simplicity, we assume both PUs and SUs experience frequency flat fading. Therefore, we focus on the amount of bandwidth allocated to PUs and SBSs instead of specifying which part of the spectrum is to be allocated. Users' locations are assumed to be static during an EST procedure. We assume the channel fading changes slowly and can be considered as a constant within the duration. Therefore, the wireless channel is modeled as a slow fading channel, which reflects the large scale fading between BSs and users.

At the beginning of an EST procedure, the kth SBS calculates its bandwidth requirements, denoted as $\phi_{k,i}$, for serving the ith PU. The calculation of $\phi_{k,i}$ consists of two steps. First, the kth SBS calculates the required bandwidth, $\phi_{k,i}^P$, to satisfy the ith PU's minimum data rate, $r_i^{\min}$. Assuming the kth SBS's transmit power-spectral density is p^s and the channel fading between the kth SBS and the ith PU is $h_{k,i}^s$, $\phi_{k,i}^P$ can be derived by solving

$$r_i^{\min} = \phi_{k,i}^P \log\left(1 + \frac{p^s |h_{k,i}^s|^2}{\mathcal{N}_0}\right). \tag{5.1}$$

Second, the kth SBS calculates the required bandwidth, $\phi_{k,i}^S$, to compensate for its cost in serving the ith PU. The kth SBS's cost includes the SBS's energy consumption and backhaul usages for serving the ith PU. The cost may be different for different ISPs. For example, in Figure 5.1, the second ISP utilizes a green energy powered access point, which may reduce the energy cost. Thus, as compared with other ISPs, the second ISP may incur a smaller cost in serving one PU. However, the calculation of $\phi_{k,i}^S$ is beyond the scope of the chatper. We assume $\phi_{k,i}^S$ is a constant. Then,

$$\phi_{k,i} = \phi_{k,i}^P + \phi_{k,i}^S. \tag{5.2}$$

The energy spectrum trading server collects $\phi_{k,i}, \forall k \in S, \forall i \in \mathcal{U}$, and optimizes the user–BS associations and bandwidth allocations to minimize the energy consumption of the PBS.

5.1.1.2 Energy Consumption Model

The PBS's power consumption consists of two parts: static power consumption and dynamic power consumption [140]. The static power consumption is the power consumption of a BS without any traffic load. Dynamic power consumption refers to the additional power consumption caused by traffic load on the BS. We consider the PBS's static power consumption, p^{fix}, as a constant, and focus on reducing the dynamic power consumption of a PBS by offloading its traffic to SBSs. The dynamic power consumption of a MBS depends on the traffic load on the BS and can be expressed as a linear function of the BS's transmit power [3]. Therefore, we model the PBS's power consumption as

$$C = \sum_{i \in \mathcal{U}} \alpha \mu_i p_i w_i + p^{\text{fix}}. \tag{5.3}$$

Here, α is a coefficient which reflects the relationship between the PBS's dynamic energy consumption and the summation of the PBS's transmit power toward its associated PUs. The value of α depends on the characteristic of the BS [3]. μ_i is an indicator function. If PU is associated with the PBS, $\mu_i = 1$; otherwise, $\mu_i = 0$. w_i is the amount of bandwidth allocated to the ith PU, and p_i is the transmit power-spectral density in w_i.

5.1.1.3 Problem Formulation and Analysis

In the EST scheme, the PBS aims to minimize its power consumption by offloading data traffic to SBSs. Therefore, the PCM problem can be formulated as follows:

$$\min_{(\mu_i, \beta_{k,i}, w_i, p_i)} \sum_{i \in \mathcal{U}} \alpha \mu_i p_i w_i + p^{\text{fix}} \tag{5.4}$$

$$\text{subject to: } \sum_{i \in \mathcal{U}} \left(\mu_i w_i + \sum_{k \in S} \beta_{k,i} \phi_{k,i} \right) = W,$$

$$r_i \geq r_i^{\min}, \ \forall i \in \mathcal{U},$$

$$\mu_i p_i \leq p^{\max}, \ \forall i \in \mathcal{U}$$

$$\mu_i + \sum_{k \in S} \beta_{k,i} = 1, \ \forall i \in \mathcal{U}. \tag{5.5}$$

Here, $\beta_{k,i}$ is an indicator function. If the ith PU is associated with the kth SBS, $\beta_{k,i} = 1$; otherwise, $\beta_{k,i} = 0$. $p^{\max}$ is the PBS's maximum transmit power-spectral density. If a PU is offloaded to an SBS, the SBS should satisfy the PU's minimum data rates. Thus, $r_i = r_i^{\min}$ when $\mu_i = 0$. Therefore,

$$r_i = \begin{cases} w_i \log \left(1 + \dfrac{p_i |h_i^p|^2}{\mathcal{N}_0} \right), & \mu_i = 1; \\ r_i^{\min}, & \mu_i = 0. \end{cases} \tag{5.6}$$

Here, h_i^p is the channel fading between the PBS and the ith PU. The PCM problem consists of four constraints. The first constraint is that the sum of the allocated bandwidth should not be larger than the total amount of bandwidth. The second constraint is that the PU's minimum data rate should be satisfied. The third constraint is that the PBS's transmit power should not be larger than its maximum transmit power. The fourth constraint is that a PU can only access either the PBS or one of the SBSs.

When $\mu_i = 1$,

$$p_i = \frac{\mathcal{N}_0(2^{\frac{r_i}{w_i}} - 1)}{|h_i^P|^2}. \tag{5.7}$$

Given the amount of bandwidth, w_i, the derivative of p_i with respect to r_i can be expressed as

$$\frac{\partial p_i}{\partial r_i} = \frac{\mathcal{N}_0 2^{r_i/w_i} \ln 2}{w_i |h_i^P|^2} > 0. \tag{5.8}$$

Since $\frac{\partial p_i}{\partial r_i} > 0$, given the amount of bandwidth, the power consumption increases as the data rate increases. Therefore, to minimize the PBS's energy consumption, PUs are served at the minimum data rate. Thus, $r_i = r_i^{\min}$ in the PCM problem. If the ith PU is associated with the PBS, the minimum required bandwidth, $w_i^{\min}$, is derived by solving

$$r_i^{\min} = w_i^{\min} \log\left(1 + \frac{p^{\max}|h_i^P|^2}{\mathcal{N}_0}\right). \tag{5.9}$$

Therefore, we replace the second and third constraints with the minimum bandwidth constraint. In addition, since a PU can be associated with at most one SBS, we select the SBS with the smallest $\phi_{k,i}$, $\forall i \in \mathcal{U}$. Here, $k = \arg\min_{j \in \mathcal{S}} \phi_{j,i}$. Thus, the PCM problem can be rewritten as

$$\min_{(\mu_i, \beta_{k,i}, w_i)} \sum_{i \in \mathcal{U}} \alpha \mu_i p_i w_i + p^{\text{fix}} \tag{5.10}$$

$$\text{subject to: } \sum_{i \in \mathcal{U}} (\mu_i w_i + \beta_{k,i} \phi_{k,i}) = W,$$

$$\mu_i w_i \geq \mu_i w_i^{\min}, \ \forall i \in \mathcal{U}$$

$$\mu_i + \beta_{k,i} = 1, \ \forall i \in \mathcal{U}. \tag{5.11}$$

Lemma 5.1.1 *The optimal solution to the problem formulated in Eq. (5.10) is the optimal solution to the PCM problem in Eq. (5.4).*

Proof: The problem formulated in Eq. (5.10) has the same objective function as that in Eq. (5.4). For the constraints, the data rate constraint and the transmit power constraint in Eq. (5.4) can be translated into the minimum bandwidth constraint in Eq. (5.10). Thus, proving Lemma 5.1.1 is equivalent to proving that $k = \arg\min_{j \in \mathcal{S}} \phi_{j,i}$ is a necessary condition of the optimal solution to the problem formulated in Eq. (5.4). This can be proved by contradiction. Assume the optimal solution to the problem formulated in Eq. (5.4) offloads the ith PU to the k^*th SBS and $k^* \neq \arg\min_{j \in \mathcal{S}} \phi_{j,i}$. Let $k = \arg\min_{j \in \mathcal{S}} \phi_{j,i}$. In this case, since the kth SBS requires less bandwidth from the PBS, offloading the ith PU to the kth SBS increases the available bandwidth in the PBS. According to the Shannon–Hartley theorem [141], to achieve a given data rate, increasing the bandwidth reduces the requirement for transmitting power. Thus, offloading the ith PU to the kth SBS can further reduce the power consumption of the PBS. Therefore, offloading the ith PU to the k^*th SBS is not the optimal solution to the problem formulated in Eq. (5.4). Thus, the optimal solution to the problem formulated in Eq. (5.10) is equal to the optimal solution to the PCM problem in Eq. (5.4). □

Theorem 5.1.1 *The PCM problem is an NP-hard problem.*

Proof: We prove the theorem by transforming a simplified PCM (SPCM) problem into a knapsack problem, which is an NP-hard problem [142]. We simplify the PCM problem by setting $p_i = p^{\max}$, $\forall i \in \mathcal{U}$. Then, the SPCM problem can be expressed as

$$\min_{(\mu_i, \beta_{k,i})} \sum_{i \in \mathcal{U}} \alpha \mu_i p^{\max} w_i^{\min} + p^{\text{fix}} \tag{5.12}$$

$$\text{subject to: } \sum_{i \in \mathcal{U}} \left(\mu_i w_i^{\min} + \beta_{k,i} \phi_{k,i} \right) = W,$$
$$\mu_i + \beta_{k,i} = 1, \ \forall i \in \mathcal{U}. \tag{5.13}$$

Define $\Delta W = W - \sum_{i \in \mathcal{U}} w_i^{\min}$ as the maximum amount of bandwidth that can be utilized by the PBS as the incentives to SBSs for traffic offloading. Define $\Delta \phi_{k,i} = \max \{\phi_{k,i} - w_i^{\min}, 0\}$ as the required incentives for the kth SBS to offload the ith PU. The PBS's power savings by offloading the ith PU equals $p^{\max} w_i^{\min}$. The SPCM can be transformed to

$$\max_{\beta_{k,i}} \sum_{i \in \mathcal{U}} \beta_{k,i} p^{\max} w_i^{\min} \tag{5.14}$$

$$\text{subject to: } \sum_{i \in \mathcal{U}} \beta_{k,i} \Delta \phi_{k,i} \leq \Delta W. \tag{5.15}$$

The above formulation is actually a knapsack problem. Therefore, the SPCM problem can be transformed into a knapsack problem which is an NP-hard problem. Thus, the PCM problem is an NP-hard problem. □

5.1.2 A Heuristic Power Consumption Minimization Algorithm

In this section, we propose a heuristic power consumption minimization (HPCM) algorithm to approximate the optimal solution of the PCM problem with low computational complexity, and prove that the maximum power savings achieved by the HPCM algorithm are at least 50% of that achieved by the brute-force search.

5.1.2.1 The HPCM Algorithm

For the PCM problem, if user–BS associations are determined, then μ_i and $\beta_{k,i}$ are known. The amount of available bandwidth in the PBS can be derived as

$$W^P = W - \sum_{i \in \mathcal{U}} \sum_{k \in S} \beta_{k,i} \phi_{k,i}. \tag{5.16}$$

Define $\mathcal{U}^P = \{i | \mu_i = 1, \forall i \in \mathcal{U}\}$ as the set of PUs associated with the PBS. Then, the PCM problem becomes a bandwidth allocation (BA) problem as follows:

$$\min_{w_i} \sum_{i \in \mathcal{U}^P} \alpha p_i w_i + p^{\text{fix}} \tag{5.17}$$

$$\text{subject to:} \sum_{i \in \mathcal{U}^P} w_i = W^P$$
$$w_i \geq w_i^{\min}, \forall i \in \mathcal{U}^P. \tag{5.18}$$

Let $f(w) = \sum_{i \in \mathcal{U}^P} \alpha p_i w_i + p^{\text{fix}}$ and $w = (w_1, w_2, \cdots, w_{|\mathcal{U}^P|})$. When $w > 0$,

$$\frac{d^2 f(w)}{dw_i^2} = \frac{\alpha \mathcal{N}_0 \mu_i (r_i^{\min})^2 (\ln 2)^2 2^{r_i^{\min}/w_i}}{|h_i^P|^2 w_i^3} > 0. \tag{5.19}$$

Thus, $f(w)$ is a convex function of w. Therefore, the objective function of the BA problem is convex. The constraints of the BA problem satisfy Slater's conditions, and therefore the Karush-Kuhn-Tucher (KKT) conditions provide necessary and sufficient conditions for the optimality of the BA problem [143]. Hence, we can derive optimal bandwidth allocations by solving the KKT conditions of the BA problem.

The PCM problem, thus, can be solved in two steps. In the first step, the user–BS associations are determined. Then, the PCM problem is reduced to the BA problem. In the second step, the BA algorithm is solved by solving its KKT conditions. Since the BA problem can be easily solved, the major difficulty of solving the PCM problem is to optimize the user–BS associations.

When $\sum_{i \in \mathcal{U}} \phi_{k,i} \leq W$ and $k = \arg \min_{j \in S} \phi_{j,i}$, all the PUs are offloaded to the SBSs, where the ith PU is offloaded to the kth SBS. In this case, the PBS does not provide a data service to any PU, and its dynamic power consumption is zero.

When $\sum_{i \in \mathcal{U}} \phi_{k,i} > W$ and $k = \arg \min_{j \in S} \phi_{j,i}$, not all PUs can be offloaded to SBSs, and the user–BS associations are to be optimized to minimize the PBS's power consumption. In this case, the PUs can be classified into three categories based on their minimum data rates, their channel conditions, and the amount of compensating bandwidth required by the SBSs.

The first category of PUs pertains to the PUs which can only be associated with the PBS. For example, if the ith PU is out of the coverage area of all SBSs or $\phi_{k,i} > W$, $k = \arg \min_{j \in S} \phi_{j,i}$, then the ith PU can only be associated with the PBS. The second category of PUs involves the PUs which have to be associated with SBSs in order to achieve the optimal solution. For example, if $\phi_{k,i} < w_i^{\min}$ and $k = \arg \min_{j \in S} \phi_{j,i}$, the ith PU is associated with SBSs because by such association, the ith PU consumes less bandwidth and zero dynamic power from the PBS. The third category of PUs refers to the PUs whose user–BS associations are to be determined. These PUs, if associated with the SBSs, consume zero dynamic power from the PBS. However, the PBS has to allocate more bandwidth to SBSs in order to incentivize them to provide data services to these PUs. This results in a reduction of the amount of bandwidth that can be allocated to the PUs which are associated with the PBS, and thus the overall power consumption of the PBS may increase. Therefore, determining user–BS associations for the third category of PUs is the essential task of the HPCM algorithm.

Thus, we first present the user filtering algorithm to classify PUs into three user sets, $\mathcal{U}^P$, $\mathcal{U}^S$, and $\mathcal{U}^T$, which represent the first, the second, and the third category of PUs, respectively. The pseudo code is shown in Algorithm 1.

Algorithm 1: The User Filtering Algorithm

---

1 **for** $i=1$ to $|\mathcal{U}|$ **do**
2   **if** $\phi_{k,i} \le w_i^{\min}$, $k = \arg\min_{j \in S} \phi_{j,i}$ **then**
3   Assign the ith PU in the user set $\mathcal{U}^S$;
4   **else if** $\phi_{k,i} > W$, $k = \arg\min_{j \in S} \phi_{j,i}$ **then**
5   Assign the ith PU in the user set $\mathcal{U}^P$;
6   **else**
7   Assign the ith PU in the user set $\mathcal{U}^T$;
8 Return $\mathcal{U}^P$, $\mathcal{U}^S$, and $\mathcal{U}^T$.

The user–BS associations of the PUs belonging to $\mathcal{U}^P$ and $\mathcal{U}^S$ are associated with the PBS and SBSs, respectively. The HPCM algorithm is to determine the user–BS associations for the PUs in $\mathcal{U}^T$. When $\sum_{i \in \mathcal{U}} \phi_{k,i} > W$ and $k = \arg\min_{j \in S} \phi_{j,i}$, the PBS is unable to offload all the PUs to SBSs. Thus, the HPCM algorithm is to iteratively offload the PU that consumes the largest amount of dynamic power from the PBS to SBSs.

Assuming the PUs belonging to $\mathcal{U}^T$ are associated with the PBS, the amount of bandwidth, w_i^t, allocated to the ith PU, $i \in \mathcal{U}^T$, can be derived by solving the BA problem. If $w_i^t \ge \phi_{k,i}$ and $k = \arg\min_{j \in S} \phi_{j,i}$, this indicates that associating the ith PU with SBSs does not require more bandwidth than associating the PU with the PBS. Meanwhile, associating the ith PU with SBSs reduces the dynamic power consumption of the PBS. Therefore, the ith PU is associated with SBSs. If $w_i^t < \phi_{k,i}$ and $k = \arg\min_{j \in S} \phi_{j,i}$, by associating the ith PU with SBSs, the PBS reduces its dynamic power consumption. However, the PBS has to allocate an additional amount of bandwidth to SBSs to incentivize them to provide data services to the ith PU. This reduces the amount of available bandwidth for the PUs which are associated with the PBS, and may result in an increase in the PBS's dynamic power consumption in serving these PUs.

When $w_i^t < \phi_{k,i}$ and $k = \arg\min_{j \in S} \phi_{j,i}$, the PBS's dynamic power savings on offloading the ith PU to the kth SBS depend on two factors. The first is the PBS's dynamic power consumption in serving the ith PU. The second is the difference between w_i^t and $\phi_{k,i}$, $k = \arg\min_{j \in S} \phi_{j,i}$. We denote the difference as $\Delta w_i^t = \phi_{k,i} - w_i^t$. A smaller Δw_i^t indicates that offloading traffic of the ith PU reduces the PBS's total bandwidth by a smaller amount, and thus results in a smaller reduction in the PBS's dynamic power consumption in serving the rest of the PUs. The ratio between the two factors is utilized by the HPCM algorithm to reflect the amount of potential power savings that can be achieved by the PBS in offloading the traffic of a PU to an SBS. The larger the ratio, the more power savings may be realized by the PBS. We refer to this ratio as the power-bandwidth ratio (PBR). Define p_i^t and w_i^t as the PBS's transmit power-spectrum density and the corresponding bandwidth allocation toward the ith PU, respectively. The PBR of the ith PU can be expressed as

$$\rho_i = \frac{\alpha p_i^t w_i^t}{\triangle w_i^t}. \tag{5.20}$$

The idea of the HPCM algorithm is to iteratively find a PU with the largest PBR, and offload its traffic to SBSs if power savings can be achieved by the PBS. The HPCM algorithm terminates when the user set $\mathcal{U}^T$ is empty.

At the beginning of each iteration, the HPCM algorithm assumes all the PUs in $\mathcal{U}^T$ are associated with the PBS, and calculates the PBS's total power consumption, C, its transmit power-spectrum density toward the ith PU, p_i^t, and the corresponding bandwidth allocation, w_i^t, $\forall i \in \mathcal{U}^T$. The HPCM algorithm finds the largest ρ_i, $i \in \mathcal{U}^T$. Assuming $m = \arg \max \rho_i, i \in \mathcal{U}^T$, the HPCM algorithm associates the mth PU with the kth SBS. Here, $k = \arg \min_{j \in S} \phi_{j,m}$. Then, the HPCM algorithm calculates the total power consumption of the PBS, which is denoted as C^m. If $C^m < C$, the HPCM algorithm offloads the traffic of the mth PU to the kth SBS and assigns $C = C^m$; otherwise, the mth PU is associated with the PBS.

To ensure its performance, before the iteration begins, the HPCM algorithm associates the mth PU, $m = \arg \max_{i \in \mathcal{U}^T} \alpha p_i^t w_i^t$, with the kth SBS, $k = \arg \min_{j \in S} \phi_{j,m}$. Then, the HPCM algorithm calculates the total power consumption of the PBS, $C^{\max}$. At the end, the algorithm compares $C^{\max}$ with C, and returns the user–BS associations that achieve the minimum power consumption of the PBS.

The pseudo code of the HPCM algorithm as described above is shown in Algorithm 2.

5.2 Auction-Based Decentralized Algorithm

In designing the auction-based decentralized algorithm, we consider a graphic area consisting of one PBS and several SBSs. The PBS provides data service to the PUs within its coverage area via the licensed spectrum. SBSs are randomly located in the area and aim to opportunistically utilize the licensed spectrum to transmit data to SUs. The PUs within SBSs' coverage areas can be associated with either the PBS or SBSs. In cellular networks, the total spectrum is usually split into multiple channels which are allocated to users to fulfill their QoS requirements. In the following, we assume the licensed spectrum band can be split into channels with arbitrary amounts of bandwidth. Each channel is allocated to an individual PU as needed. The amount of bandwidth allocated in each channel is optimized to minimize the PBS's power consumption. For simplicity, we assume both PUs and SUs experience frequency flat fading. Therefore, we focus on the amount of bandwidth allocated to PUs and SBSs instead of specifying which part of the spectrum is to be allocated. The time horizon is assumed to be divided into time frames of duration T. Users' locations are assumed to be static during one time frame. We assume the channel fading changes slowly and can be considered as a constant within a time frame. Therefore, the wireless channel is modeled as a slow fading channel which reflects the large scale fading between BSs and users. Define $r_i^{\min}$ and w_i as PU i's data rate requirement and bandwidth requirement in associating with the PBS.

Algorithm 2: The HPCM Algorithm

---

1 Assign all PUs in $\mathcal{U}$;
2 Calculate $\phi_{k,i}$ and $w_i^{\min}$, $k = \arg\min_{j \in S} \phi_{j,m}$, $\forall i \in \mathcal{U}$;
3 **if** $\sum_{i \in \mathcal{U}} \phi_{k,i} \leq W$ **then**
4 $\quad \lfloor \quad C = p^{\text{fix}}$, and all PUs are associated with SBSs;
5 **else**
6 $\quad (\mathcal{U}^P, \mathcal{U}^S, \mathcal{U}^T) =$ User Filter Alg. $(\phi_{k,i}, w_i^{\min})$;
7 $\quad$ Calculate W^P and derive C, w_i^t, and p_i^t by solving the BA problem with $\mathcal{U}^P = \mathcal{U}^P \cup \mathcal{U}^T$;
8 $\quad$ **if** $\phi_{k,i} \leq w_i^t$, $\forall i \in \mathcal{U}^P$ **then**
9 $\quad \quad \lfloor$ Assign the ith PU in $\mathcal{U}^S$;
10 $\quad$ Find $m = \arg\max_{i \in \mathcal{U}^T} \alpha p_i^t w_i^t$;
11 $\quad$ Calculate $C^{\max}$ by solving the BA problem with $\mathcal{U}^P = \mathcal{U}^P \cup \mathcal{U}^T \setminus \{m\}$;
12 $\quad$ Assign $\mathcal{U}_{\max}^P = \mathcal{U}^P \cup \mathcal{U}^T \setminus \{m\}$;
13 $\quad$ **while** $\mathcal{U}^T$ *is not empty* **do**
14 $\quad \quad$ Calculate C, w_i^t, and p_i^t by solving the BA problem with $\mathcal{U}^P = \mathcal{U}^P \cup \mathcal{U}^T$;
15 $\quad \quad$ Calculate ρ_i, $\forall i \in \mathcal{U}^T$;
16 $\quad \quad$ Find $m = \arg\max_{i \in \mathcal{U}^T} \rho_i$;
17 $\quad \quad$ **if** $\sum_{i \in \mathcal{U}^P \cup \mathcal{U}^T \setminus \{m\}} w_i^t + \sum_{i \in \mathcal{U}^S \cup \{m\}} \phi_{k,i} \leq W$ **then**
18 $\quad \quad \quad$ Calculate C^m by solving the BA problem with $\mathcal{U}^P = \mathcal{U}^P \cup \mathcal{U}^T \setminus \{m\}$;
19 $\quad \quad \quad$ **if** $C^m < C$ **then**
20 $\quad \quad \quad \quad \lfloor$ Offload the mth PU to $\mathcal{U}^S$;
21 $\quad \quad \quad \quad$ Assign $C = C^m$;
22 $\quad \quad \quad$ **else**
23 $\quad \quad \quad \quad \lfloor$ Assign the mth PU into $\mathcal{U}^P$;
24 $\quad \quad$ **else**
25 $\quad \quad \quad \lfloor$ Set primary user m in $\mathcal{U}^P$;
26 $\quad \quad \lfloor$ Update $\mathcal{U}^T = \mathcal{U}^T \setminus \{m\}$;
27 $\quad$ **if** $C^{\max} < C$ **then**
28 $\quad \quad \lfloor$ Assign $\mathcal{U}^P = \mathcal{U}_{\max}^P$, and $\mathcal{U}^S = \mathcal{U} \setminus \mathcal{U}^P$
29 Derive w_i by solving the BA algorithm;
30 Return $\mathcal{U}^P$, $\mathcal{U}^S$ and w_i, $\forall i \in \mathcal{U}^P$.

Since there may exist a SBS k which is much closer to PU i, PU i, in association with SBS k, requires less bandwidth than w_i to satisfy the data rate requirement. Thus, given the amount of bandwidth w_i, if PU i is associated with SBS k, a portion of bandwidth will be underutilized. SBS k is able to exploit the underutilized spectrum for data transmission for SUs, thus enhancing the spectral efficiency of the networks. We assume one time frame is further divided into multiple time slots. At the beginning of each time slot, SBSs calculate the bandwidth requirements in serving individual PUs, and send the bids to the PBS. Upon receiving the bids, the PBS selects the bids with the least amount of bandwidth requirement on individual PUs, calculates the maximum amount of bandwidth that can be allocated to the PUs, and then determines user–BS associations and bandwidth allocations.

A PBS's power consumption consists of two parts: static power consumption and dynamic power consumption [140]. The static power consumption is the power consumption of a BS without any traffic load. The dynamic power consumption refers to the additional power consumption caused by traffic load on the BS. Although static power consumption makes up a significant part of the total power consumption of a BS, we do not model static power consumption in our problem because we focus on reducing the dynamic power consumption of a PBS by offloading its traffic to SBSs. Therefore, we model the PBS's power consumption as the summation of the PBS's transmission power toward the PUs that are associated with the PBS.

5.2.1 An Auction-Based EST Scheme

The profit of a PBS for cooperating with SBSs is in reducing its energy consumption. Therefore, the PBS aims to maximize its energy savings by sharing its spectrum with SBSs. Meanwhile, the incentive of SBSs for cooperating is to gain spectrum for their own data transmission. To compensate for the cost in providing a data service to the PUs, SBSs set a minimum data rate requirement. SBSs cooperate with PBSs only when the minimum data rate is satisfied. According to the Shannon–Hartley theorem, given the channel fading coefficient h_i, the PBS's transmit power toward PU i, p_i, is a function of the bandwidth allocation w_i and the data rate r_i, and can be expressed as

$$p_i = \frac{\mathcal{N}_0 w_i (2^{\frac{r_i}{w_i}} - 1)}{|h_i|^2}, \tag{5.21}$$

where $\mathcal{N}_0$ is the channel noise density. Let $\mathbf{U}$ be the set of PUs, $\mathbf{U}^p$ be the set of PUs that are associated with the PBS, $\mathbf{U}^s$ be the set of PUs that are associated with SBSs, $\mathbf{U}_k^s$ be the set of PUs that are associated with SBS k, and $\mathbf{S}$ be the set of SBSs. The energy savings maximization (ESM) problem can be expressed as follows:

$$\min \sum_{i \in \mathbf{U}^p} p_i \tag{5.22}$$

$$\text{subject to: } \sum_{i \in \mathbf{U}^p} w_i + \sum_{i \in \mathbf{U}^s} \bar{w}_i \leq W,$$
$$r_i \geq r_i^{\min}, \forall i \in \mathbf{U},$$
$$p_i \leq p^{\max}, \forall i \in \mathbf{U}^p,$$
$$\bar{p}_i \leq \bar{p}_k^{\max}, \forall k \in \mathbf{S}, \forall i \in \mathbf{U}_k^s,$$
$$\bar{r}_{k,i} \geq \bar{r}_k^{\min}, \forall k \in \mathbf{S}, \forall i \in \mathbf{U}_k^s. \tag{5.23}$$

Here, p_i is the PBS's transmit power to PU i, $p^{\max}$ is the PBS's maximum transmit power, $\bar{p}_i$ is the SBS's transmit power, $\bar{p}_k^{\max}$ is the maximum transmit power of SBS k, W is the total amount of bandwidth, $\bar{w}_i$ is the amount of bandwidth allocated to SBSs that provide a data service to PU i, $\bar{r}_k^{\min}$ is the minimum data rate requirement of SBS k in serving one PU, and $\bar{r}_{k,i}$ is the secondary data rate (SDR) achieved by SBS k in serving PU i. $\bar{r}_{k,i}$ can be expressed as

$$\bar{r}_{k,i} = (\bar{w}_i - \hat{w}_{k,i}) \log\left(1 + \frac{\bar{p}_i |\bar{h}_k|^2}{\mathcal{N}_0(\bar{w}_i - \hat{w}_{k,i})}\right). \tag{5.24}$$

Here, $\bar{h}_k$ is the fading coefficient between SBS k and its secondary user, and $\hat{w}_{k,i}$ is the amount of bandwidth required by SBS k to satisfy PU i's data rate requirements, which can be derived by solving

$$r_i = \hat{w}_{k,i} \log\left(1 + \frac{\bar{p}_i |\bar{h}_{k,i}|^2}{\mathcal{N}_0 \hat{w}_{k,i}}\right). \tag{5.25}$$

Here, $\bar{h}_{k,i}$ is the fading coefficient between SBS k and PU i.

Lemma 5.2.1 *The ESM problem is an NP-hard problem.*

Proof: In the ESM problem, PUs can be associated with either the PBS or SBSs. A different user–BS association results in a different solution for the ESM problem. Therefore, achieving the optimal solution of the ESM problem involves a 0-1 integer programming problem which is an NP-hard problem [142]. Therefore, the ESM problem is an NP-hard problem. □

Since the ESM problem is NP-hard, it is not easy to solve the problem efficiently, especially when the network consists of a large number of PBSs and SBSs. We thus design an auction-based EST scheme to approximate the solution of the ESM problem. In the auction-based EST scheme, SBSs bid for the spectrum usage while the PBS selects its cooperators based on the given bids to minimize its power consumption. The interactions between the PBS and SBSs can be modeled as a repeated auction game, in which the players are the PBS and SBSs, and the actions are (1) SBSs choose the bandwidth requirement in serving individual PUs within their coverage, and (2) the PBS calculates the maximum amount of bandwidth that can be allocated to individual PUs and decides the winning bids. We shall next present both SBSs' and the PBS's actions.

5.2.1.1 SBSs' Actions
At the beginning of each time slot, SBSs calculate their bids in serving individual PUs. The incentive of SBSs in cooperating with the PBS is to utilize the licensed bandwidth to transmit data for SUs. Define the data rate achieved by individual SBSs in cooperating with the PBS as the SDR. Each SBS has its own valuation in term of the SDR in serving individual PUs. For example, if SBS k is serving PU i, SBS k expects to achieve a SDR of $\bar{r}_{k,i}$. In order to achieve the SDR, SBS k requires an amount of bandwidth equal to

$\bar{w}_k^s$. Assuming SBSs always transmit with their maximal transmission power, $\bar{r}_{k,i}$ can be expressed as

$$\bar{r}_{k,i} = \bar{w}_k^s \log\left(1 + \frac{\bar{p}_k|\bar{h}_k|^2}{\mathcal{N}_0 \bar{w}_k^s}\right). \tag{5.26}$$

Note that $\bar{w}_k^s$ can be derived by solving Eq. (5.26). Then, the amount of bandwidth required by SBS k in serving PU i is $\bar{w}_{k,i}^{\min} = \bar{w}_k^s + \hat{w}_{k,i}$. Therefore, SBS k's bid for serving PU i is $(i, \bar{w}_{k,i}^{\min})$. The first element of the bid indicates which PU the SBS is bidding for while the second element presents the amount of bandwidth required in serving the PU. Since multiple SBSs may bid for the same PU, SBSs have to learn and adapt their expected SDR in serving individual PUs, and update the bandwidth requirements in their bids. If SBS k's bid for PU i is selected, at the next time slot, SBS k increases its expected SDR in serving PU i. Otherwise, SBS k decreases its expected SDR. However, SBS k requires the minimum SDR $\bar{r}_k^{\min}$ in serving PU i. Therefore, $\bar{r}_{k,i}$ should be larger than $\bar{r}_k^{\min}$ in order to incentivize SBS k to cooperate with the PBS. Define $\Delta\beta$ as the SBS's data rate adaptation step. The individual SBS's learning and adaptation algorithm is illustrated in Algorithm 3.

Algorithm 3: The SBS's Rate Adaptation Algorithm

---

for $i = 1$ to $|U_k^s|$ **do**
  if $(\bar{w}_i > \bar{w}_{k,i}^{\min})$ **then**
    $\bar{r}_{k,i} = \bar{r}_{k,i} + \Delta\beta$;
  else
    $\bar{r}_{k,i} = \max(\bar{r}_{k,i} - \Delta\beta, \bar{r}_k^{\min})$;
  end if
end for
Return $\bar{r}_{k,i}, i \in U_k^s$.

5.2.1.2 The PBS's Actions

Since the total amount of bandwidth is limited, the PBS sets the maximum amount of bandwidth that can be allocated to individual PUs. Define $w_i^{\max}$ as the maximum bandwidth that the PBS is willing to allocate to PU i at a given time slot. The PBS's bidding selection includes two steps. Define the highest bid as the bid that requires the least amount of bandwidth in serving individual PUs. The first step is to find the highest bid for individual PUs. The second step is to decide the user–BS associations and bandwidth allocations. Let $(i, \bar{w}_{k,i}^{\min})$ and $(i, \bar{w}_{m,i}^{\min})$ be the highest bid and second highest bid for PU i, respectively. If there is only one bid for a PU, we set $\bar{w}_{m,i}^{\min} = \min(\bar{w}_{k,i}^{\min}, w_i^{\max})$. If $\bar{w}_{k,i}^{\min} > w_i^{\max}$, PU i is associated with the PBS at the current time slot. Otherwise, PU i is associated with SBS k. If $\bar{w}_{m,i}^{\min} < w_i^{\max}$, the amount of bandwidth allocated to SBS k equals $\bar{w}_{m,i}^{\min}$; otherwise, it equals $w_i^{\max}$. Let $\mathcal{B}_i$ denote the set of SBSs that bid for serving PU i. The pseudo code of the winning bid selection algorithm is illustrated in Algorithm 4.

Algorithm 4: The Winning Bid Selection Algorithm

---

for (i=1 to $|U|$) **do**

    $k = \arg\min_{j \in B_i} \bar{w}_{j,i}^{\min}$;

    if ($\bar{w}_{k,i}^{\min} \leq w_i^{\max}$) **then**

        $m = \arg\min_{j \in B, j \neq k} \bar{w}_{j,i}^{\min}$;

        if ($\bar{w}_{m,i}^{\min} \leq w_i^{\max}$) **then**

            The granted bandwidth $\bar{w}_{k,i} = \bar{w}_{m,i}^{\min}$;

        else

            The granted bandwidth $\bar{w}_{k,i} = w_i^{\max}$;

        end if

        Associates PU i with SBS k;

    else

        Associates PU i with the PBS;

    end if

end for

Return the user–BS associations and $\bar{w}_{k,i}$, $i \in U$.

Theorem 5.2.1 *The winning bid selection algorithm enforces truth telling to be individual SBSs' dominant strategy, which maximizes individual SBSs' profits.*

Proof: Let $\bar{w}_{k,i}^*$ be the actual amount of bandwidth required by SBS k in serving PU i, $\bar{w}_{k,i}^{\min}$ be the amount of bandwidth presented in the bid, and $\bar{w}_{m,i}^{\min}$ be the amount of bandwidth presented in the second highest-bid. On the one hand, assume $\bar{w}_{k,i}^{\min} > \bar{w}_{k,i}^*$, which indicates that SBS k requires more bandwidth in its bid. If $\bar{w}_{k,i}^{\min} < \min(w_i^{\max}, \bar{w}_{m,i}^{\min})$, SBS k wins the bid, and the amount of bandwidth allocated to SBS k equals $\min(w_i^{\max}, \bar{w}_{m,i}^{\min})$. However, if $\bar{w}_{k,i}^* < \min(w_i^{\max}, \bar{w}_{m,i}^{\min}) < \bar{w}_{k,i}^{\min}$, SBS k loses the bid because of its presenting a larger bandwidth demand in its bid. In this case, if SBS k bids its valuation, $\bar{w}_{k,i}^*$, then it wins the bid and is allocated the amount of bandwidth equaling $\min(w_i^{\max}, \bar{w}_{m,i}^{\min})$. Therefore, asking for more than the actual required amount of bandwidth in the bid does not gain more bandwidth but has the probability to lose the bid. On the other hand, assume $\bar{w}_{k,i}^{\min} < \bar{w}_{k,i}^*$, which indicates that SBS k asks for less than its actual required amount of bandwidth in its bid. If $\bar{w}_{k,i}^{\min} < \min(w_i^{\max}, \bar{w}_{m,i}^{\min})$, SBS k wins the bid, and the amount of bandwidth allocated to SBS k equals $\min(w_i^{\max}, \bar{w}_{m,i}^{\min})$. If $\bar{w}_{m,i}^{\min} < \min(w_i^{\max}, \bar{w}_{m,i}^{\min}) < \bar{w}_{k,i}^*$, although SBS k wins the bid, the amount of bandwidth allocated to SBS k is less than the amount of bandwidth required to achieve its expected SDR. Thus, SBS k achieves negative profits in the cooperation. Instead, if SBS k bids for its actual bandwidth requirements, $\bar{w}_{k,i}^*$, it will lose the bid and has a zero profit. Therefore, there is no incentive for SBSs not to reveal their actual bandwidth requirements in their bids. Hence, truth telling is the dominant strategy for SBSs. □

To minimize its power consumption, the PBS wants to maximize the number of PUs associating with SBSs while minimizing the amount of bandwidth allocated to SBSs. Therefore, if PU i is successfully associated with SBSs at one time slot, the PBS should reduce $w_i^{\max}$ at the next time slot; otherwise, the PBS should increase $w_i^{\max}$. To adjust $w_i^{\max}$, the PBS assigns PU i a power consumption weight η_i which is a positive number that indicates the importance of the PU in minimizing the PBS's power consumption. Given η_i, $i \in \mathbf{U}$, $w_i^{\max}$ can be calculated by the weighted power consumption minimization (WPCM) problem:

$$\min \sum_{i \in \mathbf{U}} \eta_i p_i \qquad (5.27)$$

$$\textit{subject to:} \sum_{i \in \mathbf{U}} w_i^{\max} \leq W,$$
$$r_i \geq r_i^{\min}, \forall i \in \mathbf{U},$$
$$w_i^{\max} \geq w_i^{\min}. \qquad (5.28)$$

When $w_i > 0$, $\frac{d^2 p_i}{dw_i^2} > 0$. Thus, p_i is a convex function of w_i. Therefore, the objective function is convex. The constraints of the WPCM problem satisfy Slater's conditions, and therefore the KKT conditions provide necessary and sufficient conditions for the optimality of the WPCM problem. The optimal solution of the WPCM problem can be achieved by solving its KKT conditions. If the PBS increases η_i, $w_i^{\max}$ will be increased, and vice versa. Therefore, at the beginning of each time slot, the PBS updates η_i, $i \in \mathbf{U}$ to adapt $w_i^{\max}$, $i \in \mathbf{U}$ according to the user–BS associations of the previous time slot. The power consumption weight adaptation algorithm works as shown in Algorithm 5.

Algorithm 5: Power Consumption Weight Adaptation

---

    for $i = 1$ to $|\mathbf{U}|$ **do**
      if $w_i^{\max} > \min w_{k,i}^{\min}$, $k \in \mathbf{B}_i$ **then**
        $\eta_i = min(\eta_i - \Delta\eta, \eta^{\min})$;
      else
        $\eta_i = max(\eta_i + \Delta\eta, \eta^{\max})$;
      end if
    end for
    Return $\eta_i, i \in \mathbf{U}$.

Here, $\Delta\eta$ is the PBS's power consumption weight adaptation step, and $\eta^{\min}$ is a positive number that avoids η_i going negative. $\eta^{\max}$ is an upper bound of η_i. If PU i is out of the coverage area of all SBSs, increasing η_i does not enable PU i to be associated with SBSs. If this is the case, $\eta_i = \eta^{\max}$.

Since the user–BS associations of PUs in both $\mathcal{U}^P$ and $\mathcal{U}^S$ are determined, the HPCM algorithm optimizes the user–BS associations for the PUs in $\mathcal{U}^T$. If brute-force search is applied, the total number of possible combinations of the user–BS associations for the PUs in $\mathcal{U}^T$ is $2^{|\mathcal{U}^T|}$. The computational complexity of solving the KKT conditions of the BA problem is $O(|\mathcal{U}|)$. Therefore, the computational complexity of the brute-force search, in the worst case, is $O(|\mathcal{U}|2^{|\mathcal{U}^T|})$. When $|\mathcal{U}^T|$ is large, brute-force search is very inefficient, it can even be impossible to solve the PCM problem within a reasonable time. As compared with brute-force search, the HPCM algorithm incurs significantly less computational complexity. The *while* loop at most requires $|\mathcal{U}^T|$ iterations. In the *while* loop, finding the PU with the largest PBR requires at most $|\mathcal{U}^T| \log |\mathcal{U}^T|$ iterations. The complexity of solving the BA problem is $O(|\mathcal{U}|)$. Therefore, the worst case computational complexity of the HPCM algorithm is $O(|\mathcal{U}^T|(|\mathcal{U}| + |\mathcal{U}^T| \log |\mathcal{U}^T|))$.

Although having significantly less computational complexity, the HPCM algorithm's performance in terms of minimizing the PBS's power consumption is still significantly beneficial. In fact, the PBS's power savings achieved by the HPCM algorithm are at least 50% of that achieved by brute-force search when the PBS experiences heavy traffic load from cell edge users.

Lemma 5.2.2 *If $\rho_m > \rho_j$, $m, j \in \mathcal{U}^T$, the mth PU's user–BS association does not depend on the jth PU's user–BS association.*

Proof: Proving Lemma 5.2.2 is to prove that the order of PBRs of users in $\mathcal{U}^T$ does not change during the iterations. $\phi_{k,i}$ is determined by the channel condition between the ith PU and the kth SBS, the ith PU's data rate requirement, and the kth SBS's compensating bandwidth. These parameters do not change during the iterations. Thus, ρ_i is determined by p_i^t and w_i^t, which are derived by solving the BA problem. The BA problem is solved by solving its KKT conditions. Then, w_i^t can be derived by solving the following equation array:

$$\begin{cases} \alpha(2^{r_i/w_i^*} - 1 - \frac{r_i}{w_i^*}2^{r_i/w_i^*}\ln 2)\frac{\mathcal{N}_0}{|h_i^P|^2} = v^* \\ \sum_{i\in\mathcal{U}^P} w_i^* = W^P. \end{cases} \tag{5.29}$$

Here, w_i^* and v^* are the primal and dual optimal points for the BA problem, respectively. Although there is no closed form solution for the above equation array, we can derive the structure of the optimal solutions, based on which we prove the lemma. Let

$$\psi(w_i) = \alpha \left(2^{r_i/w_i} - 1 - \frac{r_i}{w_i}2^{r_i/w_i}\ln 2 \right) \frac{\mathcal{N}_0}{|h_i^P|^2}. \tag{5.30}$$

Since $w_i \geq w_i^{\min}$ and $\frac{d\psi(w)}{dw} > 0$,

$$\psi(w_i^{\min}) = \min_{w_i} \psi(w_i). \tag{5.31}$$

Based on the first equation in the equation array, w_i^* can be expressed as a function of v^*:

$$w_i^* = \begin{cases} \varphi(r_i, |h_i^P|^2, v^*), & v^* > \psi(w_i^{\min}), \\ w_i^{\min}, & v^* \leq \psi(w_i^{\min}). \end{cases} \tag{5.32}$$

Here, $\varphi(r_i, |h_i^P|^2, v^*)$ is derived based on the first equation in the equation array (5.29). Since

$$r_i = w_i^{\min} \log \left(1 + \frac{p^{\max}|h_i^P|^2}{\mathcal{N}_0} \right), \tag{5.33}$$

$\psi(w_i^{\min})$ can be expressed as

$$\psi(w_i^{\min}) = \alpha(p^{\max} - \\ \log \left(1 + \frac{p^{\max}|h_i^P|^2}{\mathcal{N}_0} \right) \left(\frac{\mathcal{N}_0}{|h_i^P|^2} + p^{\max} \right) ln2). \tag{5.34}$$

$\psi(w_i^{\min})$ can be considered as a function of $|h_i^P|^2$. When $p^{\max}|h_i^P|^2 \gg \mathcal{N}_0$, $\frac{d\psi(w_i^{\min})}{d|h_i^P|^2} < 0$. Therefore, a small $|h_i^P|^2$ leads to a large $\psi(w_i^{\min})$. When the PBS experiences heavy traffic from both cell edge users and inner cell users, the optimal bandwidth allocations toward cell edge users are equal to their minimum required bandwidths. When the proposed scheme is applied to offload cell edge users, it is reasonable to assume that the users in $\mathcal{U}^T$ are the cell edge users. For these users, their optimal bandwidth allocations derived by solving the BA problem are their minimum required bandwidths, which do not change during each iteration. Hence, the order of PBR of users in $\mathcal{U}^T$ does not change during the iterations. $\square$

Lemma 5.2.2 is important to guarantee the correctness of the HPCM algorithm. According to the HPCM algorithm, when $\rho_m > \rho_j$, the mth PU's user–BS association is determined prior to the jth PU's. In this case, if the mth PU's user–BS association depends on the jth PU's user–BS association, then the HPCM algorithm cannot determine the mth PU's user–BS association before determining the jth PU's, which contradicts the procedure of the HPCM algorithm. However, Lemma 5.2.2 proves that the mth PU's user–BS association does not depend on the jth PU's, which ensures the correctness of the HPCM algorithm.

Let $m = \arg\max_{j \in \mathcal{U}^T} \rho_j$. Define ρ_i and ρ_i^m as the ith PU's PBR before and after determining the mth PU's serving BS, respectively.

Lemma 5.2.3 *When the PBS experiences heavy traffic from both cell edge users and inner cell users, if $\rho_i \geq \rho_k, \forall i, k \in \mathcal{U}^T \setminus \{m\}$, $\rho_i^m \geq \rho_k^m$.*

Proof: Based on the HPCM algorithm, when $\rho_m > \rho_j$, $m, j \in \mathcal{U}^T$, the mth PU's user–BS association is determined prior to the jth PU's. On the one hand, if the mth PU is associated with a SBS by the HPCM algorithm, the jth PU's user–BS association does not change the mth PU's user–BS association.

On the other hand, if the mth PU is not associated with an SBS by the HPCM algorithm, it is because either (1) $\sum_{i \in \mathcal{U}^P \cup \mathcal{U}^T \setminus \{m\}} w_i^t + \sum_{i \in \mathcal{U}^S \cup \{m\}} \phi_{k,i} > W, k = \arg\min_{j \in \mathcal{S}} \phi_{j,i}$, or (2) offloading the mth PU increases the PBS's dynamic power consumption.

For the first case, if the jth PU is associated with the kth SBS, because $\phi_{k,j} - w_j^t > 0$,

$$\sum_{i \in \mathcal{U}^P \cup \mathcal{U}^T \setminus \{m, j\}} w_i^t + \sum_{i \in \mathcal{U}^S \cup \{m\}} \phi_{k,i} + \phi_{k,j} - w_j^t > W. \tag{5.35}$$

Therefore, the mth PU still cannot be associated with an SBS.

For the second case, the jth PU's user–BS association does not change the fact that the PBS's power savings in offloading the mth PU to the SBS is negative. Given user–BS associations, in order to minimize the PBS's power consumption, the bandwidth allocations are optimized by solving the BA problem. Therefore, any bandwidth allocation solution which is different from the solution obtained by solving the BA problem will not reduce the PBS's power consumption. Based on this observation, we can prove that even if the jth PU is associated with a SBS, the PBS's power savings in offloading the mth PU to the SBS is still negative.

Assume both the mth PU and the jth PU are associated with the PBS. Let $W^0 = W - \sum_{i \in \mathcal{U}^S} \phi_{k,i}$, $k = \arg\min_{j \in \mathcal{S}} \phi_{j,i}$. Define w_i^0 and p_i^0 as the bandwidth allocation and the PBS's transmit power density for the ith PU, $i \in \mathcal{U}^P \cup \mathcal{U}^T$, derived by solving the BA problem with $W^P = W^0$ and $\mathcal{U}^P = \mathcal{U}^P \cup \mathcal{U}^T$. Define

$$C_m^0 = \sum_{i \in \mathcal{U}^P \cup \mathcal{U}^T} \alpha w_i^0 p_i^0 + p^{\text{fix}}. \tag{5.36}$$

When the mth PU is associated with the kth SBS and the jth PU is associated with the PBS, the total amount of bandwidth available for the PUs in $\mathcal{U}^P \cup \mathcal{U}^T \setminus \{m\}$ is reduced by $\Delta w_m^0 = \phi_{k,m} - w_m^0$. The bandwidth reduction results in an increase of the PBS's dynamic power consumption in serving its associated PUs. Define w_i^1 and p_i^1 as the bandwidth allocation and the PBS's transmit power density for the ith PU, $i \in \mathcal{U}^P \cup \mathcal{U}^T \setminus \{m\}$, derived by solving the BA problem with $W^P = W^0 - \Delta w_m^0$ and $\mathcal{U}^P = \mathcal{U}^P \cup \mathcal{U}^T \setminus \{m\}$. Define

$$C_m^1 = \sum_{i \in \mathcal{U}^P \cup \mathcal{U}^T} \alpha w_i^1 p_i^1 + p^{\text{fix}}; \tag{5.37}$$

since both w_i^1 and p_i^1 are derived by solving the BA problem, C_m^1 is minimized. Define $\Delta C = C_m^1 - C_m^0$ as the increased power consumption owing to offloading the mth PU to the SBS. In the process of minimization, if the bandwidth allocation and the transmit power density toward the jth PU do not change, the power consumption increases do not come from the jth PU. Therefore, whether the jth PU is offloaded to the SBS does not change the mth PU's user–BS association. On the other hand, if the jth PU's power consumption increases owing to bandwidth reduction, the bandwidth reduction is derived by solving the BA problem to minimize the overall energy consumption. In other words, in this case, if we keep the jth PU's power consumption unchanged during the process of solving the BA algorithm, the PBS's power consumption will be larger than C_m^1. As a result, even if the jth PU is offloaded to the SBS, the mth PU still cannot be offloaded to

SBSs. Thus, when the PBS experiences heavy traffic from both cell edge users and inner cell users, if $\rho_i \geq \rho_k, \forall i, k \in \mathcal{U}^T \setminus \{m\}, \rho_i^m \geq \rho_k^m$. $\qquad \square$

Theorem 5.2.2 *When $\sum_{i \in \mathcal{U}} \phi_{k,i} \leq W$ and $k = \arg\min_{j \in S} \phi_{j,i}$, both the HPCM algorithm and brute-force search achieve the same solution.*

Proof: If $\sum_{i \in \mathcal{U}} \phi_{m,i} \leq W$ and $m = \arg\min_{j \in S} \phi_{j,i}$, then all the PUs can be offloaded to SBSs, and the PBS's dynamic power consumption is zero. For this scenario, both the HPCM algorithm and brute-force search achieve the same solution. $\qquad \square$

Define a relaxed PCM problem where the PUs can be partially associated with SBSs as the RPCM problem. In this case, if the mth PU is partially associated with the kth SBS, $\gamma_m\%$ of the mth PU's data service is provided by the kth SBS, and the other portion of data service is provided by the PBS, then the kth SBS's bandwidth requirement to partially serve the mth PU is defined as $\phi_{m,k}^P = \gamma_m \phi_{m,k}$. We assume $(m-1)$ PUs are fully offloaded to SBSs. The mth PU is defined as the PU with the largest PBR after offloading the $(m-1)$ PUs to the SBS. Define E^P as the PBS's power savings. We assume that the mth PU is partially offloaded to the SBS. Define the PBS's power savings achieved by solving the PCM problem with brute-force search as E^B. We assume the PBS experiences heavy traffic load from cell edge users and the SBS's compensated bandwidth is properly selected. In this case, there always exists a PU that can be offloaded to the SBS, thus resulting in the reduction of energy consumption of the PBS.

Lemma 5.2.4 *When $\sum_{i \in \mathcal{U}} \phi_{k,i} > W$ and $k = \arg\min_{j \in S} \phi_{j,i}, E^P >= E^B$.*

Proof: Since the RPCM problem is a relaxed version of the PCM problem, every solution for the PCM problem is feasible for the RPCM problem. Since we assume the PBS experiences heavy traffic load from cell edge users and the SBS's compensating bandwidth is properly selected, a PU always exists which can be offloaded to the SBS, thus resulting in the reduction of energy consumption of the PBS. In other words, we assume that given the maximum transmit power constraint is satisfied, offloading traffic load to the SBS can reduce the PBS's power consumption. Therefore, $E^P >= E^B$. $\qquad \square$

Theorem 5.2.3 *When $\sum_{i \in \mathcal{U}} \phi_{k,i} > W$ and $k = \arg\min_{j \in S} \phi_{j,i}$, the maximum power savings achieved by the HPCM algorithm is at least 50% of that achieved by brute-force search.*

Proof: When $\sum_{i \in \mathcal{U}} \phi_{k,i} > W$ and $k = \arg\min_{j \in S} \phi_{j,i}$, not all the PUs can be offloaded to SBSs. In this case, we show that the maximum power savings achieved by the HPCM algorithm is at least 50% of that achieved by brute-force search.

Define E^H, E^M, and $E^{\max}$ as the PBS's power savings by offloading the first $(m-1)$th PUs, the mth PU, and the PU with the maximum power consumption, respectively. Then, the power saving achieved by the HPCM algorithm is $\max\{E^H, E^{\max}\}$. Since E^P is the power saving from partially offloading the mth PU, $E^H + E^M \geq E^P$.

$$\frac{\max\{E^H, E^{\max}\}}{E^B} \geq \frac{\max\{E^H, E^M\}}{E^P}$$

$$\geq \frac{\max\{E^H, E^M\}}{E^H + E^M}$$

$$\geq 0.5. \tag{5.38}$$

$\square$

5.3 Performance Evaluation

In this section, we set up simulations to evaluate the centralized and auction-based energy spectrum trading schemes presented here.

5.3.1 Centralized Energy Spectrum Trading Algorithm

Two simulation scenarios are set up to evaluate the performance of the proposed EST scheme and the HPCM algorithm. In the simulations, we adopt COST 231 Walfisch-Ikegami [144] as the propagation model with 9 dB Rayleigh fading and 5 dB shadowing fading for both the PBS and SBSs. The carrier frequency is 2110 MHz, the antenna feeder loss is 3 dB, the transmitter gain is 1 dB, the noise density is 10^{-10} $W\,Hz^{-1}$, and the receiver sensitivity is -97 dB. The total amount of licensed bandwidth is 20 MHz and the PBS's maximum transmit power is 20 W (43 dBm) [3]. Thus, the PBS's maximum transmit power-spectral density is 1 $\mu W\,Hz^{-1}$. Based on the measurement results in [3], we set $\alpha = 25$ and $p^{\text{fix}} = 700$ W. SBSs are assumed to have the same energy consumption model as that of the PBS. The SBSs' static power consumption is 14 W, and the SBSs' coefficients between the dynamic power consumption and their transmit power are the same, 2. We assume the SBS's transmit power-spectral density is 20 $\mu W\,kHz^{-1}$ [145].

The simulation typologies are shown in Figure 5.2. A radio cell, which is covered by the PBS, is divided into three sectors. The radius of the radio cell is 1.5 km, and the PBS is located at the center of the radio cell. The PU's minimum data rate is 500 $kbps$. In the simulations, the energy efficiency (EE) is calculated by dividing PUs' total data rate by the sum of the PBS's power consumption and the SBSs' dynamic power consumption in serving PUs; the spectrum efficiency (SE) is calculated by dividing PUs' total data rate by the sum of the bandwidth allocated to PUs by both the PBS and SBSs.

5.3.1.1 Simulation Scenario One

In this simulation scenario, we consider a radio cell with one PBS and one SBS in each sector, as shown in Figure 5.2(a). The SBSs have the same operational parameters such as the transmit spectral-power density, the per-PU compensated bandwidth, and the distance between the SBS and the PBS. We define the cell edge users as users whose distances from the PBS are larger than 0.9 km.

Figure 5.3 shows the EE and SE of the network versus the percentage of cell edge users. In this simulation, the total number of mobile users in each sector follows a Poisson distribution with mean 20. As the percentage of cell edge users increases, the EE of the traditional scheme decreases because serving cell edge users usually requires more energy

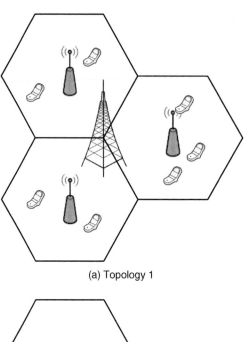

Figure 5.2 Simulation topology.
Source: Han 2014 [139]. Reproduced with permission of IEEE.

(a) Topology 1

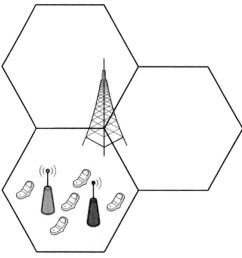

(b) Topology 2

consumption. The EE of the EST scheme increases because more users are offloaded to the SBS. For the same reason, the SE of the EST scheme also increases.

Since the EST scheme aims to offload cell edge users to enhance network efficiency, we assume that PUs are randomly distributed at the edge of each sector in the following simulations. The number of mobile users in each sector follows a Poisson distribution with mean 20.

Figure 5.4 compares the EE and SE of the traditional scheme and the EST scheme versus the distance between the PBS and SBSs. Here, the traditional scheme refers to the scheme in which all PUs are served by the PBS. In this simulation, we assume the distances between the PBS and three SBSs are the same. For all the SBSs, the per-PU

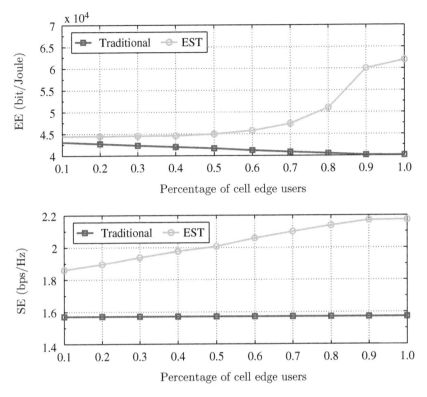

Figure 5.3 Performance of the EST vs. the percentage of cell edge users (topology 1).
Source: Han 2014 [139]. Reproduced with permission of IEEE.

compensated bandwidth is 100 *kHz*. As shown in Figure 5.4, the EST scheme enhances the EE and the SE by 43.60% and 38.79%, respectively. The EE and SE achieve their maximum values when the distance between the PBS and SBS is 1.05 *km* and 1.125 *km*, respectively. As shown in Figure 5.5, as the distance between the PBS and SBSs increases, the average distance between the mobile users and the SBS decreases until reaching its minimum, and then increases. When the average distance between the mobile users and the SBS decreases, the SBSs may offload more PUs from the PBS and may require less bandwidth for offloading the same PU. Thus, both the EE and SE increase. After the EE and SE reach their peaks, they decrease as the average distance between the mobile users and the SBSs increases. The EE decreases faster than the SE. As the average distance between the mobile users and the SBS increases, the SBS requires more bandwidth for offloading the same PU. As a result, less bandwidth is available in the PBS, and the PBS's power consumption increases.

Remark 5.3.1 *The above observation indicates that the location of the SBS significantly impacts the performance of the mobile network under the EST scheme, in terms of the EE and SE. For network planning, the locations of SBSs deployed by ISPs may not be optimized for the purpose of enhancing the EE and SE of mobile networks. However, if the EST scheme is considered, in order to maximize their profits from utilizing the licensed*

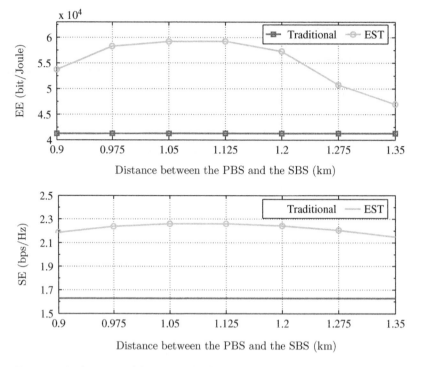

Figure 5.4 Performance of the EST vs. the distance between the PBS the SBSs (topology 1).
Source: Han 2014 [139]. Reproduced with permission of IEEE.

bandwidth, the ISPs desire to maximize their traffic loading by optimizing the locations of the SBSs increases. Furthermore, if the EST scheme is considered, the mobile network operators and ISPs can jointly optimize their network planning to maximize their profits.

Figure 5.6 shows the EE and the SE of the network versus the SBS's transmit power-spectral density. In this simulation, the distance between the PBS and SBSs is 1.125 *km*,

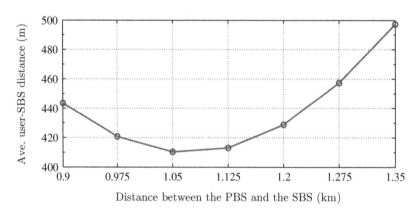

Figure 5.5 The average distance between mobile users and the SBSs (topology 1).
Source: Han 2014 [139]. Reproduced with permission of IEEE.

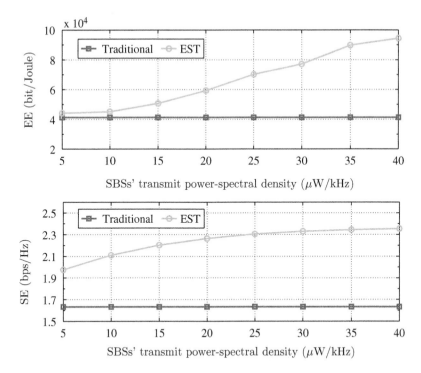

Figure 5.6 Performance of the EST vs. the SBSs' power-spectral density (topology 1). *Source:* Han 2014 [139]. Reproduced with permission of IEEE.

and the SBS's per-PU compensated bandwidth is 100 *kHz*. When the SBS's transmit power-spectral density increases, more PUs fall into the coverage area of SBSs and the SBSs' bandwidth requirements in serving the PUs are reduced. As a result, more PUs are offloaded to SBSs. Therefore, both EE and SE improve as the SBSs' transmit power-spectral density increases.

Figure 5.7 shows the performance of the EST versus the SBSs' compensating bandwidth. In this simulation, the distance between the PBS and SBSs is 1.125 *km* and the SBS's transmit power-spectral density is 20 $\mu W \ kHz^{-1}$. As the SBS's per-PU compensated bandwidth increases, the EE of the network decreases because less PUs are offloaded to SBSs. When the SBS's per-PU compensated bandwidth increases, the SE increases, peaks when the per-PU compensated bandwidth is equal to 280 *kHz*, and then decreases. This is because when the SBS's per-PU compensated bandwidth increases, although the number of offloaded PUs decreases, the total amount of the bandwidth obtained by SBSs increases due to the larger per-PU compensated bandwidth. Thus, the SE shows the concavity. This observation indicates that the per-PU compensated bandwidth should be properly selected to optimize the trade-off between the EE and the SE of the network. In addition, when the SBSs' compensating bandwidth equals 360 *kHz*, the EE of the network under EST is only slightly larger than that under the traditional scheme, while the SE of the network under EST is significantly larger than that under the traditional scheme. The improvement in SE indicates that some PUs are still offloaded to the SBSs even when the compensating bandwidth is as large as 360 *kHz*. The

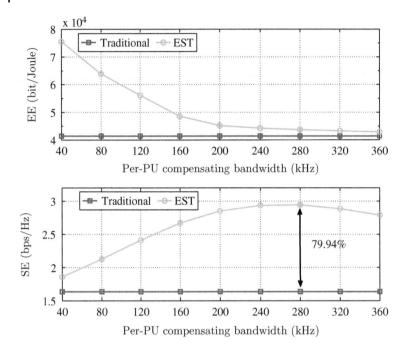

Figure 5.7 Performance of the EST vs. the SBSs' compensating bandwidth (topology 1).
Source: Han 2014 [139]. Reproduced with permission of IEEE.

traditional scheme and the EST show similar EE because a large compensating bandwidth leads to a reduced available bandwidth in the PBS. As a result, the PBS has to increase its transmit power density to satisfy users' data rate requirements. Therefore, the PBS's power consumption increases and the EE of the network under EST decreases. As the compensating bandwidth increases, the EE of the network under EST keeps decreasing until it converges to the EE of the traditional scheme. When the traditional scheme and EST have the same EE, the performance of the SE reflects whether PUs are offloaded to the SBS.

5.3.1.2 Simulation Scenario Two

As shown in Figure 5.2(b), we consider only one sector of the radio cell in this simulation scenario. PUs are randomly distributed at the edge of each sector. The distances between the PUs and the PBS are larger than $0.9\ km$. The number of mobile users in the sector follows a Poisson distribution with mean 40. Two SBSs are deployed in the sector. The distance between the PBS and SBSs is $1.125\ km$, and the distance between the two SBSs is also $1.125\ km$. Both the SBSs' transmit power-spectral density is $20\ \mu W\ kHz^{-1}$. The per-PU compensated bandwidth of one of the SBSs, SBS 1, is set to $100\ kHz$. We vary the per-PU compensated bandwidth of the other SBS to show the interactions between the PBS and the SBSs.

Figure 5.8 shows the network's EE and SE when SBS 2 varies its per-PU compensated bandwidth. When SBS 2 increases its per-PU compensated bandwidth, the EE of the network decreases because less PUs are being offloaded to SBSs. Meanwhile, the SE of the network shows concavity for the same as is shown in Figure 5.7.

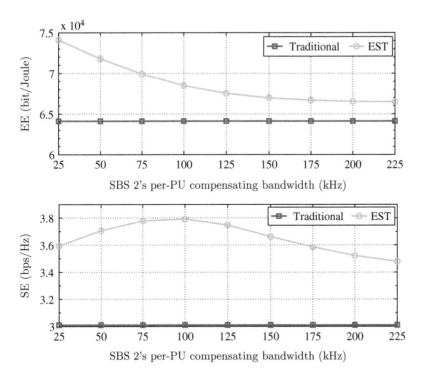

Figure 5.8 Performance of the EST vs. the SBSs' compensating bandwidth (topology 2).
Source: Han 2014 [139]. Reproduced with permission of IEEE.

Figure 5.9 shows the interaction between two SBSs. As SBS 2's per-PU compensated bandwidth increases, its dynamic power cost decreases because less PUs are offloaded to it. At the same time, SBS 1's power cost increases because more PUs are associated with SBS 1. This indicates that as SBS 2's per-PU compensated bandwidth increases, the PUs that were originally offloaded to SBS 2 are associated with SBS 1. On the other hand, as SBS 2's per-PU compensated bandwidth increases, the amount of bandwidth obtained by SBS 1 increases because it is serving more PUs. Meanwhile, the amount of bandwidth obtained by SBS 2 shows concavity for the same reason as explained before. We can observe from the simulation result that given SBS 1's strategies in terms of transmit power-spectral density and the per-PU compensated bandwidth, the profits of SBS 2 in terms of the amount of obtained bandwidth and the power cost can be maximized by selecting an optimal per-PU compensated bandwidth.

5.3.2 Auction-Based Decentralized Algorithm

In the simulations, the PBS's maximum transmit power toward individual PU is 2 W. We adopt COST 231 Walfisch-Ikegami [144] as the propagation model with 9 dB Rayleigh fading and 5 dB shadowing fading. The carrier frequency is 2110 MHz, the antenna feeder loss is 3 dB, the transmitter gain is 1 dB, the noise density is 10^{-10} $W\,Hz^{-1}$, and the receiver sensitivity is -97 dB. We deploy one PBS and two SBSs in the network. The SBSs are located at the cell edge and close to each other. Thirty PUs are randomly distributed at the cell edge.

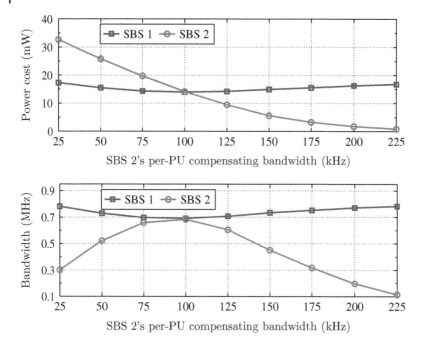

Figure 5.9 The SBSs' power cost and obtained bandwidth vs. the SBSs' compensating bandwidth (topology 2). *Source:* Han 2014 [139]. Reproduced with permission of IEEE.

Figure 5.10 compares the PBS's power consumption in traditional cellular networks versus that in cellular networks enabled with the EST scheme, and shows the total SDR achieved by the secondary networks. In the figure, the plot reflects the PBS's power consumption and the SDR in five time frames. In each time frame, there are thirty time slots. As compared with traditional cellular networks, the EST scheme significantly reduces the PBS's power consumption. Meanwhile, the data rate achieved by SBSs is up to 53 *Mbps*, which indicates that the EST scheme enhances the spectral efficiency of cellular networks. Because both SBSs and the PBS are considered to be greedy, and they always try to increase their profits, the PBS's power consumption and the SBSs' data rates fluctuate owing to the interactions between the PBS and SBSs. For example, if a PU is close to only one of the SBSs, the PU is associated to that SBS. According to the EST scheme, at the next time slot, the SBS will increase its minimum SDR in serving this PU. Meanwhile, the PBS will reduce the power consumption weight on the PU. When meeting the SBS's minimum SDR requires more bandwidth than the maximum bandwidth that the PBS is willing to spend, the PU is associated with the PBS, and thus the PBS's power consumption will increase and the SBSs' data rates will decrease. At the next time slot, the SBS will reduce its minimum SDR while the PBS will increase the power consumption weight on the PU. As a result, the PU's user–BS association may switch between the PBS and SBS from one time slot to another, and thus the PBS's power consumption and the SBSs' data rates fluctuate. The amplitude of fluctuation is determined by the adaptation step sizes of the PBS and the SBSs.

Figure 5.11 shows the performance of the EST enabled cellular networks versus the number of users in the network, in terms of power and spectral efficiency. As shown

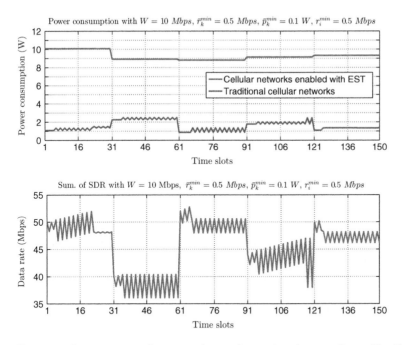

Figure 5.10 Power consumption comparison and secondary data rate. *Source:* Han 2014 [139]. Reproduced with permission of IEEE.

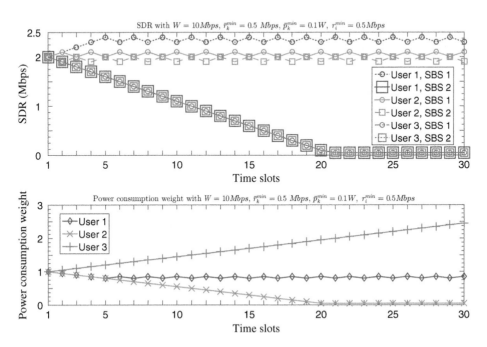

Figure 5.11 Power consumption comparison and secondary data rate. *Source:* Han 2014 [139]. Reproduced with permission of IEEE.

in the figure, the gap between the PBS's power consumption in the traditional cellular network and in the EST enabled cellular network increases with the number of users. This indicates that the EST enabled cellular network saves a significant amount of energy when a considerable number of cell edge users are offloading traffic data to secondary networks. Meanwhile, by participating in energy spectrum trading, the secondary networks achieve a data rate of up to seventy *Mbps* in the simulation. That is to say, with the same bandwidth, the EST enabled cellular networks can accommodate more users' traffic demands. Therefore, enabled with EST, both energy efficiency and spectral efficiency of the cellular networks are enhanced.

5.4 Summary

In this chapter, we have proposed a novel EST scheme that enables mobile traffic offloading between mobile networks and ISP networks by leveraging CR techniques. We have shown that achieving optimal mobile traffic offloading is NP-hard. We have proposed a HPCM algorithm to approximate the optimal solution with low computational complexity. The HPCM algorithm enables mobile traffic offloading and significantly enhances the energy and spectral efficiency of mobile networks. We have also designed an auction-based decentralized solution to approximate the solution in a distributed manner.

5.5 Questions

5.1 How does the energy spectrum trading scheme improve the energy efficiency of mobile networks?

5.2 What are the advantages and disadvantages of the centralized and decentralized algorithms?

5.3 How does the auction-based algorithm work?

5.4 Is the power minimization problem in the energy spectrum trading scheme NP-hard? If so, please prove it.

5.5 Describe the auction procedures in the auction-based decentralized solution.

6

Optimizing Green Energy Utilization for Mobile Networks with Hybrid Energy Supplies

For a network powered by green energy, the fundamental design issue is to utilize the harvested energy to sustain traffic demands of users in the network [146]. The optimal utilization of green energy over a period of time depends on the characteristics of energy arrival and energy consumption in the current period as well as in future periods. Solving this optimization problem involves at least two aspects. The first aspect is the multi-stage energy allocation problem, which determines how much energy should be used at the current time and how much energy is reserved for the future. To solve the multi-stage energy allocation problem, parameters such as the current energy arrival and consumption profile and estimations of the future energy arrival and consumption profile should be considered. The energy arrival profile depends on the renewable resources, and the energy consumption depends on user traffic demands. The second aspect is to maximize the utilization of the allocated green energy at each stage.

In this chapter, we present green energy optimization (GEO) schemes, which optimize energy utilization in cellular networks with hybrid energy supplies. The GEO scheme decomposes the problem into two sub-problems: the multi-stage energy allocation problem and the multi-BSs energy balancing problem. We propose algorithms to solve these sub-problems, and hence solve the green energy optimization problem.

We also introduce and investigate the green energy provisioning (GEP) problem, which aims to minimize the CAPEX of deploying green energy systems in BSs while satisfying the QoS requirements of cellular networks. The GEP problem is challenging because it involves optimization over multiple time slots and across multiple BSs. We decompose the GEP problem into the weighted energy minimization problem and the green energy system sizing problem, and propose a green energy provisioning solution consisting of a provision-cost aware traffic load balancing algorithm and a binary energy system sizing algorithm to solve the sub-problems and hence solve the GEP problem.

6.1 Green Energy Optimization Scheme for Mobile Networks With Hybrid Energy Supplies

Owing to the dynamics of green energy generation and the limited capacity of energy storage, green energy may not guarantee sufficient power supplies for BSs. Thus, it is

Green Mobile Networks: A Networking Perspective, First Edition. Nirwan Ansari and Tao Han.
© 2017 John Wiley & Sons Ltd. Published 2017 by John Wiley & Sons Ltd.

assumed that future BSs will be powered by multiple types of energy sources, for example the grid, solar energy, and wind energy. In such mobile networks, BSs are powered by green energy if they have enough green energy stored in their batteries; otherwise, the BSs switch to on-grid energy to serve mobile users. It is desirable to know how to optimize the utilization of green energy in order to reduce the on-grid energy consumption of mobile networks. In this chapter, algorithms are proposed to optimize the utilization of green energy during peak traffic hours. In other words, the algorithms leverage green energy to reduce the on-grid energy consumption of mobile networks during peak traffic hours.

Solar energy is considered as a green energy source. Solar energy generation depends on many factors such as temperature, sunlight intensity, the geolocation of the solar panel, and so on. However, the daily solar energy generation in a given area exhibits temporal dynamics that peaks around noon and reaches the lowest point during the night. The BSs' energy consumption depends on mobile traffic demand, which shows both temporal and spatial diversities [147]. Temporal traffic diversity indicates that the traffic demands at individual BSs are highly variable over time while spatial traffic diversity means that neighboring BSs may experience different traffic intensities at the same time of day, and therefore experience different energy consumption. Thus, in order to reduce the on-grid energy consumption of mobile networks during peak traffic hours, the GEO problem needs to balance the energy consumption among BSs.

The GEO problem involves optimization in two dimensions: the time dimension and the spatial dimension. The GEO problem is thus decomposed into two sub-problems: the multi-stage energy allocation (MEA) problem and the multi-BSs energy balancing (MEB) problem based on the characteristics of green energy generation and mobile traffic demand. The MEA problem is to optimize the green energy allocation at individual BSs to accommodate the temporal dynamics of both green energy generation and mobile traffic demand. An MEA algorithm is proposed to solve the MEA problem. By leveraging the spatial diversity of mobile traffic, the MEB problem is to balance the green energy consumption among BSs so as to reduce the on-grid energy consumption of the mobile network. The energy consumption of BSs is adjusted by adapting the cell size of the BSs. A BS adapts its cell size by varying its pilot signal strength. However, in order to maintain coverage, it may not be practical to vary the pilot signal strength arbitrarily. Therefore, the MEB algorithm selects the best pilot signal strengths for individual BSs from a few alternatives. When a BS increases its cell size, the energy efficiency of the BS may decrease [147]. Hence, the MEB algorithm only increases a BS's cell size when it has a sufficient amount of green energy. In other words, the MEB algorithm may reduce the energy efficiency of the BSs powered by green energy in order to increase the energy efficiency of the BSs powered by on-grid energy. Generally, BSs with a large amount of green energy increase their cell sizes to accommodate more mobile traffic while BSs with a small amount of green energy shrink their cell sizes to offload mobile traffic to their neighboring BSs. By adjusting the energy consumption of individual BSs, the energy consumption among BSs is balanced. As a result, the green energy shortage for BSs is reduced. Therefore, after having applied the energy allocation (EA) algorithm, the number of BSs that can be powered by green energy is increased, and thus the on-grid energy consumption is reduced.

6.1.1 System Model and Problem Formulation

6.1.1.1 Network Scenario

Consider a mobile network whose BSs can be powered by either on-grid energy or green energy. For a given time slot, if a BS's stored green energy is larger than its energy demand, the BS is powered by green energy; otherwise, the BS is powered by on-grid energy.[1] Assuming that the mobile network experiences high traffic volumes, the proposed algorithms aim to reduce the main grid energy consumption of the mobile network. Owing to this assumption, our algorithm will not enable the sleeping mode of BSs, which is usually applied to improve the energy efficiency of mobile networks during off-peak traffic hours. This assumption is valid because solar energy is considered as a green energy source, and solar energy can only be generated during the daytime, which is also the time period when the mobile networks usually experience high traffic volumes. The BSs are assumed to be able to adapt their coverage areas by changing their pilot signal power levels, and the maximum power level of a BS is Q. With a large coverage area, the BS may serve more users, and thus the BS may consume more energy.

6.1.1.2 Green Energy Model

Solar panels are considered as green energy generators. Solar panels generate electrical power by converting solar radiation into direct current electricity using semiconductors that exhibit the photovoltaic effect. Solar energy generation depends on various factors, such as the temperature, the solar intensity, and the geolocation of the solar panels. However, the hourly solar energy generation can be estimated by using typical annual meteorological weather data for a given geolocation. In this chapter, the System Advisor Model (SAM) [148] and PV Watts model [149] are adopted to estimate the hourly solar energy generation. Figure 6.1 shows the estimation of the hourly solar energy generation for four different months in New York City. In this estimation, the nameplate capacity, ϕ, and the DC-to-AC derate factor, η, are set to 1 $kWdc$ and 0.77, respectively. From the estimation, the solar panels start to generate energy from around 6:00 a.m. The solar energy generation keeps increasing and peaks at around 1:00 p.m., and ends at about 7:00 p.m. The time period is divided into time slots, and the energy generation rate at the ith time slot, α_i, is derived by using the SAM and PV Watts models [148, 149]. Here, the solar energy generation is considered to change at a time scale of several minutes rather than to vary according to the instantaneous solar energy fluctuations because the proposed algorithm optimizes the cell size of BSs at an interval of several minutes, and the instantaneous energy generation changes may not significantly affect the performance of our algorithm. Since solar energy generation exhibits temporal dynamics, the available solar energy cannot always guarantee sufficient energy supplies to the BSs. BSs located in the same geographical region are assumed to experience almost the same weather environment including solar intensity and temperature. Thus, in this chapter, it is assumed that the solar panels of all the BSs yield the same green energy generation rate.

---

[1] The case of utilizing multiple energy sources simultaneously complicates the system design (which requires multiple input source converter and controller) and is beyond the scope of this chapter.

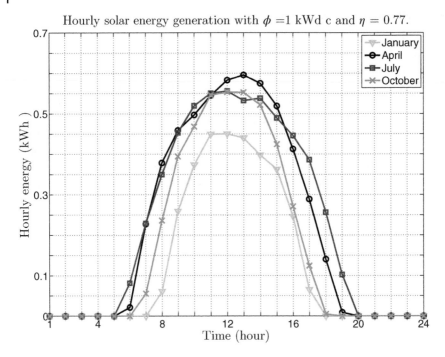

Figure 6.1 Hourly solar energy generation. *Source:* Han 2013 [18]. Reproduced with permission of IEEE.

6.1.1.3 Energy Consumption Model

According to network measurement studies, the energy consumption of BSs is directly related to the traffic loads on the BSs [150], and linear models have been proposed to express the influence of traffic load on the instantaneous energy consumption of the BSs [3]. Therefore, in this chapter, the energy consumption of BSs is modeled by two parts: the static energy consumption which is the energy consumption of the BSs without any traffic load, and the dynamic energy consumption which is related to the traffic volume of the BSs [140].

6.1.1.4 Network Traffic Model

The mobile traffic volume exhibits both temporal and spatial diversity [147]. In terms of temporal diversity, traffic volume at individual BSs is highly dynamic over time, and typically peaks between 10:00 a.m. and 6:00 p.m., and bottoms between 1:00 a.m. and 5:00 a.m. However, the traffic volume is almost stable at the same time on consecutive days. Peng *et al.* [147] showed that the traffic load difference on two consecutive days is less than 20% for 70% of BSs in their network measurement studies. Thus, it is assumed that the traffic load at individual BSs can be estimated using historical mobile traffic statistics, and hence the energy consumption of individual BSs can be estimated. Regarding spatial diversity, traffic load intensity is quite diverse among closely located BSs, and the diversity is more evident during peak times [147]. We assume that the BSs always have data transmission to mobile users during each time slot, and the BSs transmit data to all users with the same data rate. Therefore, the traffic volume at an individual BS is determined by the number of users associated with that BS. From mobile traffic statistics, the total number of mobile users in the system varies at different time slots—the temporal diversity of mobile traffic. On the other hand, it is assumed that mobile users are randomly distributed in the area. Thus, the number of users connected to individual BSs are different—the spatial diversity of mobile traffic.

6.1.2 Problem Formulation

Consider a mobile network with N BSs and M users. The duration of time is divided into L time slots, and the length of each time slot is τ seconds. Let $\vec{P_i^0} = (p_{i,1}^0, p_{i,2}^0, \cdots, p_{i,n}^0)$ be the pilot signal power of BSs at the ith time slot. Then, the user–BS association matrix at the ith time slot, X_i, is determined by $\vec{P_i^0}$. Let $X_i(k,j) = 1$ when user k is associated with BS j; otherwise, $X_i(k,j) = 0$. Assume the BSs always have data transmission to mobile users during the τ seconds. The energy consumption of BS j during the ith interval can be expressed as

$$C_{i,j} = \sum_{k=1}^{M} X_i(k,j) P_{k,j} \tau + P_{i,j}^{\text{fix}} \tau. \tag{6.1}$$

Here, $P_{k,j}$ is the dynamic power consumption of BS j for serving user k, and $P_{i,j}^{\text{fix}}$ is the static power consumption when the BS is in the active status. Since it is assumed that the mobile network experiences high traffic volume, $\sum_{k=1}^{M} X_i(k,j) > 0$. Therefore, all the BSs are in active status.

At the ith time slot, the stored green energy at BS j is $E_{i,j}$. The amount of stored green energy depends on the energy consumption and generation of the previous time slots. Hence, $E_{i,j}$ equals to

$$E_{i,j} = \begin{cases} E_{0,j} + \alpha_i \tau, & i = 1; \\ E_{i-1,j} - \beta_{i-1,j} C_{i-1,j} + \alpha_i \tau, & i \geq 2. \end{cases} \tag{6.2}$$

Here, $E_{0,j}$ is the initial green energy stored at BS j. It is assumed that the capacity of the battery is sufficiently large to store green energy at individual BSs. Therefore, energy overflow is not considered. $\beta_{i,j}$ is the energy source indicator function. Let $E_{i,j}^A$ be the amount of allocated green energy at the ith time slot in BS j, and $E_{i,j}^A \leq E_{i,j}$. If $E_{i,j}^A \geq C_{i,j}$, then BS j is powered by green energy at the ith time slot, and then $\beta_{i,j}$ equals to 1; otherwise, $\beta_{i,j}$ equals to zero. The on-grid energy consumed by BS j during the ith interval is

$$G_{i,j} = (1 - \beta_{i,j}) C_{i,j}. \tag{6.3}$$

The GEO problem can be formulated as:

$$\min_{(\vec{P_1^0}, \vec{P_2^0}, \cdots, \vec{P_i^0}, \cdots, \vec{P_L^0})} \sum_{i=1}^{L} \sum_{j=1}^{N} G_{i,j} \tag{6.4}$$

$$\text{subject to: } \lambda_{k,i} \geq \gamma, \\ k \in (1, 2, \cdots, M). \tag{6.5}$$

Here, γ is the minimum SINR requirement. It is assumed that all users have the same SINR requirement. $\lambda_{k,i}$ is the receiving SINR of user k at the ith time slot, and it can be expressed as

$$\lambda_{k,i} = \frac{P_{k,j}\Theta_{k,j}}{\mathcal{N}_0 w_k + \sum_{m\in(1,2,\cdots,N),m\neq j} P_{m,j}\Theta_{m,j}}. \tag{6.6}$$

Here, $\Theta_{k,j}$ is the channel fading between BS j and user k, $\mathcal{N}_0$ is the noise density, and w_k is the bandwidth allocated to user k.

The GEO problem is to find the optimal pilot signal power of each BS, which determines its coverage area. Given the user distribution and the coverage area of the BSs, the energy consumption of the BSs can be calculated according to Eqs. (6.1) and (6.6). In order to minimize the overall on-grid energy consumption of the mobile network, the BSs' optimal pilot signal power depends on the BSs' green energy generation and energy consumption. When $\beta_{i,j} = 1, \forall i \in (1, 2, \cdots, L), \forall j \in (1, 2, \cdots, N)$, the on-grid energy consumption is zero. To reduce the on-grid energy consumption, the green energy utilization is to be optimized among L time slots and among N BSs to make $\beta_{i,j} = 1, \forall i \in (1, 2, \cdots, L), \forall j \in (1, 2, \cdots, N)$. In other words, for individual BSs, the green energy allocation at each time slot should be optimized; for the network, the energy consumption among BSs should be balanced: the BSs with a larger amount of green energy should increase their coverage areas to absorb traffic from the BSs with less green energy. However, owing to the dynamics of green energy and mobile traffic, the energy consumption in the BSs exhibits temporal and spatial dynamics. Therefore, it is difficult to optimize the green energy utilization because the optimization involves two dimensions: the time dimension and the space dimension.

Hence, the GEO problem is decomposed into two sub-problems. The first sub-problem is the MEA problem, which aims to optimize green energy usage at different time slots to accommodate the temporal dynamics of both the green energy generation and the mobile traffic. The second sub-problem is the MEB problem, which accommodates the spatial dynamics of the mobile traffic and seeks to maximize the utilization of green energy by balancing the green energy consumption among BSs.

6.1.2.1 The MEA Problem

Define $\delta_{i,j} = \frac{C_{i,j}}{E_{i,j}^A}$ as the energy depleting ratio (EDR) which is derived by dividing the energy consumption by the allocated green energy. If $\delta_{i,j} > 1$, the allocated green energy is not sufficient to meet the energy demand. The larger the $\delta_{i,j}$, the larger the gap between energy demand and the allocated green energy. If $\delta_{i,j} \leq 1$, the gap is zero. It is desirable to optimize the green energy allocation such that $\delta_{i,j} \leq 1, \forall i \in (1, 2, \cdots, L), \forall j \in (1, 2, \cdots, N)$, which leads to zero on-grid energy consumption. For individual BSs, the available green energy and the energy consumption vary at different time slots. As a result, the energy gap varies at different time slots. The MEA problem aims to optimize the green energy allocation over time slots to reduce the energy gap.

The MEA problem can be expressed as:

$$\min_{\left(E_{1,j}^A, \cdots, E_{i,j}^A, \cdots, E_{L,j}^A\right)} (\delta_{1,j}, \delta_{2,j}, \cdots, \delta_{i,j}, \cdots, \delta_{L,j}) \tag{6.7}$$

$$\text{subject to: } E_{i,j}^S = E_{i-1,j}^S + \alpha_{i-1}\tau - E_{i-1,j}^A,$$
$$E_{i,j}^A \leq E_{i,j}^S + \alpha_i\tau. \tag{6.8}$$

Here, $E_{i,j}^S$ is the amount of residual green energy at the beginning of the ith time slot on BS j, and $E_{0,j}^S = E_{0,j}$. Sort $\delta_{i,j}, \forall i \in (1, 2, \cdots, L)$ from the largest to the smallest, and define $\vec{Y}_j = (\delta_j^1, \delta_j^2, \cdots, \delta_j^L)$ as the sorted energy ratio vector for BS j. $(\delta_{1,j}, \delta_{2,j}, \cdots, \delta_{i,j}, \cdots, \delta_{L,j})$ is minimized if its corresponding sorted energy ratio vector $\vec{Y}_j$ has the lowest lexicographical value. The MEA problem is then to balance the green energy utilization among time slots by minimizing $(\delta_{1,j}, \delta_{2,j}, \cdots, \delta_{i,j}, \cdots, \delta_{L,j})$. Solving the MEA problem has two benefits for minimizing on-grid energy consumption. The first is that solving the MEA problem may reduce the number of the time slots in which individual BSs do not have sufficient green energy. The second is that solving the MEA problem narrows the energy gaps at individual time slots. With a narrowed energy gap, the probability of filling the gap by solving the MEB problem increases. Therefore, the probability of consuming on-grid energy is reduced.

6.1.2.2 The MEB Problem

Owing to the spatial diversity of the mobile traffic, the energy consumption of closely located BSs may exhibit large differences. The unbalanced energy demand may result in an under utilization of green energy. To maximize the utilization of green energy, green energy consumption is balanced among BSs by adapting their cell sizes. The optimal cell sizes are chosen according to both the amount of green energy and mobile traffic demands. In general, BSs with a larger amount of green energy are enforced to have larger cell sizes. BSs adapt their cell sizes by changing the power of their pilot signals. Mobile users select BSs based on the strength of pilot signals from BSs. Thus, if a BS increases its cell size by increasing its pilot signal strength, the number of mobile users associated with the BS may increase, and thus the energy consumption of the BS may increase. In this way, the energy consumption among BSs is balanced. The solution to the MEB problem is to optimize BSs' cell sizes at individual time slots, and thus to balance energy consumption among the BSs.

The MEB problem can be formulated as

$$\min_{P_i^0}(\delta_{i,1}, \delta_{i,2}, \cdots, \delta_{i,j}, \cdots, \delta_{i,N}) \tag{6.9}$$

$$\text{subject to: } \lambda_{k,i} \geq \gamma,$$
$$\beta_{i,j}C_{i,j} \leq E_{i,j}^A, \tag{6.10}$$
$$k \in (1, 2, \cdots, M).$$

Sort $\delta_{i,j}, \forall j \in (1, 2, \cdots, N)$ from the largest to the smallest, and denote

$$\vec{X}_i = \left(\delta_i^1, \delta_i^2, \cdots, \delta_i^N \right) \tag{6.11}$$

as the sorted energy ratio vector in the ith time slot.

$$(\delta_{i,1}, \delta_{i,2}, \cdots, \delta_{i,j}, \cdots, \delta_{i,N}) \tag{6.12}$$

is minimized if its corresponding sorted energy ratio vector $\vec{X}_i$ has the lowest lexicographical value.

Theorem 6.1.1 *The MEB problem is NP-hard.*

Proof: The theorem can be proved by reducing any instance of the partition problem [142] to the MEB problem. The detailed proof is left as an exercise. □

6.1.2.3 Rationale of the Decomposition

The GEO problem involves optimization in two dimensions: the time dimension and the space dimension. The optimization in the time dimension, the MEA problem, is to optimize the green energy utilization at each time slot for individual BSs while the optimization in the space dimension, the MEB problem, is to balance the energy consumption among BSs. For illustrative purposes, consider the network scenario shown in Figure 6.2. BS 1 and BS 2 are neighboring BSs but experience different traffic demands, and thus they consume different amounts of energy. The green energy generation is the same in both BSs, which is 5 units in the first time slot and 3 units in the second time slot. In the first time slot, there are three users in the network: user 1, user 2, and user 3. User 1 and user 2 are associated with BS 1, and consume 2 units and 3 units of energy from BS 1, respectively. User 3 is associated with BS 2, and consumes 1 unit of energy. In the second time slot, there are four users in the network, and the new user, user 4, is associated with BS 1, and consumes 3 units of energy. Three network operation strategies are compared: (1) with no optimization, (2) with optimization in the space dimension

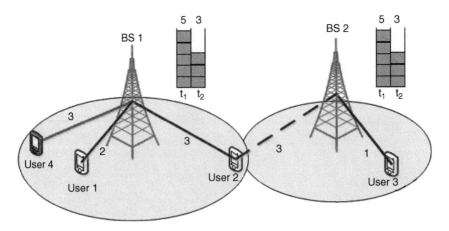

Figure 6.2 The rationale of decomposition. *Source:* Han 2013 [18]. Reproduced with permission of IEEE.

only, and (3) with optimization in both the time and space dimensions. For the first network operational strategy, BS 1 consumes zero units of on-grid energy in the first time slot. In the second time slot, three users are associated with BS 1 and the total energy consumption is 8 units. Since BS 1 only has 3 units of green energy, which are less than the energy consumption, BS 1 is powered by on-grid energy and consumes 8 units of on-grid energy. BS 2 consumes zero units of on-grid energy in both time slots. For the second operational strategy, BS 1 consumes zero units of on-grid energy in the first time slot. In the second time slot, since BS 1 does not have sufficient green energy, it reduces the coverage area and offloads user 2 to BS 2. After the offloading, the energy consumption of BS 1 is 5 units, which is larger than the amount of green energy in BS 1. As a result, BS 1 is powered by on-grid energy and consumes 5 units of on-grid energy in the second time slot. Since BS 2 only consumes 1 unit of green energy in the first time slot, the available green energy in BS 2 in the second time slot is 7 units. Therefore, BS 2 has sufficient green energy to provide service to both user 2 and user 3, and thus BS 2 consumes zero units of on-grid energy in both time slots. For the third operational strategy, BS 1 optimizes the green energy utilization in the time dimension. As a result, BS 1 allocates 3 units of green energy in the first time slot, and allocate 5 units of green energy in the second time slot. Then, in the first time slot, BS 1 reduces its coverage area and offloads user 2 to BS 2. The energy consumption of BS 1 becomes 2 units, which is less than the green energy allocation. Thus, BS 2 is powered by green energy, and consumes zero units of on-grid energy. In the second time slot, user 2 is still associated with BS 2. The available green energy and the energy consumption in BS 1 are 6 units and 5 units, respectively. Hence, BS 1 can be powered by green energy. Thus, by optimizing green energy utilization in both the time dimension and the space dimension, BS 1 consumes zero units of on-grid energy in both time slots. Although user 2 is offloaded to BS 2, BS 2 has sufficient green energy to provide service to both user 2 and user 3. Then, the network consumes zero units of on-grid energy. Hence, optimizing green energy utilization in both the time dimension and the space dimension reduces on-grid energy consumption. Therefore, the GEO problem can be decomposed into the MEA problem and the MEB problem.

6.1.3 The GEO Algorithm

In this section, a GEO algorithm is proposed to solve the GEO problem with low computational complexity. Since the GEO problem is decomposed into two sub-problems (the MEA problem and the MEB problem), the GEO problem is addressed by solving the sub-problems. The solution of the MEA problem estimates the amount of green energy allocated at individual BSs during each time slot. Based on this solution, optimal cell size adaptation is achieved by solving the MEB problem. Given the cell sizes, individual BSs apply the EA^2 algorithm to recalculate their green energy usage, and determine whether to consume on-grid energy at the current time slot. Hence, the GEO algorithm consists of the MEA algorithm, the MEB algorithm, and the EA algorithm. As shown in Figure 6.3, the MEA algorithm and the EA algorithm are implemented in individual BSs, and the MEB algorithm is implemented in the network controller which coordinates the

---

^{2} This abbreviation of EA only applies in this chapter. That is, EA may represent other meanings in other chapters.

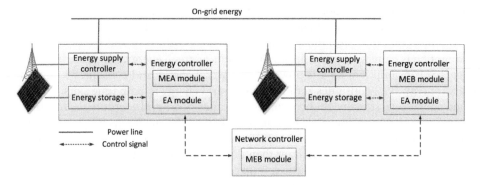

Figure 6.3 Illustration of the GEO algorithm. *Source:* Han 2013 [18]. Reproduced with permission of IEEE.

BSs in the network. Individual BSs apply the MEA algorithm to optimize green energy allocation for individual time slots based on the estimation of the mobile traffic and the green energy generation, and feed back the green energy allocation to the network controller. The network controller balances the green energy usage among the BSs by applying the MEB algorithm, and determines the BSs' cell sizes. Based on the cell sizes, individual BSs calculate their energy consumption and apply the EA algorithm to calculate the green energy allocation, and determine whether to utilize green energy for the current time slot.

6.1.3.1 The MEA Algorithm

Since the mobile traffic volume is almost stable at the same time on two consecutive days [147], energy consumption can be estimated based on historical mobile traffic statistics. Given the same weather conditions, the solar energy generation at an individual BS on two consecutive days is also stable. Hence, the MEA problem is solved based on the estimated energy generation and consumption. The solution of the MEA problem provides a good estimation of the optimal energy usage at individual BSs during each time slot. Therefore, the idea of the GEO algorithm is to derive an initial energy allocation by solving the MEA problem based on the estimated energy generation and energy demands, and then solve the MEB problem based on the solution of the MEA problem. Solving the MEA problem requires balancing the green energy usage among time slots. There are two major difficulties in solving the MEA problem. The first one is that the amount of available green energy at one time slot depends not only on the amount of energy generated in the current time slot but also on the residual green energy from previous time slots. This couples the energy usages in individual time slots, implying that changing the green energy usage in one time slot will affect the green energy allocations of later time slots. The proposed MEA algorithm decouples the dependency of the energy allocations at individual time slots by simply breaking the total green energy into that generated in the current time slot and the residual green energy from previous time slots. At the beginning of the algorithm, green energy is allocated to each time slot according to Eq. (6.13):

$$
E_{i,j}^A = \begin{cases} E_{0,j}^S + \alpha_{i,j}\tau, & i = 1; \\ \alpha_{i,j}\tau, & i > 1. \end{cases} \tag{6.13}
$$

When the MEA algorithm reduces the green energy allocation in one time slot, the saved green energy is stored as the residual energy. If there is sufficient residual energy from previous time slots, the MEA algorithm is allowed to increase the green energy allocation in the current time slot. In this way, changing the green energy allocation in one time slot will not immediately affect the green energy allocation in other time slots. The other difficulty is the constraint that the energy generated in one time slot cannot be used in previous time slots. To accommodate this constraint, the MEA algorithm optimizes the green energy allocation to individual time slots sequentially. The MEA algorithm calculates the energy allocation of the first time slot, and then iteratively adds the next time slot into the energy allocation optimization. If the EDR of the newly added time slot is larger than that of the previous time slot, the MEA algorithm reduces the green energy allocation in the previous time slots, and allocates the saved energy to the current time slot. The pseudo code of the MEA algorithm is illustrated in Algorithm 6.

Algorithm 6: The MEA Algorithm

Input : $\hat{C}_{i,j}$, $E^S_{0,j}$, α_i $i \in (1, 2, \cdots, L)$;

Output: $E^A_{i,j}, i \in (1, 2, \cdots, L)$;

1 Initialize $E^A_{i,j}$ and calculate $\hat{\delta}_{i,j} = \frac{\hat{C}_{i,j}}{E^A_{i,j}}$;

2 **for** $i = 2$ *to* L **do**

3   **if** $\hat{\delta}_{i,j} > \hat{\delta}_{i-1,j}$ **then**

4   **for** $m = 1$ *to* $i - 1$ **do**

5   Calculate $\bar{\delta} = \frac{\sum_{k=m}^{i} \hat{C}_{k,j}}{\sum_{k=m}^{i} E^A_{i,j}}$;

6   **if** $\hat{\delta}_{m,j} < \bar{\delta}$ **then**

7          $h = m$, and break;

8   **for** $k = h$ *to* $i - 1$ **do**

9   **if** $\hat{\delta}_{k,j} < \bar{\delta}$ **then**

10   Reduce $E^A_{k,j}$ to let $\hat{\delta}_{k,j} = \bar{\delta}$;

11   **for** $k = h$ *to* i **do**

12   **if** $\hat{\delta}_{k,j} > \bar{\delta}$ **then**

13   Increase $E^A_{k,j}$ to let $\hat{\delta}_{k,j} = \bar{\delta}$.

Theorem 6.1.2 *The proposed MEA algorithm achieves the optimal solution to the MEA problem.*

Proof: Since green energy cannot be utilized until it is generated, the BS's energy ratio for one time slot can be adjusted only by changing the green energy allocation of prior time slots. For example, if a BS wants to decrease the energy ratio of the *i*th time slot, the BS has to increase $E^S_{i,j}$ by reducing the energy allocation for the previous time slots. If $\hat{\delta}_{i,j} > \hat{\delta}_{i-1,j}$, the proposed MEA algorithm decreases the energy allocations of the time slots prior to the ith time slot, and thus increases the green energy allocation at the ith time slot to ensure $\hat{\delta}_{i,j} \leq \hat{\delta}_{i-1,j}$. As a result, $\delta^*_{i,j} \leq \delta^*_{i-1,j}$, $\forall i \in (2, 3, \cdots, L)$. Here, $\delta^*_{i,j}$ is the solution achieved by the MEA algorithm. When minimizing $\hat{\delta}_{i,j}$, the MEA algorithm, from the previous time slots, finds the hth time slot such that $\hat{\delta}_{h,j} < \frac{\sum_{k=h}^{i} \hat{C}_{k,j}}{\sum_{k=h}^{i} E^A_{i,j}}$. Then, the MEA algorithm reduces the energy allocation of the hth to the $(i-1)$th time slot, and let $\hat{\delta}_{m,j} = \frac{\sum_{k=h}^{i} \hat{C}_{k,j}}{\sum_{k=h}^{i} E^A_{i,j}}$, $m \in (h, h+1, \cdots, i)$. Assume $\hat{\delta}_{i,j}$ is the nth largest energy ratio among the time slots till the ith time slot. Since $\hat{\delta}_{i,j} \leq \hat{\delta}_{m,j}, m \in (1, 2, \cdots, i-1)$, $\hat{\delta}_{i,j}$ cannot be further decreased without increasing the largest to the $(n-1)$th largest energy ratio. Hence, the proposed MEA algorithm achieves the optimal solution to the MEA problem. □

The computational complexity of the MEA algorithm is $O(n^2)$ in the worst case, where n is the total number of time slots. The MEA algorithm is intended to optimize green energy allocation based on estimates of green energy generation and consumption. Therefore, the MEA algorithm can run offline, and is executed only once per day. Thus, the computational complexity of $O(n^2)$ is acceptable even when the number of time slots is very large.

6.1.3.2 The MEB Algorithm

Starting with the initial energy allocation derived from the MEA algorithm, the MEB problem is tackled to balance the energy consumption among BSs. The MEB problem is solved in two steps. First, an algorithm is designed to minimize the largest EDR among BSs. Then, the MEB algorithm iteratively minimizes the mth largest EDR among BSs. Here, $2 \leq m \leq N$. Therefore, the proposed algorithm first finds the BS with the largest EDR, and reduces the cell size of the BS by reducing its pilot signal power in order to reduce its energy demand. While reducing the cell size of a BS, the user–BS associations are updated: users originally associated with the BS may be offloaded to its neighboring BSs. Based on the new user–BS associations, the MEB algorithm re-calculates BSs' energy demands and derives their EDRs. On calculating BSs' EDRs, the MEB algorithm first derives the minimum required transmit power, $P_{k,j}$, to satisfy user k's minimum SINR requirement according to Eq. (6.6), and then calculates the energy demand of BS j according to Eq. (6.1), and eventually updates BS j's EDR.

While reducing the energy demand of a BS, EDRs of the other BSs may increase beyond or equal to the largest EDR. For example, in Figure 6.4, both BSs have 10 units of power storage, and user 2 is currently associated with BS 1. The energy demand of BS 1 is 11, which makes the largest EDR 1.1. Then, BS 1 reduces its pilot power to enable user 2 to switch to BS 2. As a result, the EDR of BS 2 will be 1.1 which is the largest EDR, and BS 2 will reduce its pilot power. Then, user 1 will switch back to BS 1. This ping-pong process is unlikely to reach the optimal solution.

To address this problem, the concept of the energy dependent set (EDS) is introduced.

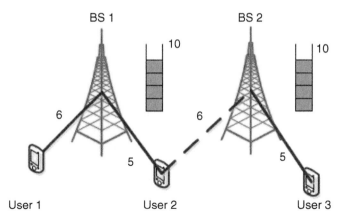

Figure 6.4 Illustration of the problem in cell size adaptation. *Source:* Han 2013 [18]. Reproduced with permission of IEEE.

Definition 6.1.1   *Let $\delta_{i,j}$ be the EDR of BS j at the ith time slot. $\delta_{i,\kappa} = \max_{1 \le j \le N} \delta_{i,j}$. Let $\delta'_{i,j}$ be the EDR of BSs j after the pilot power reduction of BS κ. Then, EDS $\mathbf{D} = \{j | \delta'_{i,j} > \delta_{i,\kappa}, j \in (1, 2, \cdots, N)\}$.*

In order to minimize the largest EDR, the pilot signal power of the BSs in the EDS should be reduced together to enable users to be offloaded to the BSs outside the EDS. Define **A** as the set of all BSs. The pseudo code of the proposed largest EDR minimization (LEM) algorithm is executed as Algorithm 7 below.

Algorithm 7: The LEM Algorithm

Input : $\vec{P}_i^0$ and $E_{i,j}^A, j \in (1, 2, \cdots, N)$;

Output: $\vec{P}_i^0$ and $\delta_i^{\max}$;

1 Set OPT = FALSE;
2 **while** *OPT = FALSE* **do**
3 $\quad$ Find the largest EDR, $\delta_i^{\max}$;
4 $\quad$ Find the set **S** that $\delta_{i,j} = \delta_i^{\max}, j \in$ **S**;
5 $\quad$ Update EDS **D** and calculate the pilot power reduction, $\vec{w}$;
6 $\quad$ **if D** == **A** $\| \exists j \in$ **D** *such that* $w_j \ge p_{i,j}^0$ **then**
7 $\quad\quad \lfloor$ Set OPT = TRUE;
$\quad$ **else**
8 $\quad\quad \lfloor$ Reduce $p_{i,j}^0$ by $w_j, j \in$ **D**.

Lemma 6.1.1 *The largest EDR is minimized if either of the following conditions is satisfied:*
1) *The energy dependent set* **D** *equals to the BS set* **A**;
2) $w_j \geq p_{i,j}^0, \exists j \in \mathbf{D}$.

Proof: For the first condition, reducing the pilot power level of all BSs does not change the user–BS association. Thus, the largest EDR cannot be reduced further. The second condition corresponds to the scenario that there exists at least one BS in EDS having its pilot power level being less than its pilot power level decrement. For the second condition, there exist BSs in EDS whose pilot power level cannot be further reduced. Reducing the pilot power level of partial BSs result in the EDRs of certain BSs in **D** being larger than the largest EDR, and therefore the largest EDR cannot be reduced further. □

The LEM algorithm minimizes the largest EDR. While further minimizing the *m*th largest EDR, the algorithm should avoid increasing the smaller EDRs. Therefore, the definition of energy distance is introduced:

Definition 6.1.2 *Given two subsets of BSs,* **A, B**, $\mathbf{A} \cap \mathbf{B} = \emptyset$, $a \in \mathbf{A}$, *and* $b \in \mathbf{B}$. *The energy distance from a to b is* $g_{a,b}$ *if reducing BS a's pilot power by up to g power levels does not increase the EDR of BS b. The energy distance from* **A** *to* **B** *is* $g_{\mathbf{A},\mathbf{B}}^s = \arg\min g_{a,b}, a \in \mathbf{A}, b \in \mathbf{B}$.

Assume the EDS derived from the *m*th largest EDR is $\mathbf{D^m}$, and the BSs with the 1st to $(m-1)$th largest EDR are in set $\mathbf{D^o}$. From Definition 6.1.2, Lemma 6.1.2 is derived.

Lemma 6.1.2 *The mth maximal EDR,* $2 \leq m \leq N$, *is minimized if* $\exists d^m \in \mathbf{D^m}$, *whose pilot power level decrement,* w_{b^m}, *is larger than* $g_{\mathbf{D^m},\mathbf{D^o}}^s$.

Proof: Lemma 6.1.2 can be readily derived by Definition 6.1.2. □

Then, the MEB algorithm balances green energy consumption among BSs by minimizing the *m*th maximal EDR until any condition in Lemma 6.1.2 or 6.1.2 is satisfied. The pseudo code of the MEB algorithm is summarized in Algorithm 8.

The computational complexity of MEB in the worst case is $O(QN^5M)$. Here, Q is the number of power levels of the pilot signal. Theoretically, MEB is a pseudo polynomial time algorithm. However, if any upper bound signal is imposed on the number of power levels of the pilot signal, the MEB algorithm becomes a polynomial time algorithm [142]. Although scaling with N^5, the MEB algorithm does not restrict the real time responsiveness of the GEO for two reasons. On the one hand, for a given coverage area, the number of BSs is fixed. If the number of power levels is given, the MEB algorithm only scales with M. Since the MEB algorithm is implemented in the network controller, the computational resources required for executing the MEB algorithm with large N can be pre-allocated to enable real time responsiveness. On the other hand, the time interval between two consecutive cell size adaptations (the duration of one time slot) depends on the dynamics of the green energy generation and the mobile traffic intensity. For green energy generation, the granularity for solar energy generation prediction is usually an hour [48]. For mobile traffic intensity, the hourly mobile traffic profile can represent

---

Algorithm 8: The MEB Algorithm

Input : $E_{i,j}^A, j \in (1, 2, \cdots, N)$;

Output: $\vec{P}_i^0$;

1 Initialize $\vec{P}_i^0$ and $\mathbf{D^o} = \emptyset$;

2 $[\vec{P}_i^0, \delta] = LEM(\vec{P}_i^0, E_{i,j}^A, \mathbf{A})$;

3 **while** $\mathbf{D^o} \neq \mathbf{A}$ **do**

4   Find the set $\mathbf{C^*}$ such that $\delta_{i,j} \geq \delta, j \in \mathbf{C^*}$;

5   Set $\mathbf{D^o} = \mathbf{D^o} \cup \mathbf{C^*}$;

6   **if** $\mathbf{D^o} \neq \mathbf{A}$ **then**

7   Set OPT = FALSE;

8   **while** *OPT == FALSE* **do**

9   Find the largest EDR $\delta = \max \delta_{i,j}, j \in \mathbf{A} \setminus \mathbf{D^o}$;

10   Find the set $\mathbf{C}$ such that $\delta_{i,j} = \delta, j \in \mathbf{A} \setminus \mathbf{D^o}$;

11   Update EDS, $\mathbf{D}$, and calculate pilot power decrements, $\vec{w}$;

12   **if** $\mathbf{D} == \mathbf{A} \setminus \mathbf{D^o} \parallel \exists j \in \mathbf{D}$ *such that* $w_j \geq p_{i,j}^0$ **then**

13   Set OPT = TRUE;

        else if $\mathbf{D^o} \cap \mathbf{D} \neq \emptyset \parallel \arg\max w_j \geq g_{\mathbf{D},\mathbf{D^o}}^s$ **then**

14   Set OPT = TRUE;

        else

15   Reduce $p_{i,j}^0$ by $w_j, j \in \mathbf{D}$, update δ.

---

mobile traffic characteristics sufficiently to guide the operation of the BSs [147]. Therefore, the time slot duration could be tens of minutes, which are long enough to execute the MEB algorithm.

6.1.3.3 The EA Algorithm

Based on the solution of the MEB algorithm, individual BSs apply the EA algorithm to adjust their green energy allocations. At the ith time slot, if $C_{i,j} \leq E_{i,j}^A$, BS j is powered by green energy; the superfluous green energy, which equals $(E_{i,j}^A - C_{i,j})$, will be evenly allocated to subsequent time slots. If $E_{i,j}^A < C_{i,j} \leq E_{i,j}$, this indicates that the energy demand is larger than the allocated green energy, but less than the actual available green energy in the batteries. In this case, BS j increases $E_{i,j}^A$ to $C_{i,j}$. As a result, the green energy allocation of the subsequent time slots will be decreased. The energy decrements are proportional to the amount of allocated green energy at each time slot. If $C_{i,j} > E_{i,j}$, BS j is unable to be powered by green energy, and the green energy allocated at the ith time slot will be evenly allocated to the subsequent time slots. The pseudo code of the EA algorithm is shown in Algorithm 9.

Algorithm 9: The EA Algorithm

Input : $\vec{P}_i^0$;
Output: $E_{k,j}^A$, $\beta_{i,j}$, $k \in (i+1, \cdots, L)$;

1 Calculate energy cost $C_{i,j}$ based on $\vec{P}_i^0$;

2 **if** $C_{i,j} \le E_{i,j}^A$ **then**

3   Set $\beta_{i,j} = 1$ and $E_{k,j}^A = E_{k,j}^A + \frac{(E_{i,j}^A - C_{i,j})}{(L-k+1)}$, $k \in (i+1, \cdots, L)$;

    else if $E_{i,j}^A < C_{i,j} \le E_{i,j}$ **then**

4   Set $\beta_{i,j} = 1$ and $E_{k,j}^A = E_{k,j}^A (1 - \frac{C_{i,j} - E_{i,j}^A}{\sum_{k=i+1}^{L} E_{k,j}^A})$, $k \in (i+1, \cdots, L)$;

    else

5   Set $\beta_{i,j} = 0$ and $E_{k,j}^A = E_{k,j}^A + E_{i,j}^A/(L - k + 1)$, $k \in (i+1, \cdots, L)$.

6.1.4 Performance Evaluation

A total of 36 BSs are located in a six by six grid. The distance between two adjacent BSs is 400 meters. The carrier frequency is 2110 *MHz* and the bandwidth is 5 *MHz*. COST 231 Walfisch-Ikegami [144] is adopted as the propagation model with 9 *dB* Rayleigh fading and 5 *dB* Shadow fading. For the BS power model, it is assumed that the antenna feeder loss is 3 *dB*, the transmitter gain is 10 *dB*, and the power amplifier efficiency is 50%. The SAM [148] and PVWatts [149] models are adopted to estimate the solar energy generation at individual time slots. Since the SAM and PVWatts models only provide hourly energy generation estimations, the linear interpolation method is adopted to derive the solar energy generation at each time slot. The solar energy generation profile reflects the solar energy generation in New York City during summer time.

To simulate the temporal and spatial diversities of mobile traffic, it is assumed that mobile users are uniformly distributed in the area, and the total number of users in the system peaks between 10:00 a.m. and 7:00 p.m., and bottoms between 1:00 a.m. and 5:00 a.m. During the peak hours, the total number of users is up to 500 while during the off-peak hours, the total number of users is up to 20. In order to reflect the mobile traffic characteristic that the traffic volume is almost stable at the same time on two consecutive days, it is assumed that the number of users within BS *j*'s coverage area at the *i*th time slot of a day is the same as that of the previous day. However, the mobile users are randomly distributed in BS *j*'s coverage area, and thus the energy consumption may vary. The data rate of all mobile users is set to 384 *kbps*. For simplicity, it is assumed that BSs have complete knowledge of the users' distances from BSs. The mobile network starts consuming green energy at 8:00 a.m. and ends at 7:00 p.m. The time slot duration is 600 seconds. In the simulation, the GEO algorithm is compared with the best-effort approach, in which the BSs consume green energy as long as the in-stored green energy is larger than the energy demand. The best-effort algorithm neither optimizes the energy allocation among time slots nor adapts BSs' cell sizes.

Figures 6.5 and 6.6 show the performance of the MEA algorithm for two selected BSs. Figure 6.5 shows the EDR for BS 1 whose total green energy generation during the

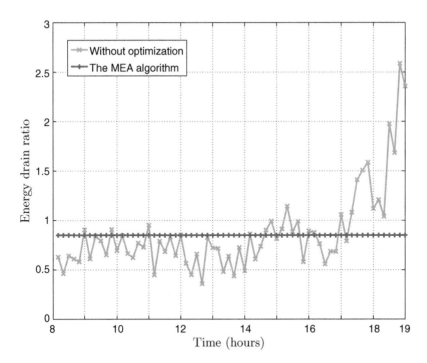

Figure 6.5 The energy drainage rate of BS 1. *Source:* Han 2013 [18]. Reproduced with permission of IEEE.

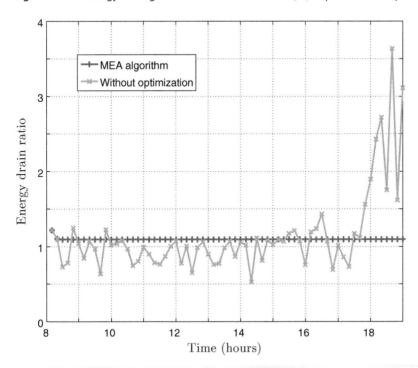

Figure 6.6 The energy drainage rate of BS 2. *Source:* Han 2013 [18]. Reproduced with permission of IEEE.

simulated time slots is larger than its total energy consumption. The y-axis is the value of EDR, and the x-axis is the time. From Figure 6.5, the MEA algorithm optimizes the green energy allocation over time slots. As a result, the EDR of BS 1 is less than 1 for all the time slots, implying that BS 1 can be powered by green energy for all the time slots. Without the MEA algorithm, the EDRs of BS 1 after 5:00 p.m. increase dramatically and become larger than 1, implying that the BS has to consume on-grid energy.

Figure 6.6 shows the EDR for BS 2 whose total green energy generation during the simulated time slots is smaller than its total energy consumption. In this case, the MEA algorithm balances the energy allocation among time slots to make the EDRs of the BS at individual time slots slightly larger than 1. However, it does not indicate that the BS consumes on-grid energy at every time slot. Since all the EDRs are slightly larger than 1, the energy gap is likely to be filled by applying the MEB algorithm. Hence, the overall on-grid energy consumption will be reduced. Without the MEA algorithm, the EDRs of the BS at the end of the day are extremely large, and the energy gap is less likely to be filled by the MEB algorithm.

Figure 6.7 shows the effectiveness of the MEB algorithm. In the figure, the EDRs of the BSs are sorted from the largest to the smallest, and the x-axis is the BS index while the y-axis is the EDR of the BSs in one time slot. As shown in the figure, the MEB algorithm balances the energy consumption among BSs. In the network, when the EDR of the BS is less than 1, the BS is powered by green energy; otherwise, it consumes on-grid energy. For both the best-effort approach and the MEB algorithm, the EDRs of the BSs indexed from 16 to 36 are less than 1. It indicates that these BSs are powered by green energy. The EDRs of the BSs indexed from 1 to 15 are larger than 1, implying that these BSs may

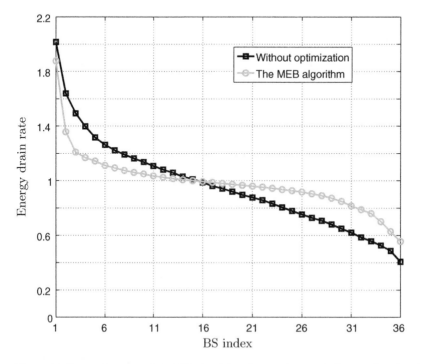

Figure 6.7 The performance of the MEB algorithm. *Source:* Han 2013 [18]. Reproduced with permission of IEEE.

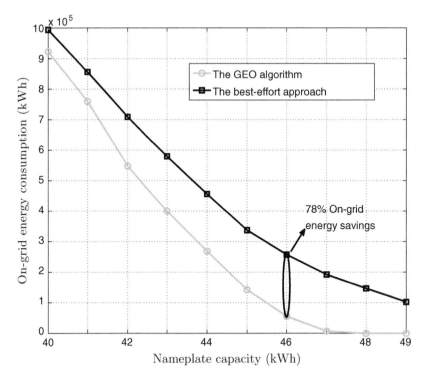

Figure 6.8 The on-grid energy consumption with different generation rates. *Source:* Han 2013 [18]. Reproduced with permission of IEEE.

consume on-grid energy. However, the MEB algorithm reduces the EDRs of these BSs by balancing the energy consumption among the BSs. In other words, the energy gaps are reduced by applying the MEB algorithm. With a reduced energy gap, individual BSs may be more likely to adjust their green energy allocation to fill the energy gap. Hence, the probability of these BSs being powered by green energy is increased. Thus, applying the MEB algorithm balances energy consumption among BSs, and therefore reduces the energy gaps of the BSs which are short of green energy. As a result, with reduced energy gaps, the BSs may be enabled to utilize green energy by adjusting the energy allocation using the EA algorithm.

Figures 6.8 and 6.9 show the mobile network's on-grid energy consumption and green energy consumption, respectively, at different green energy generation rates. The nameplate capacity of the solar panel indicates the green energy generation rate. The larger the nameplate capacity, the larger the green energy generation rate. As shown in Figure 6.8, as the green energy generation rate increases, the on-grid energy consumption of the mobile network reduces. When the green energy generation rate is larger, more electricity is generated from green energy. Then, more BSs can serve mobile users using electricity generated by green energy instead of consuming on-grid energy. When the nameplate capacity of the solar panel is larger than 48 kWh, the GEO algorithm achieves zero on-grid energy consumption while the best-effort algorithm still consumes a significant amount of on-grid energy. As compared with the best-effort approach, the GEO algorithm can save up to 78% of on-grid energy.

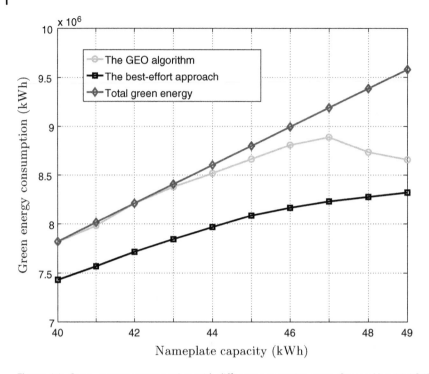

Figure 6.9 Green energy consumption with different generation rates. *Source:* Han 2013 [18]. Reproduced with permission of IEEE.

Figure 6.9 shows the mobile network's green energy utilization versus different green energy generation rates. As compared with the best-effort algorithm, the GEO algorithm enables the BSs to maximize the utilization of green energy. While the green energy generation rate increases, the green energy utilization of the GEO algorithm increases. When the nameplate capacity of the solar panel is larger than 47 kWh, the green energy utilization of the GEO algorithm decreases because the number of BSs which are short of green energy is reduced. In other words, an increasing number of BSs is able to generate sufficient green energy to serve the mobile users under their coverage. Therefore, fewer BSs are required to increase their cell sizes to cover the area of their neighboring BSs. Hence, the total green energy consumption decreases. As shown in Figures 6.8 and 6.9, the GEO algorithm is able to maximize the utilization of green energy. As a result, the GEO algorithm requires a smaller green energy generation rate to achieve zero on-grid energy consumption.

6.2 Optimal Renewable Energy Provisioning for BSs

With the development of green energy technologies, BSs can be powered by green energy in order to reduce on-grid energy consumption, and subsequently reduce carbon footprints. However, equipping a BS with a green energy system incurs additional CAPEX that is determined by the size of the green energy generator, the battery capacity, and other installation expenses. It is preferable to minimize the CAPEX on provisioning green energy while achieving the target QoS. We refer to this problem as the GEP problem. In this section, we investigate the GEP problem. We consider solar energy as the green energy source. Given the per unit cost of the solar panel and the battery capacity, the CAPEX of a BS's green energy system is determined by three variables: the size of the solar panel, the battery capacity, and the cost weight. Here, the cost weight indicates the per unit installation expense of the green energy system on a BS. Given the solar power generation rate and the characteristics of the battery, the size of the solar panel and the battery capacity are determined by the BS's power consumption. Thus, the CAPEX of a BS's green energy system is closely related to its power consumption. A BS's power consumption consists of the static power consumption and the dynamic power consumption [3]. The dynamic power consumption is the amount of power consumed for carrying traffic loads. For SBSs, dynamic power consumption accounts for a small portion of the total power consumption [3]. Thus, in this section, we do not study green energy provisioning for SBSs.

An MBS's power consumption is, however, highly traffic load dependent. Thus, the power consumption of MBSs can be adjusted by properly balancing traffic loads among BSs. Adapting MBSs' power consumption can change the green energy provisioning costs and thus reduce the network CAPEX. In order to minimize network CAPEX, it is desirable to reduce the power consumption of the BSs which have a large cost weight by optimizing the traffic loads among BSs. Thus, we decompose the GEP problem into two sub-problems: the weighted energy minimization (WEM) problem and the green energy system sizing (GESS) problem. An MBS's power consumption depends on its traffic load. Therefore, the WEM problem is solved by designing a provisioning cost aware traffic load balancing algorithm to optimize the traffic loads among BSs. Given MBSs' traffic loads, their energy consumption can be derived. Then, the solar panel size and battery capacity are optimized by solving the GESS problem.

6.2.1 Related Work on Provisioning the Green Power System

The process of sizing a green power system involves three basic models: the load model which characterizes energy demands, the battery model which defines the battery capacity and charging characteristics, and the green power generator model which describes the generator capacity [50]. Based on these models, three methods can be utilized to determine and evaluate the size of a green power system [50]: the loss of load and energy probability method, the fixed autonomy and recharge method, and the Markov chain probabilistic method. These methods are not applicable to solve the GEP problem because they do not optimize energy demands to minimize the size of the green energy system. For the GEP problem, the energy demands of individual MBSs depend on their traffic loads which should be optimized to minimize the network CAPEX. Badawy *et al.* [136] investigated the energy provisioning problem for solar powered wireless mesh networks, and designed a generic algorithm to incorporate energy aware routing in the energy provisioning procedure. Although, by incorporating energy aware routing, this method optimizes the energy consumption of wireless nodes, it is designed for wireless mesh networks and cannot be applied to solve the GEP problem. Marsan *et al.* [151] proposed the concept of "zero grid electricity networking" in which cellular networks are powered solely by renewable energy and investigated the problem of dimensioning power generator and battery storage capacities. The authors studied the green energy provisioning problem for a single MBS based on measurement of the BS power consumption and the renewable energy generation. In this section, we focus on optimizing green energy provisioning for a collection of BSs in HetNets.

6.2.2 System Model and Problem Formulation

In this section, we consider a heterogeneous cellular network with multiple MBSs and SBSs. The MBSs are powered by both solar power and grid power while the SBSs are powered by grid power. We focus on optimizing the size of each MBS's green energy system for downlink transmission. The time horizon is divided into N time slots. In the following analysis, a BS generally refers to an MBS or an SBS.

6.2.2.1 Traffic Model

We consider the scenario in which MBSs and SBSs are deployed to provide data communications in an area. Denote the set of MBSs and SBSs as B^m and B^s, respectively. Since provisioning the green energy system is a network planning problem, the solution of the green energy provisioning depends on the traffic statistics of a target area. Therefore, in this section, we adopt a location-based traffic model. We assume that, in the kth time slot, the traffic arrives at location x according to a Poisson process with the average arrival rate equal to $\lambda(x, k)$ per second, and the traffic size (packet size) per arrival having a general distribution with an average traffic size of $v(x, k)$ bits. Here, $\lambda(x, k)$ and $v(x, k)$ can be derived based on the statistics of traffic data from traffic measurements. For presentation simplicity, we assume there is only one user at location x. Assuming that the mobile user at location x is associated with the jth BS, then the user's data rate $r_j(x)$ bits per second can be generally expressed as a logarithmic function of the perceived signal to interference plus noise ratio, $SINR_j(x)$, according to the Shannon–Hartley theorem [19]:

$$r_j(x) = W_j log_2(1 + SINR_j(x)). \tag{6.14}$$

Here, W_j is the total bandwidth in the jth BS,

$$SINR_j(x) = \frac{P_j g_j(x)}{\sigma^2 + \sum_{h \in B, h \neq j} P_h g_h(x)}, \tag{6.15}$$

where P_j is the transmit power of the jth BS, σ^2 denotes the noise power level, and $g_j(x)$ is the channel gain between the jth BS and the user at location x. Here, the channel gain reflects only slow fading including path loss and shadowing. For the purposes of energy provisioning, the channel gain is measured at a large time scale, and thus fast fading is not considered. The average traffic load density at location x in the jth BS is

$$\varrho_j(x, k) = \frac{\lambda(x, k) v(x, k) \eta_j(x)}{r_j(x)}. \tag{6.16}$$

Here, $\eta_j(x)$ is an indicator function. If $\eta_j(x) = 1$, the user at location x is served by the jth BS; otherwise, the user is not served by the BS. Assuming mobile users are uniformly distributed in the area and denoting $\mathcal{A}$ as the coverage area of all the BSs, the traffic load on the jth BS can be expressed as

$$\rho_j(k) = \int_{x \in \mathcal{A}} \varrho_j(x, k) dx. \tag{6.17}$$

This value of $\rho_j(k)$ indicates the fraction of time BS j is busy in the kth time slot.

6.2.2.2 Energy Model

In the network, MBSs are powered by both green energy and on-grid energy. Since we aim to investigate the green energy provisioning for MBSs, we assume SBSs are powered by on-grid power. The MBS's power consumption consists of two parts: the static power consumption and the dynamic power consumption [140]. The static power consumption is the power consumption of an MBS without any traffic load. The dynamic power consumption refers to the additional power consumption incurred by carrying traffic loads in the MBS, which can be well approximated by a linear function of the traffic load [140]. Define p_j^s as the static power consumption of the jth MBS. Then, the jth MBS's power consumption in the kth time slot can be expressed as

$$p_j(k) = \beta_j \rho_j(k) + p_j^s. \tag{6.18}$$

Here, β_j is a linear coefficient which reflects the relationship between the traffic load and the dynamic power consumption in the jth MBS.

Define $e_j(k)$ as the green energy capacity per unit area of solar panel in the jth MBS in the kth time slot. We define S_j, B_j^{max}, and B_j^{min} as the solar panel size, the battery capacity, and the minimum permitted battery energy of the jth MBS's green power system, respectively. We adopt the linear charge model for the solar power system [136]. Then, the jth MBS's battery energy in the kth time slot can be expressed as

$$b_j(k) = \min \left\{ B_j^s(k), B_j^{max} \right\}, \tag{6.19}$$

where

$$b_j^s(k) = \max \left\{ b_j(k-1) + e_j(k)S_j - p_j(k), B_j^{min} \right\}. \tag{6.20}$$

Note that $b_j(k)$ depends on the battery energy in the $(k-1)$th time slot and the energy generation and consumption in the current time slot. For reasons of safety and battery life, the battery is not allowed to be discharged below B_j^{min}. In other words, if $b_j(k) \leq B_j^{min}$, the charge controller disconnects the jth MBS from the battery and pulls power from the power grid. For simplicity, we assume $B_j^{min} = 0$ in this section. The battery also cannot be charged beyond its capacity, B_j^{max}.

The cost of the green energy system is determined by the solar panel size and the battery capacity. In this section, we adopt a simple linear model to reflect the cost of the green energy system versus the solar panel size and battery capacity as follows:

$$f_j\left(S_j, B_j^{max}\right) = \phi_s S_j + \phi_b B_j^{max}. \tag{6.21}$$

Here, ϕ_s and ϕ_b indicate the cost per unit area of solar panel and per unit battery capacity, respectively. In addition, equipping an MBS with green energy also incurs installation expenses including labor costs and costs for leasing the space to install the green energy system. The per unit energy system installation cost may be different for MBSs at different locations. In this section, we assume the locations of MBSs are predetermined. We define w_j as the cost weight of installation per unit green energy system in the jth MBS. The CAPEX of the jth MBS's green energy system is $w_j f_j(S_j, B_j^{max})$.

6.2.2.3 Problem Formulation

The CAPEX of MBSs' green energy systems depends on the power consumption of the MBSs. In order to minimize the CAPEX, it is desirable to offload as much traffic load from MBSs to SBSs as possible. Aggressive traffic offloading may lead to traffic congestion in SBSs, and thus downgrading the QoS of the network. Therefore, when optimizing green energy provision, traffic offloading should be properly considered to ensure the QoS of the network.

We assume that traffic arrival processes at individual users are independent and follow Poisson distributions. Then, the traffic arrival in the jth BS, which is the sum of the traffic arrivals of all users in its coverage area, is also a Poisson process. The required service time for a user at location x in the jth BS is

$$\gamma_j = \frac{v(x,k)}{r_j(x)}. \tag{6.22}$$

Since $v(x,k)$ follows a general distribution, the user's required service time is also a general distribution. Hence, a BS's service rate follows a general distribution. Therefore, a BS's downlink transmission process realizes an M/G/1 processor sharing queue,[3] in which multiple users share the BS's downlink radio resource [152].

In mobile networks, various downlink scheduling algorithms have been proposed to enable proper sharing of the limited radio resource in a BS [153]. These algorithms are designed to achieve different network optimization objectives such as increasing network capacity, enhancing fairness among users, or improving network QoS. According to the scheduling algorithm, users are assigned different priorities on sharing the downlink radio resource. We assume that during the traffic balancing process, users' data rates do not change. As a result, users in different priority groups perceive different average waiting times. Since traffic arrives at a BS according to Possion arrival statistics, the allowed variation in the average waiting times among different priority groups is constrained by the Conservation Law [152]. The integral constraint on the average waiting time in the jth BS in the kth time slot can be expressed as

$$\mu(\rho_j(k)) = \frac{\rho_j(k)E(\gamma_j^2)}{2(1-\rho_j(k))}. \tag{6.23}$$

This indicates that given the users' required service time in the jth BS, if the scheduling algorithm gives some users higher priority and reduces their average waiting time, it will increase the average waiting time of other users. Therefore, $\mu(\rho_j(k))$ generally reflects the jth BS's performance in terms of users' average waiting time. Since $E(\gamma_j^2)$ mainly

---

[3] Markovian arrival; general distribution of service times; single server.

reflects the traffic characteristics, we assume that $E(\gamma_j^2)$ is roughly constant during a user association process and define

$$\vartheta_j = \frac{E(\gamma_j^2)}{2}.$$ (6.24)

Thus, we adopt

$$\mu(\rho_j(k)) = \frac{\vartheta_j \rho_j(k)}{1 - \rho_j(k)}$$ (6.25)

as a general latency indicator for the jth BS. A smaller $\mu(\rho_j(k))$ indicates that the jth BS introduces less latency to its associated users. For simplicity, we use $\mu_j(k)$ to represent $\mu(\rho_j(k))$. We utilize $\mu_j(k)$ as the QoS indicator for the jth BS in the kth time slot. To ensure the QoS of the network, $\mu_j(k)$ should be less than a threshold, ζ. Then, the GEP problem can be formulated as:

$$\min_{(S_j, B_j^{max}, \forall j \in B^m)} \sum_{j \in B^m} w_j f_j\left(S_j, B_j^{max}\right)$$ (6.26)

subject to: $\mu_j(k) \leq \zeta, \ \forall j \in B^m \cup B^s;$ (6.27)

$$b_j(k-1) + e_j(k)S_j \geq$$
$$\alpha(k)p_j(k), \ \forall j \in B^m;$$ (6.28)

$$0 \leq \rho_j(k) \leq 1 - \epsilon,$$
$$\forall j \in B^m \cup B^s, \forall k.$$ (6.29)

There are three constraints for the GEP problem. The first constraint, Eq. (6.27), requires the latency ratios of all BSs to be no larger than ζ. Here, ζ should be properly selected to ensure the feasibility of the GEP problem. The second constraint, Eq. (6.28), requires individual MBSs' green power supplies not to be less than its green power demand. Here, green power is defined as power generated from green energy. $0 \leq \alpha(k) \leq 1$ is a system parameter that defines the percentage of the power consumption that should be pulled from the MBS's green energy system. Then, $\alpha(k)$ can be selected by the mobile network operator in provisioning the green energy system. A larger $\alpha(k)$ usually results in a higher CAPEX and a lower OPEX in terms of the energy cost. The third constraint, Eq. (6.27), is to ensure the queuing system is stable by restricting the traffic load in individual BSs to be less than 1. Here, ϵ is an arbitrarily small positive real number.

Given the BS deployment and the traffic load statistics, the lower bound of ζ can be derived in two steps. First, by solving the QoS bound (QB) problem expressed as

$$\min_{\rho_j(k)} \max_{j \in B^m \cup B^s} \mu_j(k)$$ (6.30)

subject to: $0 \leq \rho_j(k) \leq 1 - \epsilon,$ (6.31)

we can derive the lower bound of ζ in the kth time slot, which is denoted as

$$\mu^*(k) = \frac{\vartheta_j \rho_j^*(k)}{1 - \rho_j^*(k)}, \ j = \arg \max_{l \in B^m \cup B^s} \mu_l(k).$$ (6.32)

Here, $\rho_j^*(k)$ is the jth BS's optimal traffic load derived by solving the QB problem in the kth time slot. Then, ζ's lower bound is $\zeta^* = \max_{k\in\{1,2,\cdots,N\}} \mu^*(k)$. To ensure the GEP problem is feasible, $\zeta \geq \zeta^*$. Similar to the parameter $\alpha(k)$, ζ is predetermined by the mobile network operator for green energy provision.

6.2.3 The Green Energy Provisioning Solution

Solving the GEP problem is equivalent to determining the optimal solar panel sizes and battery capacities for MBSs. Since the green power systems are provisioned to operate the MBSs during a certain time period (multiple time slots), the solar panel sizes and battery capacities are determined to satisfy the MBSs' green power demands over the time slots. Since battery energy in a time slot depends on that in the previous slots, the optimal solar panel size and battery capacity for an MBS is determined by the MBS's green power demands in multiple time slots. Within a time slot, say the kth time slot, the green power demand in an MBS, say the jth MBS, depends on its traffic load $\rho_j(k)$ and the parameter $\alpha(k)$. Thus, solving the GEP problem involves optimizing traffic load in multiple time slots. Owing to the complex coupling of network optimization in multiple time slots, it is very challenging to solve the GEP problem.

6.2.3.1 Problem Decomposition

In order to solve the GEP problem, we decompose it into two sub-problems: the WEM problem and the GESS problem. In this way, we decouple the interdependence of network optimization in multiple time slots. The WEM problem optimizes the network's weighted energy cost in individual time slots while the GESS problem optimizes solar panel sizes and battery capacities for individual MBSs according to their energy demands over multiple time slots.

For optimizing the solar panel size, we assume that the jth MBS's initial battery energy, $b_j(0)$, is zero, and that the green energy consumed by the jth MBS is all generated from its solar panel. Thus,

$$\sum_{k\in\{1,2,\cdots,N\}} e_j(k)S_j \geq \sum_{k\in\{1,2,\cdots,N\}} \alpha(k)p_j(k). \tag{6.33}$$

Considering all MBSs and their weights,

$$\sum_{k\in\{1,2,\cdots,N\}}\sum_{j\in B^m} w_j e_j(k)S_j \geq$$
$$\sum_{k\in\{1,2,\cdots,N\}}\sum_{j\in B^m} w_j \alpha(k)p_j(k). \tag{6.34}$$

E.q. (6.34) can be rewritten as

$$\sum_{j\in B^m}\left(w_j S_j \sum_{k\in\{1,2,\cdots,N\}} e_j(k)\right) \geq$$
$$\sum_{k\in\{1,2,\cdots,N\}}\sum_{j\in B^m} w_j \alpha(k)p_j(k). \tag{6.35}$$

Here, $e_j(k)$, $\forall j \in B^m$, $\forall k \in \{1, 2, \cdots, N\}$ is derived based on the statistical solar power data and is considered as a constant for the jth MBS. We assume all MBSs have the similar geolocations. Thus,

$$\sum_{k\in\{1,2,\cdots,N\}} e_j(k) = \sum_{k\in\{1,2,\cdots,N\}} e_i(k), \forall i,j \in B^m. \tag{6.36}$$

Therefore, minimizing

$$\sum_{k\in\{1,2,\cdots,N\}} \sum_{j\in B^m} w_j \alpha(k) p_j(k) \tag{6.37}$$

is necessary in order to minimize $\sum_{j\in B^m} w_j S_j$. Since the traffic arrival is a Poisson process, an MBS's traffic loads in different time slots are independent. Thus, the MBS's energy consumptions in different time slots are independent. Therefore, minimizing Eq. (6.37) is equivalent to minimizing

$$\sum_{j\in B^m} w_j \alpha(k) p_j(k), \ \forall k \in \{1, 2, \cdots, N\}. \tag{6.38}$$

Since the MBSs share a similar geolocation, the time slots in which the solar panels do not generate green energy are the same for all the MBSs. Define $\mathcal{K}^b$ as the set of these time slots. During these time slots, battery energy is utilized to satisfy the MBSs' demands for green energy. Thus,

$$B_j^{\max} \geq \sum_{k\in\mathcal{K}^b} \alpha(k) p_j(k). \tag{6.39}$$

Considering all the MBSs and their cost weights,

$$\sum_{j\in B^m} w_j B_j^{\max} \geq \sum_{k\in\mathcal{K}^b} \sum_{j\in B^m} w_j \alpha(k) p_j(k). \tag{6.40}$$

Since MBSs' energy consumption in different time slots is independent, minimizing

$$\sum_{j\in B^m} w_j \alpha(k) p_j(k), \ \forall k \in \mathcal{K}^b \tag{6.41}$$

is necessary in order to minimize

$$\sum_{j\in B^m} w_j B_j^{\max}. \tag{6.42}$$

Based on the above analysis, the WEM problem can be expressed as

$$\min_{(\rho_j(k),\forall j\in B^m)} \sum_{j\in B^m} w_j \alpha(k) p_j(k) \tag{6.43}$$

subject to: $\mu_j(k) \leq \zeta, \ \forall j \in B^m \cup B^s,$

$$0 \leq \rho_j(k) \leq 1 - \epsilon, \ \forall j \in B^m \cup B^s. \tag{6.44}$$

Given the jth MBS's energy consumption in all time slots, the GESS problem can be expressed as

$$\min_{\left(S_j,B_j^{\max}\right)} f_j\left(S_j, B_j^{\max}\right) \tag{6.45}$$

subject to: $b_j(k-1) + e_j(k)S_j \geq \alpha(k) p_j(k),$

$$\forall j \in B^m, \forall k \in \{1, 2, \cdots, N\}. \tag{6.46}$$

6.2.3.2 Provisioning Cost Aware (PCA) Traffic Load Balancing

The PCA traffic load balancing scheme is to determine the traffic load of individual BSs according to location-based traffic load density derived from traffic statistics rather than from the real time user distribution. Since the WEM problem minimizes MBSs' weighted power consumption within a time slot, we use μ_j, ρ_j, and α instead of $\mu_j(k)$, $\rho_j(k)$, and $\alpha(k)$. Since ϑ_j is a constant within a time slot, we assume $\vartheta_j = 1$ for presentation simplicity. For the WEM problem, since $0 < \zeta < \infty$, when $\mu_j \leq \zeta$,

$$\rho_j \leq \frac{\zeta}{1+\zeta} \leq 1 - \epsilon. \tag{6.47}$$

Since $\rho_j > 0$, $\mu_j(k) \leq \zeta$ indicates $0 \leq \rho_j \leq 1 - \epsilon$. Therefore, the second inequality constraint of the WEM problem can be eliminated. We then apply Lagrangian dual decomposition to design a provisioning cost aware traffic load balancing algorithm to solve the WEM problem.

Let $\mathcal{B} = \mathcal{B}^m \cup \mathcal{B}^s$ and $w_j = 0$, $\forall j \in \mathcal{B}^s$. We introduce a Lagrangian multiplier vector, $\upsilon = (\upsilon_1, \cdots, \upsilon_j, \cdots, \upsilon_{|\mathcal{B}|})$. The Lagrangian function of the WEM problem is

$$L(\rho, \upsilon) = \sum_{j \in \mathcal{B}} \alpha w_j p_j - \upsilon_j \left(\frac{\zeta}{1+\zeta} - \rho_j \right). \tag{6.48}$$

According to Eq. (6.18),

$$p_j = \beta_j \rho_j + p_j^s \tag{6.49}$$

$$L(\rho, \upsilon) = \sum_{j \in \mathcal{B}} (\alpha w_j \beta_j + \upsilon_j) \rho_j + \alpha w_j \beta_j p_j^s - \upsilon_j \frac{\zeta}{1+\zeta} \tag{6.50}$$

Here, $\rho = (\rho_1, \cdots, \rho_j, \cdots, \rho_{|\mathcal{B}|})$. Since

$$\rho_j = \sum_{x \in \mathcal{A}} \frac{\lambda_i \upsilon_i \eta_j(x)}{r_j(x)}, \tag{6.51}$$

the dual function is given as

$$g(\upsilon) = \inf_{\eta} h(\upsilon, \eta) + \sum_{j \in \mathcal{B}} \alpha w_j \beta_j p_j^s - \upsilon_j \frac{\zeta}{1+\zeta}, \tag{6.52}$$

where

$$h(\upsilon, \eta) = \sum_{x \in \mathcal{A}} \sum_{j \in \mathcal{B}} (\alpha w_j \beta_j + \upsilon_j) \frac{\lambda_i \upsilon_i \eta_j(x)}{r_j(x)} \tag{6.53}$$

and

$$\eta = \{\eta_j(x) | j \in \mathcal{B}, x \in \mathcal{A}\}. \tag{6.54}$$

The dual problem is

$$\max_{\upsilon} g(\upsilon) \tag{6.55}$$

$$\textit{subject to: } \upsilon_j \geq 0, \forall j \in \mathcal{B}. \tag{6.56}$$

The provisioning cost aware traffic load balancing algorithm solves the dual problem and thus addresses the WEM problem. The proposed algorithm includes two parts: the traffic redirect algorithm and the traffic load update algorithm.

The traffic redirect algorithm derives $\eta_j(x)$ that minimizes $h(v, \eta)$ while the traffic load update algorithm finds the optimal v that maximizes $g(v)$. The provisioning cost aware traffic load balancing algorithm involves multiple iterations. We denote the Lagrangian multiplier in the tth iteration as $v^t = \{v_j^t | i \in B\}$.

The traffic redirect algorithm: this algorithm calculates the downlink data rates from all BSs based on the SINR measurements for a user at a location. The traffic to the user at location x is redirected to the j^*th BS according to the following traffic redirect rule:

$$j^* = \arg \min_{j \in B} \left(\alpha w_j \beta_j + v_j^t \right) \frac{\lambda_i v_i \eta_j(x)}{r_j(x)}. \tag{6.57}$$

Lemma 6.2.1 *Given v^t, the traffic redirect algorithm minimizes $h(v^t, \eta)$.*

Proof: Since a user can only be associated with one BS, if $\eta_{j^*}(x) = 1$, $\forall j \in B$ and $j \neq j^*$, $\eta_j(x) = 0$. Define η^* as the traffic redirection derived by traffic redirect algorithm. Assume η is an arbitrary traffic redirection such that $\eta \neq \eta^*$.

$$h(v^t, \eta^*) - h(v^t, \eta)$$
$$= \sum_{x \in A} \left[\left(\alpha w_{j^*} \beta_{j^*} + v_{j^*}^t \right) \frac{\lambda_i v_i \eta_{j^*}(x)}{r_{j^*}(x)} \right.$$
$$\left. - \left(\alpha w_j \beta_j + v_j^t \right) \frac{\lambda_i v_i \eta_j(x)}{r_j(x)} \right] \tag{6.58}$$

because

$$j^* = \arg \min_{j \in B} \left(\alpha w_j \beta_j + v_j^t \right) \frac{\lambda_i v_i \eta_j(x)}{r_j(x)}. \tag{6.59}$$

$h(v, \eta^*) - h(v^t, \eta) \leq 0$. Thus, the lemma is proved. □

Traffic load update algorithm: given the traffic redirection, η^*, the traffic load update algorithm measures BSs' traffic load and updates the Lagrangian multiplier to maximize $g(v)$. Define ρ_j^t as the jth BS's traffic load after the tth iteration. The multiplier in the jth BS in the $(t + 1)$th iteration is updated as

$$v^{t+1} = v^t + \delta^t \left(\rho_j^t - \frac{\zeta}{1 + \zeta} \right). \tag{6.60}$$

Here, $\delta^t > 0$ is a dynamically selected step size that ensures the convergence of the iterations between users and BSs. δ^t is chosen based on

$$\delta^t = \gamma^k \frac{g(v^t) - g(\hat{v}) + \varepsilon^t}{\left\| \rho_j^t - \frac{\zeta}{1+\zeta} \right\|^2}, \tag{6.61}$$

where $0 < \underline{\gamma} \le \gamma^k \le \overline{\gamma} < 2$, $\underline{\gamma}$ and $\overline{\gamma}$ are some scalars [154] and ε^t is updated according to

$$\varepsilon^t = \begin{cases} a\varepsilon^t, & g(v^{t+1}) \le g(v^t), \\ \max(b\varepsilon^t, \varepsilon), & g(v^{t+1}) > g(v^t), \end{cases} \tag{6.62}$$

where a, b and ε are fixed positive constants with $a \ge 1$ and $b < 1$. In Eq. (6.61), $\hat{v} = \{\hat{v}_j | j \in B\}$ is an estimation of the optimal Lagrangian multiplier as

$$\hat{v} = \arg\min_{(v^m, 0 \le m \le t)} g(v^m). \tag{6.63}$$

Proposition 6.2.1 *There exists some scalar c such that*

$$\sup\{\|q(v)\| \mid q(v) \in \partial g(v^t), \forall t \ge 0\} \le c. \tag{6.64}$$

Proof: It is known that

$$\partial g(v^t) = \inf_\eta \sum_{x \in A} \frac{\lambda_i v_i \eta_j(x)}{r_j(x)} - \frac{\zeta}{1+\zeta}. \tag{6.65}$$

Because $\eta_j(x) = \{0,1\}$,

$$\sum_{x \in A} \frac{\lambda_i v_i \eta_j(x)}{r_j(x)} \tag{6.66}$$

is bounded. Thus, the subgradient of the dual problem is bounded:

$$\sup\{\|q(v)\| \mid q(v) \in \partial g(v^t), \forall t \ge 0\} \le c. \tag{6.67}$$
□

Theorem 6.2.1 *Assume that δ^t is determined by the dynamic step size rule in Eq. (6.61) with the adjustment procedures in Eqs. (6.62) and (6.63). If $g(v^*) < \infty$,*

$$\sup_{t \ge 0} g(v^t) \ge g(v^*) - \varepsilon, \tag{6.68}$$

where v^* denotes the optimal Lagrangian multiplier.

Proof: Based on Proposition 6.2.1, the dual problem satisfies the necessary condition of Proposition 6.3.6 in [154]. The theorem is proved by applying this proposition. □

After v converges, optimal traffic load balancing is derived according to the traffic redirect algorithm, based on which we obtain the optimal traffic load ρ^* and thus calculate the BSs' energy consumption in the time slot.

6.2.3.3 Green Energy System Sizing
After solving the WEM problem for all the time slots, we obtain individual MBSs' energy consumption in each time slot. Based on the energy consumption, we solve the GESS problem to derive the optimal solar panel size and battery capacity for MBSs.

Figure 6.10 shows an example of solar power generation and green power demand in an MBS. The solar power generation starts in the k_1th time slot and ends in the k_5th

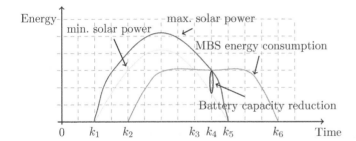

Figure 6.10 An illustration of solar panel size and battery capacity. *Source:* Han 2016 [155]. Reproduced with permission of IEEE.

time slot. The MBS is activated in the k_2th time slot and turned off in the k_6th time slot. In order to power the MBS, the solar power generation should be at least equal to the MBS's green power consumption. An MBS's green power consumption equals the MBS's total power consumption multiplied by the percentage of power pulled from green energy. We define the minimum solar panel size as the solar panel size with which the solar power generation is equal to the MBS's green power consumption. As shown in Fig. 6.10, when the solar power is generated using the minimum panel size (min. solar power), the MBS's green power comes from both the solar panel and the battery in the k_4th time slot, while the battery is responsible for the power supplies in the k_5th and the k_6th time slot. The battery capacity should be at least equal to the MBS's green power consumption from the k_4th to the k_6th time slot minus the solar power generation in the k_4th time slot.

Lemma 6.2.2 *On powering the SBS, increasing the solar panel size does not increase the required battery capacity.*

Proof: The battery is responsible for the power supplies during the time slots in which solar power is less than the MBS's power consumption. Given the solar energy generation rate, increasing the solar panel size does not increase the MBS's energy consumption, and thus does not increase the required battery capacity. □

In some cases, increasing the solar panel size enables a reduction in the required battery capacity. As shown in Fig. 6.10, an increase in the solar panel size increases the solar power generation. As a result, the MBS's energy consumption in the k_4th time slot is fully covered by solar power. The battery is only responsible for the power supplies from the k_5th to k_6th time slot. Thus, the battery capacity can be reduced. However, when the solar panel size is large enough, a further increase of the solar panel size does not decrease the required battery capacity because a sufficient battery capacity is required to power the network after the solar power generation rate reaching zero. We define the maximum solar panel size as the solar panel size for which a further increase of the panel size does not decrease the required battery capacity. Define $S_j^{\max}$ as the jth MBS's maximum solar panel size.

$$S_j^{\max} = \left\lceil \max_{k \in \{l \mid e_j(l) > 0,\, l \in \{1,2,\cdots,N\}\}} \frac{\alpha(k)p_j(k)}{e_j(k)} \right\rceil. \tag{6.69}$$

Here, $\lceil x \rceil$ denotes the smallest integer that is greater than or equal to x. Define S_j^{min} as the jth SBS's minimum solar panel size. Let $m_j = \arg\max_{k \in \{l | p_j(l) > 0,\, l \in \{1,2,\cdots,N\}\}} k$.

$$S_j^{min} = \left\lceil \frac{\sum_{k \in \{1,2,\cdots,m_j\}} p_j(k)}{\sum_{k \in \{1,2,\cdots,m_j\}} e_j(k)} \right\rceil. \qquad (6.70)$$

Solving the GESS problem involves a trade-off between solar panel size and battery capacity. We apply the binary search method to find the optimal solar panel size, and then derive the corresponding battery capacity. Given the jth MBS's solar panel size and solar energy generation rates, the jth MBS's solar power generation in an individual time slot is calculated. Given the MBS's green power consumption, the battery capacity is derived to guarantee that sufficient energy is stored to satisfy the MBS's energy demand in each time slot. If the current solar panel size cannot sufficiently charge the battery to power the MBSs during the time slots in which the solar power generation is less than the power consumption, we set the battery capacity to be infinite. As a result, the cost of the green energy system will be infinite. Thus, the binary energy system sizing (BESS) algorithm increases the solar panel size. Denote the jth MBS's intermediate solar panel size and battery capacity as S_j^{tmp} and B_j^{tmp}, respectively. The pseudo code of the (BESS) algorithm is shown in Algorithm 10.

Algorithm 10: The BESS algorithm

Input : $e_j(k), p_j(k), \forall k \in \{1,2,\cdots,N\}$;
Output: S_j, B_j^{max};Calculate S_j^{min} and S_j^{max};
2 Assign $S_j = S_j^{min}$, derive B_j^{max} and $f_j(S_j, B_j^{max})$;
3 **while** $S_j^{max} \neq S_j^{min}$ **do**
4 $\quad$ Assign $S_j^{tmp} = \lceil 1/2(S_j^{max} + S_j^{min}) \rceil$;
5 $\quad$ Calculate B_j^{tmp} and $f_j(S_j^{tmp}, B_j^{tmp})$;
6 $\quad$ **if** $f_j(S_j^{tmp}, B_j^{tmp}) \leq f_j(S_j, B_j^{max})$ *or* $f_j(S_j^{tmp}, B_j^{tmp}) = \inf$ **then**
7 $\quad\quad$ Assign $S_j = S_j^{tmp}$, $B_j^{max} = B_j^{tmp}$;
8 $\quad\quad$ Assign $S_j^{min} = S_j^{tmp}$;
9 $\quad$ **else**
10 $\quad\quad$ Assign $S_j^{max} = S_j^{tmp}$;
11 Assign $S_j = S_j^{min}$, derive B_j^{max} and $f_j(S_j, B_j^{max})$.

6.2.3.4 Computational Complexity

The solution to the GEP problem includes two parts: the PCA traffic load balancing algorithm and the BESS algorithm for sizing the energy system. The PCA traffic load balancing algorithm is designed based on the subgradient method [154]. The computational complexity of the subgradient method is $O(1/\varepsilon^2)$ to achieve the ε-optimal solution [154]. The binary search based algorithm requires at most $log_2(S_j^{max} - S_j^{min})$ iterations to

Figure 6.11 Average traffic rate. *Source:* Han 2016 [155]. Reproduced with permission of IEEE.

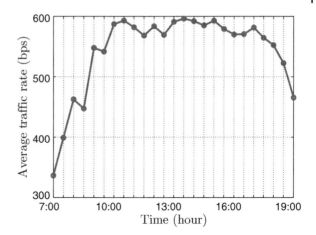

find the optimal solar panel size. Within each iteration, calculating the battery capacity with a given solar panel size requires N iterations. Thus, the complexity of the BESS algorithm is $O(N \log_2(S_j^{max} - S_j^{min}))$. Hence, the computational complexity of the proposed solution is $O(N \log_2(S_j^{max} - S_j^{min})/\varepsilon^2)$.

6.2.4 Performance Evaluation

We set up simulations to evaluate the performance of the proposed heuristic green energy provisioning solution in HetNets. In the simulation, we consider a HetNet with five MBSs and fifteen SBSs deployed in a $2km \times 2km$ area. The area is divided into 160 000 locations with each location representing a $5m \times 5m$ area. The average traffic load in each location is shown in Figure 6.11. The MBS's transmit power is 43 *dBm* while the SBS's transmit power is 33 *dBm*. The channel propagation model is based on COST 231 Walfisch-Ikegami [144]. The model and parameters are summarized in Table 6.1. Here, PL_{MBS} and PL_{SCBS} are the path losses between the users and MBSs and between the users and SCBSs, respectively. d is the distance between users and BSs. All these parameters are utilized to calculate the perceived SINR at individual locations. The total bandwidth is 10 *MHz* and the frequency reuse factor is one.

Table 6.1 Channel model and parameters

Parameters	Value
PL_{MBS} (dB)	$PL_{MBS} = 113.1 + 37.6 \log_{10}(d)$
PL_{SCBS} (dB)	$PL_{SCBS} = 38 + 10 \log_{10}(d)$
Rayleigh fading	9 *dB*
Shadowing fading	5 *dB*
Noise power level	−174 *dBm*
Receiver sensitivity	−123 *dBm*

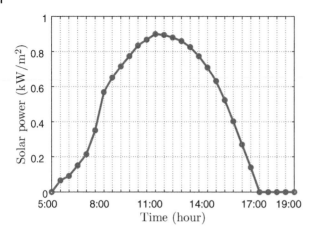

Figure 6.12 Solar power rate. *Source:* Han 2016 [155]. Reproduced with permission of IEEE.

The static power consumption and the load-power coefficient of the MBS are 750 *W* and 500, respectively [3]. Here, we assume all MBSs have the same static power consumption and the same linear coefficient, $\beta_j = \beta, \forall j \in B^m$. The duration of a time slot for the energy provisioning is 30 minutes. Solar power is utilized from 7:00 a.m. to 7:00 p.m. The solar power generation rate, which is shown in Figure 6.12, is obtained from the University of California – San Diego solar resource web application [156]. We generate the mobile traffic rates based on the mobile traffic pattern [147]. We assume the green power percentage α is the same in all time slots. The approximate threshold, ε, indicates that the algorithm achieves the ε-optimal solution. The values of parameters are shown in Table 6.2. The MBSs' provisioning cost weights are randomly selected.

Figure 6.13 shows the convergence of the PCA load balancing algorithm. The x-axis is the number of iterations while the y-axis is the value of the dual function. After about fifty iterations, the value of the dual function converges. When ζ increases, the dual function converges to a smaller value. When the dual function converges to a smaller value, the primal function also has a smaller value. This indicates that increasing ζ reduces the provisioning costs. This is because when ζ increases, the network can tolerate additional traffic latency. As a result, more traffic load will be redirected to SBSs, thus reducing the power consumption of MBSs.

Table 6.2 Simulation parameters

Parameters	Symbol	Value
Static power consumption	p_j^s	750 (W)
Load-power coefficient	β	500
Green power percentage	α	$0.5 \leq \alpha \leq 1$
Maximum latency ratio	ξ	$2 \leq \xi \leq 4$
Approximate threshold	ε	0.1

Figure 6.13 The convergence of provisioning cost aware load balancing ($\alpha = 1$). *Source:* Han 2016 [155]. Reproduced with permission of IEEE.

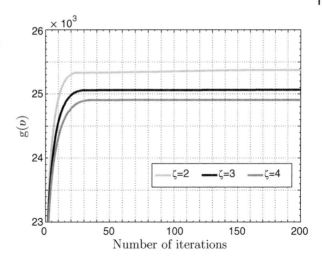

The traffic load balancing scheme is critical in minimizing green energy provisioning cost. We compare the proposed PCA traffic load balancing scheme with the data rate bias (DRB) scheme [36] and the traffic latency minimization (LM) scheme.

In the simulation, we consider a two-tier data rate bias scheme and assume that BSs in the same tier have the same cell bias. Since different data rate bias leads to different traffic load balancing results, we first evaluate the two-tier data rate bias scheme and find a proper data rate bias. In the simulation, MBSs are in the first tier while SBSs are in the second tier. The cell bias of an MBS is one. We vary the cell bias of a SBS to investigate the performance of the scheme. In the data rate bias algorithm, a user selects the BS to maximize the biased data rate.

$$b(x) = \arg \max_{j \in B^m \cup B^s} Z_j r_j(x). \tag{6.71}$$

Here, $b(x)$ and Z_j are the index of the selected BS and the cell bias of the jth BS, respectively.

Figure 6.14 shows that the maximum traffic latency ratio is a convex function of the data rate bias. The minimum value is achieved when the data rate bias is about 3.9.

Figure 6.14 The maximum latency ratio. *Source:* Han 2016 [155]. Reproduced with permission of IEEE.

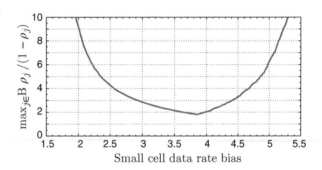

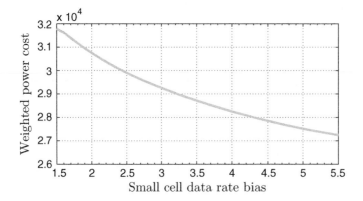

Figure 6.15 The weighted power cost. *Source:* Han 2016 [155]. Reproduced with permission of IEEE.

Meanwhile, Figure 6.15 shows that the weighted power cost reduces as the data rate bias increases. This is because increasing the data rate bias allows more traffic offloaded to SBSs and thus reduces the power consumption of MBSs. In the simulation, since $\zeta = 2$, we set the data rate bias to 4 for comparing with the PCA scheme. Note that when the data rate bias equals 4, the maximum traffic latency ratio is around 2.

The traffic latency minimization scheme solves the latency aware problem (LAP) as:

$$\min_{\rho} \sum_{j \in B^m \cup B^s} L(\rho_j) \tag{6.72}$$

$$subject\ to\text{:}\ 0 \le \rho_j \le 1 - \epsilon. \tag{6.73}$$

Figure 6.16 compares the maximum traffic latency ratio of the network under three traffic load balancing schemes. Since $\zeta = 2$, the PCA scheme maximizes traffic offloading while ensuring $\zeta \le 2$. The maximum traffic latency ratio of the data rate bias scheme depends on the traffic intensity of the networks. When traffic intensity is low (high), the DRB scheme achieves a small (large) traffic latency ratio. This is because the

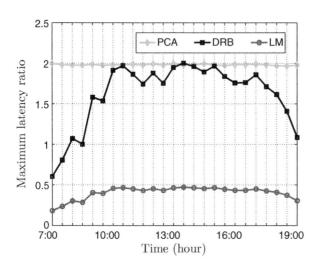

Figure 6.16 Maximum latency ratio of BSs ($\alpha = 1$). *Source:* Han 2016 [155]. Reproduced with permission of IEEE.

Figure 6.17 The weighted power cost of the network ($\alpha = 1$). *Source:* Han 2016 [155]. Reproduced with permission of IEEE.

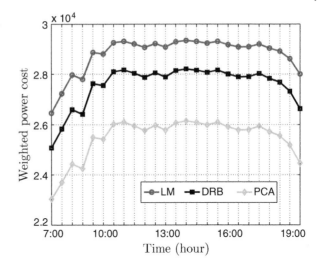

data rate bias is fixed and the traffic balancing rule does not change over time slots. The LM scheme achieves the lowest maximum traffic latency ratio.

Figure 6.17 shows the weighted power cost of the network under these traffic load balancing schemes. The PCA scheme has the lowest weighted power cost of the three schemes. This is because the PCA scheme offloads as much traffic load as allowed by the QoS constraint to SBSs. In this way, the total power consumption of MBSs is reduced. In addition, the PCA scheme also balances the traffic load among MBSs according to their provisioning weights. The MBS with a large provisioning weight serves less traffic loads than the MBS with a small provisioning weight does. Although the PCA scheme has the highest traffic latency ratio, the QoS of the network is guaranteed. Moreover, the traffic latency ratio of the algorithm can be adjusted by adapting ζ.

In Figure 6.18, we compare the total green energy provisioning costs of different solutions. The green energy provisioning solutions consist of two parts: the traffic load balancing scheme and the green energy system sizing scheme. In the simulation, the per m^2

Figure 6.18 The total provisioning cost. *Source:* Han 2016 [155]. Reproduced with permission of IEEE.

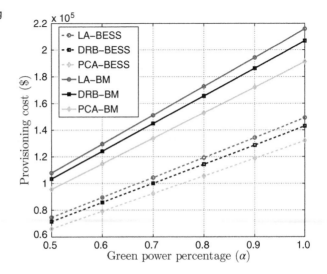

cost of the solar panel and the per Watt costs of the battery are $0.9 and $0.2, respectively. For the traffic load balancing scheme, we adopt the PCA traffic load balancing scheme, the DRB traffic load balancing scheme, and the LM traffic load balancing scheme. For the green energy system sizing scheme, we compare the proposed BESS algorithm and a battery minimization (BM) sizing algorithm that minimizes the battery capacity. In the simulation, the proposed solution consisting of the PCA load balancing scheme and the BESS algorithm incurs the smallest provisioning cost. The provisioning cost of the network increases with the green energy percentage. This is because a larger green energy percentage indicates more power should be pulled from the green energy generator, thus requiring a more powerful green energy system.

6.3 Summary

In this chapter, an optimization framework has been proposed to reduce the on-grid energy consumption of mobile networks with hybrid energy supplies. The green energy optimization problem is formulated and decomposed into two sub-problems: the multi-stage energy allocation problem and the multi-BSs energy balancing problem. The MEA algorithm, the MEB algorithm and the EA algorithm are proposed to solve these sub-problems, and thus address the GEO problem. The proposed solution has been demonstrated via extensive simulations to be able to save a significant amount of on-grid energy.

We have also investigated the green energy provisioning problem and proposed an energy provisioning solution to minimize the CAPEX of deploying the green energy system for MBSs in HetNets while achieving the targeted QoS requirement. The green energy provisioning solution consists of a provisioning cost aware traffic load balancing algorithm and a binary energy system sizing algorithm. Given a traffic load, the provisioning cost aware traffic load balancing algorithm balances the traffic load among BSs based on the QoS requirements and the provisioning costs. The energy consumption of MBSs are calculated based on their traffic loads. The BESS algorithm optimizes solar panel sizes and battery capacities for individual MBSs based on their power consumption. The simulation results have validated the performance and the viability of the proposed solution. Although various traffic load balancing algorithms may be adopted in HetNets, the energy provisioning solution based on provisioning cost aware traffic load balancing provides a lower bound on the provisioning costs of green energy systems. The results provide guidance for network planning and deployment from the perspective of provisioning green energy in cellular networks.

6.4 Questions

6.1 What is the framework for optimizing green energy utilization in a mobile network powered by hybrid energy sources?

6.2 What characteristics of mobile traffic are important for optimizing green energy utilization?

6.3 How does the GEO algorithm work?

7

Energy Aware Traffic Load Balancing in Mobile Networks

Dramatic mobile data traffic growth has spurred a dense deployment of small cell base stations (SCBSs). Small cells enhance the spectrum efficiency and thus enlarge the capacity of mobile networks. Although SCBSs consume much less power than MBSs do, the overall power consumption of a large number of SCBSs is phenomenal. As energy harvesting technology advances, BSs can be powered by green energy to alleviate on-grid power consumption. However, since electricity generated from renewable energy is not stable, green power may not be a reliable energy source for mobile networks. Therefore, future small cell networks (SCNs) are likely to adopt hybrid energy supplies: brown power and green power. Green power is utilized to reduce the brown power consumption while brown power is utilized as a backup power source [157]. For mobile networks with high BS density, traffic load balancing is critical in order to exploit the capacity of SCBSs. In order to optimize green power utilization, it is desirable to balance traffic load according to the availability of green power. For instance, mobile networks may enable BSs with sufficient green power to serve higher traffic loads while reducing the traffic loads of BSs, which consume brown power [109]. Such traffic load balancing strategies, however, may not maximize network utilities such as network capacity and traffic delivery latency. Hence, a trade-off between green power utilization and network utilities should be carefully evaluated in balancing traffic loads among BSs. In this chapter, we discuss traffic load balancing schemes that optimize network utilities, for example, average traffic delivery latency and green energy utilization.

7.1 Traffic Load Balancing in Mobile Networks

Proliferation of wireless devices and bandwidth-greedy applications are driving the exponential growth of mobile data traffic that is resulting in a continuous surge in capacity demand across mobile networks. The heterogeneous network (HetNet) is one of the key technologies for enhancing mobile network capacity to satisfy capacity demands [36]. In HetNet, low power SCBSs are densely deployed to enhance the spectrum efficiency of the network and thus increase the network capacity. Owing to disparate transmit power and BS capabilities, traditional user association metrics such as the SINR and the received-signal-strength-indication (RSSI) may lead to a severe traffic load imbalance [36]. Hence, user association algorithms should be well designed to balance traffic loads and thus to fully exploit the capacity potential of HetNet.

Green Mobile Networks: A Networking Perspective, First Edition. Nirwan Ansari and Tao Han.
© 2017 John Wiley & Sons Ltd. Published 2017 by John Wiley & Sons Ltd.

Balancing traffic loads in HetNet has been extensively studied in recent years [159]. In mobile networks, traffic loads among BSs are balanced by executing handover procedures. In the LTE system there are three types of handover procedures: Intra-LTE handover, Inter-LTE handover, and Inter-RAT (radio access technology) handover [160]. There are two ways to trigger handover procedures. The first is "Network Evaluated" in which the network triggers handover procedures and makes handover decisions. The other is "Mobile Evaluated" in which a user triggers the handover procedure and informs the network about the handover decision. The network decides whether to approve the user's handover request based on the status of radio resources. In 4G and LTE networks, a hybrid approach is usually implemented, where a user measures parameters of the neighboring cells and reports the results to the network. The network makes the handover decision based on the measurements. Here, the network can decide which parameters should be measured by users.

Aligning with the above procedures, various traffic load balancing algorithms have been proposed to optimize the network utilities [36, 20, 161, 19, 162]. The most practical traffic load balancing approach is the cell range expansion (CRE) technique, which biases users' receiving SINRs or data rates from some BSs to prioritize these BSs in associating with users [163]. Owing to the transmit power difference between MBSs and SCBSs, a large bias is usually given to SCBSs to offload users to small cells [36]. By applying CRE, a user is associated with the BS from which they receive the maximum biased SINR or data rate. Although CRE is simple, it is challenging to derive the optimal bias for BSs. Singh *et al.* [164] provided a comprehensive analysis on traffic load balancing using CRE in HetNet. The authors investigated the choice of the bias value and its impact on SINR coverage and downlink rate distribution in HetNet. Jo *et al.* [20] proposed cell biasing algorithms to balance traffic loads among MBSs and PBSs. These cell biasing algorithms perform user–BS association according to biased measured pilot signal strength, and enable traffic to be offloaded from MBSs to PBSs.

The traffic load balancing problem can also be modeled as an optimization problem and solved by convex optimization approaches. Ye *et al.* [161] modeled the traffic load balancing problem as a utility maximization problem and developed distributed user association algorithms based on primal-dual decomposition. Kim *et al.* [19] proposed an α-optimal user association algorithm to achieve flow level load balancing under spatially heterogeneous traffic distribution. The proposed algorithm may maximize different network utilities, for example, traffic latency and network throughput, by properly setting the value of α. Corroy *et al.* [26] proposed a dynamic user–BS association algorithm to maximize the total rate of network and adopted cell biasing to balance traffic loads among BSs. In addition, game theory has been exploited to model and solve the traffic load balancing problem. Aryafar *et al.* [162] modeled the traffic load balancing problem as a congestion game in which users are the players and user association decisions are the actions. Pantisano *et al.* [165] formulated the traffic load balancing problem in backhaul constrained SCNs as a one-to-many matching game between SCBSs and users, and proposed a distributed algorithm based on a deferred acceptance scheme to obtain a stable match for mobile users.

The above solutions, though they effectively balance traffic loads to maximize network utilities, do not consider energy efficiency as a performance metric in balancing traffic loads. The dense deployment of SCBSs may incur excessive energy consumption. Enhancing energy efficiency is also a critical task for next generation mobile networks [5, 1]. Although SCBSs consume less power than MBSs (MBSs), the number of SCBSs will be orders of magnitude larger than those of MBSs for a large scale network deployment. Hence, the overall power consumption of SCNs will be phenomenal. As energy harvesting technologies advance, renewable energy such as sustainable biofuels, solar, and wind energy can be utilized to power BSs [17]. Telecommunications companies such as Ericsson and Nokia Siemens have designed renewable energy powered BSs for mobile networks [34]. We define the electricity pulled from renewable energy systems and the power grid as green power and brown power, respectively. By adopting renewable energy powered BSs, mobile networks may further reduce their brown power consumption and reduce their carbon footprint [17]. Therefore, it is desirable to recognize green energy use as one of the performance metrics when balancing the traffic loads. Zhou *et al.* [10] proposed a handover parameter tuning algorithm for target cell selection and a power control algorithm for coverage optimization to guide mobile users to access BSs having renewable energy supplies. Considering a mobile network powered by multiple energy sources, we have proposed optimizing the utilization of green energy by optimizing BSs' transmit power [109]. The proposed algorithm achieves significant on-grid power savings by scheduling green energy consumption along the time domain for individual BSs, and balancing green energy consumption among BSs. We have also proposed a user association algorithm that jointly optimizes average traffic delivery latency and green energy utilization [55].

In addition, most existing solutions optimize traffic load balancing in a mobile network with the assumption that the air interface between BSs and mobile users is the bottleneck of the network. This assumption is generally correct for BSs whose deployments are well planned. However, in the case of the potentially dense deployment of SCBSs, various suboptimal backhaul solutions may be adopted, for example, xDSL, non-line-of-sight (NLOS) microwave, and wireless mesh networks, rather than the ideal backhaul approach provided by optical fiber and LOS microwave [166]. As a result, backhaul, rather than BSs, may become the bottleneck of SCNs. To alleviate backhaul constraints, content caching techniques have been exploited to enable caching popular content in BSs to reduce backhaul traffic loads [167, 168, 169, 170]. Therefore, it is desirable to optimize user association in consideration of backhaul constraints and the performance of BSs' content cache systems in SCNs.

In the following sections, we will discuss four energy efficient traffic load balancing schemes for mobile networks with different networking and energy settings.

7.2   ICE: *I*ntelligent Cell br*E*athing to Optimize the Utilization of Green Energy

In this section, we present Intelligent Cell brEathing (ICE) to minimize the maximal EDRs of LBSs, thus maximizing the utilization of green energy at each stage (every time slot of cell breathing). The EDR of a BS reflects the energy drainage speed of the BS. A large EDR indicates that energy drains faster. Here, we assume each LBS has a dedicated power generator, and the power is not shared among LBSs. Owing to limited energy storage, the energy consumption of certain LBSs under the default user-cell association algorithm may be larger than their energy storage, and thus these BSs are not able to serve all users with green energy. As a result, users under their coverage will switch to high power BSs (HBSs) that consume on-grid energy. ICE balances the users among the BSs through cell breathing, minimizes the maximal energy depleting rates of LBSs, and therefore enables more users to be served with green energy. Cell breathing techniques have recently been applied to minimizing the energy consumption of cellular networks [171]. However, unlike previous works, we are trying to maximize the utilization of green energy rather than minimizing total energy consumptions. The total energy consumption including green energy and on-grid energy under ICE may be larger than that of the default user-cell association algorithm. However, ICE reduces the on-grid energy consumption. Since green energy is renewable, ICE maximizes the utilization of green energy in order to save on-grid energy.

7.2.1 Problem Formulation

To simplify the problem formulation, we assume that LBSs update their cell size every τ seconds by changing their beacon power levels, and that LBSs always have data transmission to mobile users during the τ seconds. The LBS EDR is the normalized rate of energy consumption over the allocated green energy for the LBS; the EDR of LBS i is equal to:

$$R_i = \frac{(n_i p_i^{tx} + p_i^{static})\tau}{E_i}. \tag{7.1}$$

Here, R_i is EDR, n_i is the number of associated users, p_i^{tx} is the transmit power, p_i^{static} is the static power consumption, and E_i is the allocated green energy for LBS i at the current stage. For simplicity, we use $n_i p_i^{tx}$ to represent the dynamic power consumption of LBS i. Assume there are N LBSs in the LBS set **A**, and M mobile users in the user set **U**. Then, the problem can be formulated as follows:

$$\min_{\vec{b}} \max\{R_1, R_2, \ldots, R_N\} \tag{7.2}$$

$$subject\ to : p^r k d_{i,j}^2 \rho(i,j) \le p_i^{tx},$$
$$i \in (1, 2, \ldots, N), j \in (1, 2, \ldots, M). \tag{7.3}$$

Here, $\vec{b} = (b_1^{tx}, b_2^{tx}, \ldots, b_i^{tx}, \ldots, b_N^{tx})$ is the beacon power level vector of LBSs, $\vec{p} = (p_1^{tx}, p_2^{tx}, \ldots, p_i^{tx}, \ldots, p_N^{tx})$ is the transmit power vector of LBSs, and $d_{i,j}$ is the distance between LBS i and user j. p^r is the minimal receiving power that satisfies users' QoS requirements. $\rho(i,j)$ is an indication function which equals 1 if user j attaches to LBS i; otherwise, it equals 0. We assume the signal from LBSs experiences free space attenuation, which varies as the inverse square of distance between the users and their associated LBSs, and k is the path loss factor, which is a constant. The beacon power level vector determines the user–LBS association, and thus determines the EDRs of LBSs. Each LBS has G beacon power levels, and $b_i^{tx} \in (1, 2, \ldots, G)$. The mobile users attach to the LBS with the largest receiving signal strength.

Theorem 7.2.1   *The problem that minimizes the maximal power depleting rate is NP-hard.*

Proof: Consider the case of the problem with only two LBSs. Each user $u \in \mathbf{U}$ can be covered by both LBSs. If the user u is associated with LBS 1, the energy consumed by this user on LBS 1 is $s(u)$; if the user attaches to LBS 2, the user consumes $v(u)$ energy from LBS 2. We assume that both LBSs have the same energy storage. Minimizing the maximal power depleting rate equals to finding a subset $\mathbf{U}' \subseteq \mathbf{U}$ that satisfies

$$\sum_{u \in \mathbf{U}'} s(u) = \sum_{u \in \mathbf{U} - \mathbf{U}'} v(u). \tag{7.4}$$

By restricting the simple case of the problem to $s(u) = v(u)$ and assuming $\sum_{u \in \mathbf{U}} s(u)$ be evenly divisible by 2, the problem equals to the partition problem [142], which is a known NP-hard problem.   □

7.2.2   The ICE Algorithm

The most intuitive method to solve the above problem is the greedy algorithm that assigns each LBS with the largest transmit power, p^{max}, and then iteratively reduces the beacon power level of the LBSs with the largest EDR until the constraint given by Eq. (7.3) is violated. This greedy method may not yield the optimal solution to the min-max problem. Taking the network shown in Figure 7.1 as an example, assume both LBSs have the same energy storage, one unit, and each LBS has two transmit power levels. The energy costs of the mobile users when they are associated with different LBSs are shown in the figure. For the greedy method, LBS 1 will drop its transmit power level in the first iteration because it has the largest EDR, then in the second iteration, LBS 2 will drop its transmit power level for the same reason, and the user–LBS association returns to the original status. Then, the greedy algorithm stops since LBS 1 cannot drop its transmit power level anymore. This algorithm achieves 0.5 as its optimal value; this is clearly larger than 0.4, which is the result after the first iteration. ICE resolves this problem by introducing the dependent set (EDS). Denote the largest EDR at the current stage as δ. Let $\mathbf{D}' = \{a | R_a \geq \delta, a \in \mathbf{A}\}$. Let R'_a be the EDRs of BSs after the beacon power level reduction of LBSs in $\mathbf{D}'$. Then, the EDS $\mathbf{D} = \{a | R'_a \geq \delta, a \in \mathbf{A}\}$.

Guideline 1.   *Every LBS in EDS has to reduce its beacon power level in order to enable users switching from LBSs in EDS to those outside EDS.*

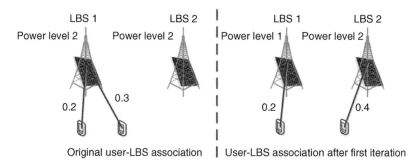

Figure 7.1 Illustration of the failure of the greedy algorithm. *Source:* Han 2012 [11]. Reproduced with permission of IEEE.

134 | *Green Mobile Networks*

Guideline 1 engineers the design of ICE from two aspects: (1) to identify which LBSs are to reduce their beacon power levels in each iteration, and (2) to determine the amount that the beacon power levels of each LBS should be reduced. Since the signals from LBSs to individual users experience different pathloss, the same amount of beacon power level reduction does not imply the same amount of receiving power reduction for individual users. Therefore, reducing the same amount of beacon power level may trigger users to switch among the LBSs in the EDS; this violates Guideline 1. Given the user distribution, the ICE algorithm, in each iteration, finds the EDS and the vector of beacon power level decrements, $\vec{w}$, for LBSs in the EDS. The vector of beacon power level decrements determines the amount of power level reduction for each LBS in EDS. Then, ICE reduces the beacon power levels of these LBSs accordingly. The algorithm iterates until the optimal solution is found. The pseudo code for ICE is shown in Algorithm 11.

Algorithm 11: The ICE Algorithm

Initialize $b_a^{tx} = G, a \in \mathbf{A}$;
OPT = FALSE;
while (OPT == FALSE) **do**
    Calculate the EDR, $R_a, a \in \mathbf{A}$;
    Find the largest EDR, δ;
    Find the set $\mathbf{D}$ such that $R_a \geq \delta, a \in \mathbf{D}$;
    while ($\mathbf{D} \neq \mathbf{D}^*$) **do**
        Initialize $w_a = 1$, and $w_a' = 0, a \in \mathbf{D}$;
        while ($\vec{w}' \neq \vec{w}$) **do**
            Reduce $b_a^{tx'} = b_a^{tx} - w_a, a \in \mathbf{D}$;
            Calculate $R_a', a \in \mathbf{D}$;
            $\vec{w}' = \vec{w}$, update $\vec{w}$ to guarantee $R_a' \leq R_a, a \in \mathbf{D}$;
        end while
        if ($\exists a \in \mathbf{D}$ such that $w_a \geq b_a^{tx}$) **then**
            $\mathbf{D} = \mathbf{A}$;
            Break;
        end if
        Calculate $R_a, a \in \mathbf{A} - \mathbf{D}$;
        Find a subset $\mathbf{T} \subseteq \mathbf{A} - \mathbf{D}$ such that $R_a \geq \delta, a \in \mathbf{T}$;
        $\mathbf{D}^* = \mathbf{D}, \mathbf{D} = \mathbf{D} \cup \mathbf{T}$;
    end while
    if ($\mathbf{D} == \mathbf{A}$) **then**
        OPT = TRUE;
    else
        Reduce b_a^{tx} by $w_a, a \in \mathbf{D}$;
    end if
end while
Return $\vec{b}$.

Theorem 7.2.2   *ICE always minimizes the maximal EDR.*

Proof: ICE initializes all the LBSs with their maximal beacon power level. Therefore, the maximal EDR, δ, of the network at the initial state is not less than that at the optimal state. At every iteration, ICE attempts to reduce the maximal EDR. Therefore, the maximal EDR at the current iteration will not be larger than that at the previous iteration. ICE finds the EDS and reduces the beacon level of all the LBSs contained in EDS. According to Guideline 1, users can only switch from LBSs within EDS to those outside EDS. As a result, the EDRs of LBSs within EDS do not increase. According to the definition of the EDS, the EDRs of LBSs outside EDS are strictly less than δ. Therefore, the iteration keeps reducing the maximal EDR. ICE stops at the optimal solution. The algorithm stops in two cases. The first is when $\mathbf{D} = \mathbf{A}$. In this case, all the LBSs are in the EDS. According to Guideline 1, reducing the beacon power level of all LBSs does not change the user–LBS association. Therefore, the maximal EDR cannot be reduced further. The second case corresponds to the scenario that there exists at least one LBS in the EDS with its current beacon power level being less than its beacon power level decrement. Thus, there exist LBSs in the EDS whose beacon power level cannot be reduced. Reducing the beacon power level of partial LBSs in the EDS violates Guideline 1, and therefore the maximal EDR cannot be reduced further. $\square$

The computational complexity of ICE in the worst case is $O(GN^4 M)$. Theoretically, ICE is a pseudo polynomial time algorithm. However, if any upper bound is imposed on the number of beacon power levels, ICE becomes a polynomial time algorithm [142].

7.2.3   ICE Algorithm Performance

A total of 25 LBS stations are located in a 5 by 5 grid. The distance between two adjacent LBSs is 400 meters. Users are uniformly distributed in the area. For simplicity, we assume the interference between LBSs is well managed by frequency planning, and the LBSs have complete knowledge of the users' locations. In the simulations, we compare our algorithm with the default user–BSs association method called strongest-signal-first (SSF), which always associates a user to the BS with the strongest received signal strength.

Figure 7.2 shows the maximal EDR achieved by ICE and SSF. In this simulation, we assume all the LBSs have the same amount of energy allocation, and the users do not move during the cell breathing interval, τ seconds. As the number of users is increasing, the maximal EDR is increasing. However, ICE outperforms SSF by up to 20% in terms of minimizing the maximal EDR. ICE with 25 power levels achieves a better solution than that with 15 power levels because when the number of beacon power levels increases, the search region for ICE becomes larger. Therefore, ICE with a larger number of power levels has more opportunities to balance energy consumption among the LBSs.

Figure 7.3 shows a comparison of the EDRs of LBSs between ICE and SSF. In the figure, we sort LBSs by their EDRs from the largest to the smallest; the x-axis is the LBS index while the y-axis is the energy depleting rate. For SSF, some LBSs experience large EDRs while the EDRs of the other LBSs are small. This indicates that the energy consumption among LBSs is not balanced, with the EDRs of the first 6 LBSs being larger than 1. This means that these LBSs cannot serve the associated users with green energy due to their limited green energy storage capacity. ICE minimizes the maximal EDR

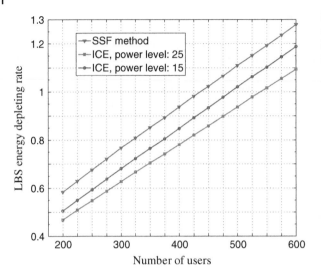

Figure 7.2 The maximal EDR comparison (N = 25). *Source:* Han 2012 [11]. Reproduced with permission of IEEE.

of LBSs by offloading some users to their neighboring LBSs. In this simulation, when applying ICE, the EDRs of all LBSs are smaller than 1, thus enabling all users to be served by green energy.

Figure 7.4 shows the green energy allocation at individual LBSs and the number of users who are associated with the corresponding LBSs. The x-axis is the LBS index, and each index represents an LBS. There are 25 LBSs in this simulation. Here, we sort LBSs by the normalized green energy storage from the largest to the smallest. The left y-axis represents the user ratio, which is defined as the number of users associated with individual LBSs determined by ICE and SSF divided by that determined by SSF. The right y-axis is the normalized green energy storage, which is derived by dividing the green

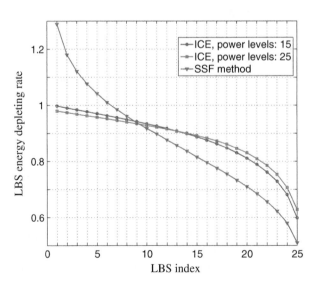

Figure 7.3 The EDR comparison (N = 25, M = 600). *Source:* Han 2012 [11]. Reproduced with permission of IEEE.

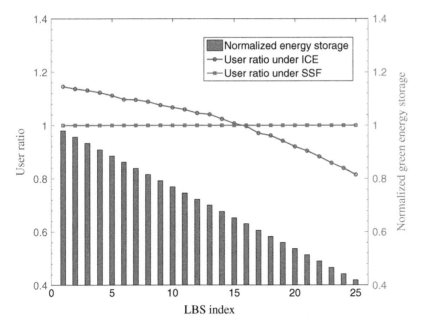

Figure 7.4 LBSs' statuses ($N = 25$, $M = 600$, $G = 25$). *Source:* Han 2012 [11]. Reproduced with permission of IEEE.

energy allocation on each LBS by the maximal green energy allocation among the LBSs, in the current period. The normalized energy storage indicates the amount of green energy allocated to each LBS, and is represented by the vertical bars in the figure. The user ratio reflects the number of users associated with individual LBSs, and is represented by the curves in the figure. We can see that ICE associates more users with the LBSs that have a larger amount of the green energy allocation. This benefits the utilization of green energy from two aspects. First, for the LBSs with a small amount of energy allocation, ICE offloads users from these LBSs, and enables the LBSs to serve users with their limited energy allocation. Second, for the LBSs with a large amount of energy allocation, ICE directs users to be associated with these LBSs, and avoids the waste of arrival energy at these LBSs because of the finite battery capacity.

Figure 7.5 shows the user outage under ICE and SSF, respectively. The y-axis represents the user outage percentage, which is defined as the percentage of users who are not served by green energy. Users may not be served by green energy if their associated LBSs have less green energy allocations than the energy demands. When the number of users is small, both ICE and SSF achieve zero user outage. As the number of users increases, user outage increases because of limited green energy allocations. However, when the number of users is less than 660, ICE incurs much less user outage than SSF does. In fact, when the number of users is less than 550, ICE achieves almost zero user outage while SSF suffers from up to 15% of outage users. When the number of users is larger than 660, SSF achieves better performance because there are too many users in the networks, and most LBSs do not have sufficient green energy to serve all their associated users. Therefore, ICE may increase the EDRs of the LBSs that already have large EDRs. This may prevent these LBSs from serving their associated users with

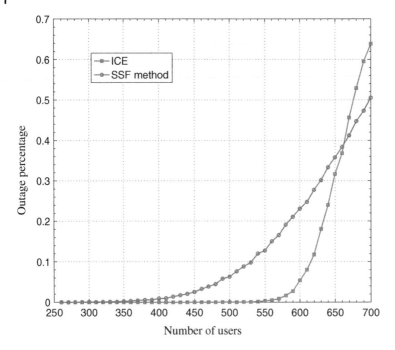

Figure 7.5 The user outage percentage. *Source:* Han 2012 [11]. Reproduced with permission of IEEE.

green energy. Thus, the user outage percentage is large. However, when the network is not overloaded and green energy is allocated properly, ICE provides better performance in terms of user outage percentage than SSF does.

Figure 7.6 shows the user outage versus different green energy arrival rates. In this simulation, we assume the amount of green energy allocation at each cell breathing time slot is equal to the amount of green energy arrival of the previous time slot. Assume the energy arrival rate, e_a, of each time slot is identical for all LBSs; then $E_i = e_a \tau$. As the energy arrival rate increases, the user outage decreases. When the energy arrival rate is larger than 800 $mW\ h^{-1}$, ICE achieves almost zero user outage while SSF still suffers from about 20% user outage, and SSF requires the energy arrival rate to be more than 1100 $mW\ h^{-1}$ in order to eliminate user outages.

7.3 Energy- and QoS-Aware Traffic Load Balancing

In this section, we consider a HetNet with multiple MBSs and SCBSs as shown in Figure 7.7. Both the MBSs and SCBSs are powered by on-grid power and green energy. We consider solar power as the green energy source. We focus on balancing the downlink traffic loads among BSs by designing a green energy and latency aware user association scheme. We adopt a software-defined radio access network (SoftRAN) architecture in which all BSs are controlled by the RAN controller (RANC). The RANC has a global view of BSs' traffic loads and green energy. The user association is optimized by the RANC. The specific design of the RANC is beyond the scope of this section.

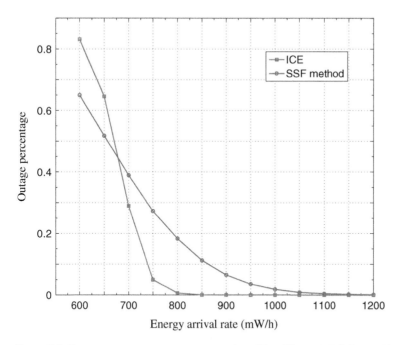

Figure 7.6 The user outage percentage comparison (M = 600, $\tau = 360s$). *Source:* Han 2012 [11]. Reproduced with permission of IEEE.

7.3.1 System Model and Problem Formulation

7.3.1.1 Traffic Model

Define $\mathcal{B}$ as a set of BSs including both the MBS and SCBSs. We assume that the traffic arrives according to a Poisson point process with the average arrival rate per unit area at location x equal to $\lambda(x)$, and the traffic size (packet size) per arrival having a general distribution with average traffic size $v(x)$. Assuming a mobile user at location x is associated with the jth BS, then the user's downlink data rate $r_j(x)$ that will end up

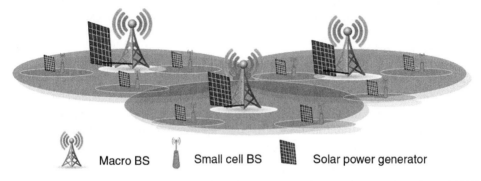

Figure 7.7 A HetNet powered by hybrid energy sources: on-grid power and green energy. *Source:* Han 2016 [157]. Reproduced with permission of IEEE.

becoming available to the user can generally be expressed as a logarithmic function of the perceived SINR, $SINR_j(x)$, according to the Shannon–Hartley theorem [19],

$$r_j(x) = W_j log_2(1 + SINR_j(x)), \tag{7.5}$$

where W_j is the total bandwidth in the jth BS, and

$$SINR_j(x) = \frac{P_j g_j(x)}{\sigma^2 + \sum_{k \in \mathcal{I}_j} I_k(x)}. \tag{7.6}$$

Here, P_j is the transmission power of the jth BS, $\mathcal{I}_j$ represents the set of interfering BSs which is defined as the set of BSs whose transmissions interfere with the jth BS's transmission toward a user at location x, $I_k(x)$ is the average interference power seen by a user at location x from the kth BS, σ^2 denotes the noise power level, and $g_j(x)$ is the channel gain between the jth BS and the user at location x. Here, the channel gain reflects only slow fading including path loss and shadowing. We assume the channel gain is measured over a large time scale, and thus fast fading is not considered.

In HetNet, the total bandwidth in a BS is determined by the network's frequency planning. Different frequency reuse strategies result in different inter-BS interference. In this section, we assume the network's frequency reuse strategy is given and static. Thus, $\mathcal{I}_j$ contains the set of BSs that share the same spectrum with the jth BS. We assume users experience roughly static interference from the interfering BSs. Although the inter-BS interference in HetNet varies depending on the activities in the interfering BSs, the interference can be well coordinated via time domain techniques, frequency domain techniques, and power control techniques [172]. Therefore, the inter-BS interference can be reasonably modeled as a static value for analytical simplicity. The static inter-BS interference model has also been adopted in previous works for modeling the user association problem [173, 19].

The average traffic load density at location x in the jth BS is

$$\varrho_j(x) = \frac{\lambda(x)v(x)\eta_j(x)}{r_j(x)}. \tag{7.7}$$

Here, $\eta_j(x)$ is an indicator function. If $\eta_j(x) = 1$, the user at location x is associated with the jth BS; otherwise, the user is not associated with the jth BS. Assuming mobile users are uniformly distributed in the area and denoting $\mathcal{A}$ as the coverage area of all the BSs, based on Eq. (7.7), we can derive the average traffic loads in the jth BS, expressed as

$$\rho_j = \int_{x \in \mathcal{A}} \varrho_j(x) dx. \tag{7.8}$$

The value of ρ_j indicates the fraction of time during which the jth BS is busy.

We assume that traffic arrival processes at individual locations are independent. Since the traffic arrival per unit area is a Poisson point process, the traffic arrival in the jth BS, which is the sum of the traffic arrivals in its coverage area, is also a Poisson process. The required service time per traffic arrival for a user at location x in the jth BS is $\gamma_j = \frac{v(x)}{r_j(x)}$.

Since $v(x)$ is the average traffic size per arrival, which follows a general distribution, the user's required service time is also a general distribution. Hence, a BS's service rate follows a general distribution. Therefore, a BS's downlink transmission process implements an M/G/1 processor sharing queue, in which multiple users share the BS's downlink radio resource [152].

In mobile networks, various downlink scheduling algorithms have been proposed to enable proper sharing of the limited radio resource in a BS [153]. These algorithms are designed to maximize network capacity, enhance fairness among users, or provision QoS services. According to the scheduling algorithm, users are assigned different priorities on sharing the downlink radio resource. As a result, users in different priority groups perceive different average waiting time. Since traffic arrives at a BS according to Possion arrival statistics, the allowed variation in the average waiting times among different priority groups is constrained by the Conservation Law [152]. The integral constraint on the average waiting time in the jth BS can be expressed as

$$\bar{L}_j = \frac{\rho_j E(\gamma_j^2)}{2(1 - \rho_j)}. \tag{7.9}$$

This indicates that, given the users' required service time in the jth BS, if the scheduling algorithm gives some users higher priority and reduces their average waiting time, it will increase the average waiting time of other users. Therefore, $\bar{L}_j$ generally reflects the jth BS's performance in terms of users' average waiting time. Since $E(\gamma_j^2)$ mainly reflects the traffic characteristics, we assume that $E(\gamma_j^2)$ is roughly constant during a user association process and thus $\vartheta_j = \frac{E(\gamma_j^2)}{2}$ can be considered as a constant. Thus, we adopt

$$L(\rho_j) = \frac{\vartheta_j \rho_j}{1 - \rho_j} \tag{7.10}$$

as a general latency indicator for the jth BS. A smaller $L(\rho_j)$ indicates that the jth BS introduces less latency to its associated users. Therefore, we use $L(\rho_j)$ to reflect the jth BS's average traffic delivery latency.

7.3.1.2 Energy Model

In the network, both MBSs and SCBSs have their own solar panels for generating green energy. Therefore, BSs are powered by hybrid energy sources: on-grid power and green energy. If the green energy generated by its solar panel is not sufficient, the BS consumes on-grid power. Since MBSs usually consume more energy than SCBSs, we assume that MBSs are equipped with larger solar panels that have a higher energy generation capacity than those of a SCBS. A reference design of a hybrid energy powered BS [17] is shown in Fig 7.8. The charge controller optimizes the green energy utilization based on the solar power intensity, the power consumption of BSs, and energy prices on the power grid. Here, green energy utilization is optimized over the time horizon. For example, the charge controller may predict the solar power intensity and mobile traffic load in a BS over a certain period of time, say 24 hours. The prediction can be based on statistical data and real time weather forecasts. The charge controller determines how much green energy should be utilized to power a BS, according to this prediction, during a specific time period, for example, the time duration between two consecutive traffic load balancing procedures.

In this section, instead of investigating how to optimize green energy utilization over a time horizon, we aim to study how to balance traffic loads among BSs to save on-grid

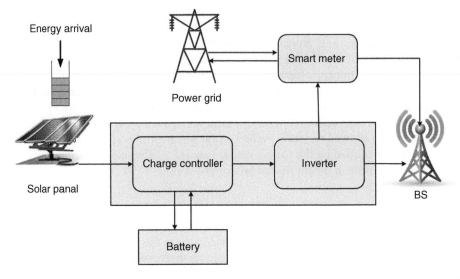

Figure 7.8 A hybrid energy powered BS. *Source:* Han 2016 [157]. Reproduced with permission of IEEE.

energy within the duration of a traffic balancing procedure. Therefore, we assume that the amount of available green energy for powering a BS is a constant within this duration as determined by the charge controller. This is a reasonable assumption because the traffic load balancing process is at a time scale of several minutes [19] while solar power generation is usually modeled at a time scale of an hour [53]. Define e_j as the amount of green energy for powering the jth BS in a traffic load balancing procedure. If the power consumption of the jth BS is larger than e_j, the BS consumes on-grid power. Otherwise, the residual green energy will be either stored in batteries for future usage or uploaded to the power grid via the smart meter. Since we are not focusing on optimizing the green energy utilization over the time horizon, we simply model the BS's on-grid energy consumption to be zero when its power consumption is less than e_j. In other words, we do not consider the redistribution of the residual green energy in our model.

A BS's power consumption consists of two parts: static power consumption and dynamic power consumption [3]. The static power consumption is the power consumption of a BS without carrying any traffic load. The dynamic power consumption refers to the additional power consumption incurred by the traffic load in the BS, which can be well approximated by a linear function of the traffic load [3]. Define p_j^s as the static power consumption of the jth BS. Then, the jth BS's power consumption can be expressed as

$$p_j = \beta_j \rho_j + p_j^s. \tag{7.11}$$

Here, β_j is the load-power coefficient that reflects the relationship between the traffic load and the dynamic power consumption in the jth BS. The BS power consumption model can be adjusted to model the power consumption of either MBSs or SCBSs by incorporating and tweaking the static power consumption and the load-power coefficient. The on-grid power consumption in the jth BS is

$$p_j^o = \max{(p_j - e_j, 0)}. \tag{7.12}$$

7.3.1.3 Problem Formulation

In determining user association, the network aims to strive for a trade-off between network utilities, for example, the average traffic delivery latency and the on-grid power consumption. In this study, we focus on designing a user association algorithm to enhance the network performance by reducing the average traffic delivery latency in BSs as well as to reduce the on-grid power consumption by optimizing green energy usage.

On the one hand, to reduce the average traffic delivery latency, the network desires to minimize the summation of the latency indicators of BSs. On the other hand, since BSs are powered by both green energy and on-grid power, the network seeks to minimize the usage of on-grid power by optimizing the utilization of green energy. According to Eq. (7.12), on-grid power is only consumed when green energy is not sufficient in the BS. When $p_j > e_j$, to alleviate on-grid power consumption, the jth BS has to reduce its traffic load. We define the green traffic capacity as the maximum traffic load that can be supported by green energy. Define $\hat{\rho}_j$ as the green traffic capacity of the jth BS. Then,

$$\hat{\rho}_j = \max\left(\epsilon, \min\left(\frac{e_j - p_j^s}{\beta_j}, 1 - \epsilon\right)\right). \tag{7.13}$$

Here, ϵ is an arbitrary small positive constant to guarantee $0 < \hat{\rho}_j < 1$. To reduce the traffic load from ρ_j to $\hat{\rho}_j$, the jth BS has to shrink its coverage area. As a result, its traffic load is offloaded to its neighboring BSs and may lead to traffic congestion in the neighboring BSs. The traffic congestion increases the average traffic delivery latency of the network. To achieve a trade-off between the average traffic delivery latency and the on-grid power consumption, we define the energy-latency coefficient in the jth BS as θ_j. We further define the desired traffic load in the jth BS after the energy–latency trade-off as

$$\tau_j = (1 - \theta_j)\rho_j + \theta_j\hat{\rho}_j. \tag{7.14}$$

Here, $0 \leq \theta_j \leq 1$. If θ_j is set to zero, the jth BS's desired traffic load is its actual traffic load without considering green energy. In this case, we consider the jth BS being latency-sensitive; otherwise, if θ_j is equal to one, the jth BS's desired traffic load is dominated by its green traffic capacity, and thus the BS is energy-sensitive. We assume θ_j remains constant within the duration of a user association process.

Since mobile devices are battery powered, it is desirable to guarantee their energy efficiency during traffic load balancing [174]. To ensure the energy efficiency of the mobile devices, we restrict a user to be only be associated with the BSs to which the user's uplink pathloss is smaller than a predefined threshold. Considering all the above factors, the user association (UA) problem can be formulated as:

$$\min_{\rho} \sum_{j \in B} w_j(\rho_j)L(\rho_j) \tag{7.15}$$

$$\text{subject to} : 0 \leq \rho_j \leq 1 - \epsilon.$$
$$(\alpha_j(x) - \alpha^*(x))\eta_j(x) \leq 0,$$
$$\forall x \in \mathcal{A}, j \in B. \tag{7.16}$$

Here, $\alpha_j(x)$ and $\alpha^*(x)$ are the uplink pathloss from the user at location x to the jth BS and the uplink pathloss threshold for the user, respectively. $0 < \epsilon < 1$ is a small real number to ensure $\rho_j < 1$. $\rho = (\rho_1, \rho_2, \ldots, \rho_{|B|})$, and

$$
\begin{aligned}
w_j(\rho_j) &= e^{\kappa(\rho_j - \tau_j)} \\
&= e^{\kappa(\rho_j - (1-\theta_j)\rho_j - \theta_j \hat{\rho}_j)} \\
&= e^{\kappa\theta_j(\rho_j - \hat{\rho}_j)}.
\end{aligned}
\tag{7.17}
$$

In the objective function, $w_j(\rho_j)$ indicates the weight of the jth BS's latency indicator. If the jth BS has sufficient green energy ($\hat{\rho}_j \geq \rho_j$), $0 < w_j(\rho_j) \leq 1$; otherwise, $w_j(\rho_j) > 1$. This is because when the amount of available green energy in the jth BS is sufficient, the green traffic capacity, $\hat{\rho}_j$, is larger than ρ_j. Then, $\tau_j > \rho_j$ and $w_j < 1$. With a large weight, the jth BS has a high priority in reducing its latency indicator while minimizing Eq. (7.15) as compared with the BSs having a small weight. Therefore, as compared with $w_j(\rho_j) \leq 1$, $w_j(\rho_j) > 1$ enables the jth BS to achieve a smaller latency indicator. Since

$$
\frac{dL(\rho_j)}{d\rho_j} = \frac{\vartheta_j}{(1-\rho_j)^2} > 0,
\tag{7.18}
$$

a smaller latency indicator means less traffic load in the jth BS, which is desirable for saving on-grid power in the jth BS. Thus, introducing the weights for BSs' latency indicators in the objective function enables the green energy aware and traffic delivery latency aware user association. κ is a parameter that further adjusts the value of the weight according to that of the traffic latency indicator and enables the network to control the trade-off between the on-grid power consumption and average traffic delivery latency.

7.3.2 vGALA: A Green Energy and Latency Aware Load Balancing Scheme

In this section, we present the virtual Green energy Aware and Latency Aware (vGALA) scheme and prove its properties. The vGALA scheme generally consists of three phases. The first phase is initial user association and network measurement, during which the RANC collects network information, for example, available green energy, traffic loads, and users' data rates. The second phase is the user association optimization, in which the RANC optimizes the user association and derives the corresponding BSs' operation statuses based on the information collected in the first phase. Here, a BS's operation status reflects the price for a user to access the BS. In the third phase, the user association is determined based on the optimized BSs' operation statuses and users' downlink data rates. The major optimization of the vGALA scheme is in the second phase. To be analytically tractable, we assume that (1) the RANC can successfully collect the network information from all BSs and users, and (2) the users' data rates do not change within one user association process. We will evaluate these assumptions in the next section where we discuss the practicality of the vGALA scheme.

Based on the collected network information, the RANC optimizes the user association and derives the optimal BS operation status. We leverage the SoftRAN architecture to facilitate the user association optimization as the RANC has a global view of the traffic loads and the availability of green energy in the network. However, owing to the large number of users and BSs, the user association algorithm if not well designed may be time consuming and incur excessive delays. In order to efficiently optimize user association, the vGALA scheme divides the user association algorithm into two parts: the user side algorithm and the BS side algorithm. The user side algorithm calculates the user's BS selection. The BS side algorithm updates the BS's operation status based on the green traffic capacity and the traffic load. Based on the updates, the user side algorithm recalculates the BS selection. The user association algorithm iterates until it converges. After the convergence, the optimal BS operation status is obtained and the optimal user association is subsequently determined.

The information exchanges over the air interface between users and BSs may introduce additional communications overhead and incur extra power consumption. The vGALA scheme, by leveraging cloud computing and virtualization, generates virtual users and virtual BSs (vBSs) in the RANC. The user side algorithm runs on virtual users while the BS side algorithm runs on vBSs. In this way, instead of exchanging information over the air interface, the virtual users and vBSs can iteratively update their information locally within the RANC. Here, the virtualization only virtualizes the computational resources for BSs and users rather than virtualizing all their functions.

7.3.2.1 The User Side Algorithm

We define the time interval between two consecutive BS selection updates as a time slot. At the beginning of the kth time slot, vBSs send their operation statuses to virtual users. The user association algorithm is designed based on the subgradient algorithm. Therefore, a vBS calculates its operation status based on the gradient of the objective function. Let

$$\psi(\rho) = \sum_{j \in B} w_j(\rho_j) L(\rho_j). \tag{7.19}$$

The jth vBS's operation status in the kth time slot is defined as

$$\begin{aligned}
\phi_j(\rho_j(k)) &= \frac{\partial \psi(\rho(k))}{\partial \rho_j(k)} \\
&= \frac{\vartheta_j e^{\kappa \theta_j(\rho_j(k) - \hat{\rho}_j)}}{(1 - \rho_j(k))^2} (\kappa \theta_j \rho_j(k) \\
&\quad - \kappa \theta_j \rho_j(k)^2 - 1).
\end{aligned} \tag{7.20}$$

Here, the jth vBS is mapped to the jth BS in the mobile network.

Let $\bar{B}(x) = \{j | a_j(x) \le a^*(x)\}$ be the set of BSs whose uplink pathloss is less than the user's pathloss threshold. Assign $r_j(x) = \zeta$, $\forall j \in B \setminus \bar{B}(x)$ where ζ is a very small positive number that approaches zero. This is equivalent to restricting the user from associating with the BSs outside $\bar{B}(x)$. Then, the BS selection rule for a user at location x can be expressed as

$$b^k(x) = \arg \max_{j \in B} \frac{r_j(x)}{\phi_j(\rho_j(k))}. \tag{7.21}$$

Here, $b^k(x)$ is the index of the vBS selected by the virtual user at location x in the kth time slot. The pseudo code of the user side algorithm is shown in Algorithm 12. The BS selection rule is derived based on the subgradient algorithm. This BS selection rule enables the network to minimize the objective function in Eq. (7.15). The computational complexity of the user side algorithm for an individual user is $O(|B|)$.

Algorithm 12: The user side algorithm

Input : BSs' operation status: $\phi_j(\rho_j(k)), j \in \mathcal{B}$;
Output: The BS selection: $b^k(x)$;
1 Estimate the uplink pathloss: $\alpha_j(x)$;
2 Find $\bar{B}(x) = \{j | \alpha_j(x) \leq \alpha^*(x)\}$;
3 Assign $r_j(x) = \zeta, \forall j \in \mathcal{B} \setminus \bar{B}(x)$;
4 Find $b^k(x) = \arg\max_{j \in \mathcal{B}} \frac{r_j(x)}{\phi_j(\rho_j(k))}$.

7.3.2.2 The BS Side Algorithm

Upon receiving vBSs' operation status updates, virtual users select vBSs according to the user side algorithm. The coverage area of the jth vBS in the kth time slot is updated as

$$\tilde{A}_j(k) = \{x | j = b^k(x), \forall x \in \mathcal{A}\}. \tag{7.22}$$

Then, given $\rho(k) = (\rho_1(k), \rho_2(k), \ldots, \rho_{|B|}(k))$, $\theta = (\theta_1, \theta_2, \ldots, \theta_{|B|})$, and $\hat{\rho} = (\hat{\rho}_1, \hat{\rho}_2, \ldots, \hat{\rho}_{|B|})$, the jth vBS's perceived traffic load in the kth time slot is

$$M_j(\rho(k), \theta, \hat{\rho}) = \min\left(\int_{x \in \tilde{A}_j(k)} \varrho_j(x)dx, 1 - \epsilon\right). \tag{7.23}$$

Here, $\int_{x \in \tilde{A}_j(k)} \varrho_j(x)dx$ is the perceived traffic load in the jth BS. The maximum traffic load which can be served in a BS is defined as $1 - \epsilon$. If $\int_{x \in \tilde{A}_j(k)} \varrho_j(x)dx$ is larger than $1 - \epsilon$, the excessive traffic load will not be served. Since θ and $\hat{\rho}$ are assumed not to change within the duration of a user association process, $M_j(\rho(k), \theta, \hat{\rho})$ evolves based only on $\rho(k)$. Thus, we use $M_j(\rho(k))$ instead of $M_j(\rho(k), \theta, \hat{\rho})$ for simplicity in the following analysis.

After having derived the perceived traffic load, the jth vBS updates its traffic load as

$$\rho_j(k+1) = \delta(k)\rho_j(k) + (1 - \delta(k))M_j(\rho(k)). \tag{7.24}$$

Here, $0 \leq \delta(k) < 1$ is a system parameter calculated by the RANC to enable

$$\psi(\rho(k+1)) \leq \psi(\rho(k))$$
$$+ \varsigma(1 - \delta(k)) \sum_{j \in \mathcal{B}} \phi_j(\rho_j(k))(M_j(\rho(k)) - \rho_j(k)). \tag{7.25}$$

Here, $0 < \varsigma < 0.5$ is a constant. In the $(k+1)$th time slot, the jth vBS's operation status is $\phi_j(\rho_j(k+1))$. The pseudo code of the BS side algorithm is presented in Algorithm 13. The computational complexity of the BS side algorithm is determined by the "while" loop, for which the runtime depends on the convergence of $\psi(\rho(k))$. When $\psi(\rho(k))$ is closer to the optimal value, it may take longer to find $\delta(k)$. In the following, we will analyze the convergence of the vGALA scheme, which reflects the computational complexity of the BS side algorithm.

Algorithm 13: The BS side algorithm

Input : Users' vBS selection: $b^k(x), \forall x \in \mathcal{A}$;
Output: vBSs'operation status, $\phi_j(\rho_j(k+1)), \forall j \in \mathcal{B}$;
1 vBSs measure their perceived traffic loads, $M_j(\rho(k))$;
2 Assign $\delta(k) = 0$;
3 **while** *Eq. (7.25) is not true* **do**
4 $\quad \lfloor \; \delta(k) = 1 - \xi(1 - \delta(k))$, here, $0 < \xi < 1$ is a real number;
5 vBSs update their traffic loads: $\rho_j(k+1) = \delta(k)\rho_j(k) + (1 - \delta(k))M_j(\rho(k))$;
6 Calculate $\phi_j(\rho_j(k+1))$ based on $\rho_j(k+1), \forall j \in \mathcal{B}$.

7.3.3   Properties of vGALA

7.3.3.1   vGALA Convergence
In order to prove the convergence of vGALA, we first prove that the vBSs' traffic load vector converges. The feasible set for the UA problem is

$$\mathcal{F} = \{\rho | \rho_j = \int_{x \in \mathcal{A}} \varrho_j(x)dx,$$

$$0 \le \rho_j \le 1 - \epsilon, \sum_{j \in \mathcal{B}} \eta_j(x) = 1,$$

$$\eta_j(x) = \{0, 1\}, \forall j \in \mathcal{B}, \forall x \in \mathcal{A}\}. \tag{7.26}$$

Here, $\rho_j = \int_{x \in \mathcal{A}} \varrho_j(x)dx$ is the traffic load in the jth vBS; $0 \le \rho_j \le 1 - \epsilon$ constrains the traffic load in a vBS to be larger than zero and smaller than $1 - \epsilon$; $\eta_j(x)$ is an indicator function which indicates that a user at location x is associated with the jth vBS if $\eta_j(x) = 1$; since a user can only be associated with one vBS, $\sum_{j \in \mathcal{B}} \eta_j(x) = 1$. Since $\eta_j(x) = \{0, 1\}$, $\mathcal{F}$ is not a convex set. Thus, the traffic update in Eq. (7.24) cannot guarantee the updated traffic load is in the feasible set. In order to show the convergence of vGALA, we first relax the constraint to let $0 \le \eta_j(x) \le 1$ and then prove the traffic load vector converges to the traffic load vector that is in the feasible set. Define

$$\tilde{\mathcal{F}} = \{\rho | \rho_j = \int_{x \in \mathcal{A}} \varrho_j(x)dx,$$

$$0 \le \rho_j \le 1 - \epsilon, \sum_{j \in \mathcal{B}} \eta_j(x) = 1,$$

$$0 \le \eta_j(x) \le 1, \forall j \in \mathcal{B}, \forall x \in \mathcal{A}\} \tag{7.27}$$

as the relaxed feasible set.

Lemma 7.3.1   *The relaxed feasible set $\tilde{\mathcal{F}}$ is a convex set.*

Proof: The lemma is proved by showing that the set $\tilde{\mathcal{F}}$ contains any convex combination of the traffic load vector ρ. □

Lemma 7.3.2 $\psi(\rho)$ *is a strong convex function of ρ when ρ is defined in $\tilde{F}$.*

Proof: The lemma is proved by showing $\nabla^2 \psi(\rho) \geq qI$ where $q = 4e^{-1}$ and I is an identity matrix. Here, $\geq$ is componentwise inequality between vectors $\nabla^2 \psi(\rho)$ and qI. □

Let $M(\rho) = \{M_1(\rho), M_2(\rho), \dots, M_{|B|}(\rho)\}$.

Lemma 7.3.3 *When $M(\rho(k)) \neq \rho(k)$, $M(\rho(k))$ provides a descent direction of $\psi(\rho)$ at $\rho(k)$.*

Proof: Since $\psi(\rho)$ is a convex function, proving this lemma is equivalent to proving

$$\langle \nabla \psi(\rho)|_{\rho=\rho(k)}, M(\rho(k)) - \rho(k) \rangle < 0. \tag{7.28}$$

Let $\hat{\eta}_j(x)$ and $\eta_j(x)$ be the user association indications of the jth BS that result in the traffic load $M_j(\rho(k))$ and $\rho_j(k)$, respectively.

$$\langle \nabla \psi(\rho)|_{\rho=\rho(k)}, M(\rho(k)) - \rho(k) \rangle \tag{7.29}$$
$$= \sum_{j \in B} (M_j(\rho(k)) - \rho_j(k)) \phi_j(\rho_j(k))$$
$$= \sum_{j \in B} \frac{\int_{x \in A} \lambda(x)v(x)(\hat{\eta}_j(x) - \eta_j(x))dx}{r_j(x)\phi_j^{-1}(\rho_j(k))}$$
$$= \int_{x \in A} \lambda(x)v(x) \sum_{j \in B} \frac{\hat{\eta}_j(x) - \eta_j(x)}{r_j(x)\phi_j^{-1}(\rho_j(k))} dx.$$

Since

$$\hat{\eta}_j(x) = \begin{cases} 1, & j = b^k(x) \\ 0, & otherwise, \end{cases} \tag{7.30}$$

$$\sum_{j \in B} \frac{\hat{\eta}_j(x) - \eta_j(x)}{r_j(x)\phi_j^{-1}(\rho_j(k))} \leq 0. \tag{7.31}$$

Because $M(\rho(k)) \neq \rho(k)$, there exists $j \in B$ such that $\hat{\eta}_j(x) \neq \eta_j(x)$, $x \in A$. Hence,

$$\sum_{j \in B} \frac{\hat{\eta}_j(x) - \eta_j(x)}{r_j(x)\phi_j^{-1}(\rho_j(k))} < 0, \tag{7.32}$$

and

$$\langle \nabla \psi(\rho)|_{\rho=\rho(k)}, M(\rho(k)) - \rho(k) \rangle < 0. \tag{7.33}$$

□

Theorem 7.3.1 *The traffic load vector ρ converges to the traffic load vector $\rho^* \in F$.*

Proof: Since $\sum_{j \in B} (M_j(\rho(k)) - \rho_j(k)) \phi_j(\rho_j(k)) < 0$ when $M(\rho(k)) \neq \rho(k)$, Algorithm 3 ensures that $\psi(\rho(k+1)) \leq \psi(\rho(k))$ in each time slot. Since $\psi(\rho) \geq 0$, $\psi(\rho)$ will converge.

Let $\psi(\rho)$ converge to $\psi(\rho^*)$. Since

$$\rho(k+1) = \delta(k)\rho(k) + (1 - \delta(k))(M\rho(k))$$
$$= \rho(k) + (1 - \delta(k))(M(\rho(k)) - \rho(k)), \tag{7.34}$$

$M(\rho)$ and ρ will converge to ρ^*. Because $M(\rho^*)$ is derived based on the user side algorithm where $\eta_j^m(x) = \{0, 1\}$, $\forall j \in \mathcal{B}$, $x \in \mathcal{A}$, ρ^* is in the feasible set $\mathcal{F}$. □

Corollary 7.3.1 *The vBSs' operation status $\phi_j(\rho_j)$, $\forall j \in \mathcal{B}$, converges to $\phi_j(\rho_j^*)$.*

Proof: Within the duration of a user association process, ϑ_j, θ_j, and $\hat{\rho}_j$ are constant. Thus, $\phi_j(\rho_j)$ is only determined by ρ_j. Since ρ_j converges to ρ_j^*, $\phi_j(\rho_j)$ converges to $\phi_j(\rho_j^*)$. □

Since $\psi(\rho)$ is a strong convex function, there exist $q > 0$ and $Q > 0$ such that $qI \preceq \nabla^2 \psi(\rho) \preceq QI$, $\rho \in \tilde{\mathcal{F}}$ [143]. Denote the optimal solution as $\psi(\rho^*)$. $\psi(\rho(k+1))$ is said to be the ϵ sub-optimal solution if $\psi(\rho(k+1)) - \psi(\rho^*) \leq \epsilon$ where $\epsilon > 0$ is a small real number.

Lemma 7.3.4 *The number of iterations required to ensure $\psi(\rho(k+1)) - \psi(\rho^*) \leq \epsilon$ is at most equal to*

$$\frac{\log((\psi(\rho(1)) - \psi(\rho^*))/\epsilon)}{\log 1/z}, \tag{7.35}$$

where $z = 1 - \min\{2q\varsigma, 2q\varsigma\xi/Q\} < 1$ and $\rho(1)$ is the initial traffic load vector.

Proof: Let $\Delta\rho(k) = M(\rho(k)) - \rho(k)$. The termination condition of the BS side algorithm (Algorithm 3) can be expressed as

$$\psi(\rho(k+1))$$
$$\leq \psi(\rho(k)) + \varsigma(1 - \delta(k))\nabla\psi(\rho)^\top \Delta\rho(k). \tag{7.36}$$

Since $\Delta\rho(k)$ is a descent direction of $\psi(\rho(k))$, $\Delta\rho(k)$ can be replaced by $-\nabla\psi(\rho)$. Thus, the termination condition of Algorithm 13 can be rewritten as

$$\psi(\rho(k+1)) \leq \psi(\rho(k)) - \varsigma(1 - \delta(k))\|\nabla\psi(\rho)\|_2^2. \tag{7.37}$$

We will next prove that the termination condition is satisfied whenever $0 \leq 1 - \delta(k) \leq 1/Q$. Since $\psi(\rho) \preceq QI$, we can derive, according to [143],

$$\psi(\rho(k+1)) \leq$$
$$\psi(\rho(k)) + \left(\frac{(1 - \delta(k))Q}{2} - 1\right)(1 - \delta(k))\|\nabla\psi(\rho)\|_2^2. \tag{7.38}$$

When $0 \leq 1 - \delta(k) \leq 1/Q$, $\frac{(1-\delta(k))Q}{2} - 1 \leq -1/2$. Therefore,

$$\psi(\rho(k+1)) \leq$$
$$\psi(\rho(k)) - \frac{(1 - \delta(k))}{2}\|\nabla\psi(\rho)\|_2^2. \tag{7.39}$$

Since $0 < \varsigma < 0.5$, $-\frac{(1-\delta(k))}{2} \leq -(1 - \delta(k))\varsigma$. Thus, we have

$$\psi(\rho(k+1)) \leq \psi(\rho(k)) - (1 - \delta(k))\varsigma\|\nabla\psi(\rho)\|_2^2, \tag{7.40}$$

which satisfies the termination condition of Algorithm 13. Therefore, Algorithm 13 terminates either with $\delta(k) = 0$ or $(1 - \delta(k))$ equaling a value that is larger than ξ/Q.

In the first case $(\delta(k) = 0)$, we have

$$\psi(\rho(k+1)) \leq \psi(\rho(k)) - \varsigma\|\nabla\psi(\rho)\|_2^2. \tag{7.41}$$

In the second case $((1 - \delta(k)) \geq \xi/Q)$, we can derive that

$$\psi(\rho(k+1)) \leq \psi(\rho(k)) - \varsigma\xi/Q\|\nabla\psi(\rho)\|_2^2. \tag{7.42}$$

Thus,

$$\psi(\rho(k+1)) \leq \psi(\rho(k)) - \min\{\varsigma, \varsigma\xi/Q\}\|\nabla\psi(\rho)\|_2^2. \tag{7.43}$$

Subtracting $\psi(\rho^*)$ from both sides, we have

$$\psi(\rho(k+1)) - \psi(\rho^*) \leq$$
$$\psi(\rho(k)) - \psi(\rho^*) - \min\{\varsigma, \varsigma\xi/Q\}\|\nabla\psi(\rho)\|_2^2. \tag{7.44}$$

Since $qI \leq \nabla^2\psi(\rho)$, according to [143],

$$\|\nabla\psi(\rho(k))\|_2^2 \geq 2q(\psi(\rho(k)) - \psi(\rho^*)). \tag{7.45}$$

Combining these, we can derive that

$$\psi(\rho(k+1)) - \psi(\rho^*) \leq$$
$$(1 - \min\{2q\varsigma, 2q\varsigma\xi/Q\})(\psi(\rho(k)) - \psi(\rho^*)). \tag{7.46}$$

Let $z = 1 - \min\{2q\varsigma, 2q\varsigma\xi/Q\}$, and by applying the inequality recursively, we find that

$$\psi(\rho(k+1)) - \psi(\rho^*) \leq z^k(\psi(\rho(1)) - \psi(\rho^*)). \tag{7.47}$$

Let $z^k(\psi(\rho(1)) - \psi(\rho^*)) = \epsilon$; we can derive the number of iterations required to achieve ϵ optimality:

$$k = \frac{\log((\psi(\rho(1)) - \psi(\rho^*))/\epsilon)}{\log 1/z}. \tag{7.48}$$

$\square$

Eq. (7.35) indicates that $\psi(\rho)$ converges at least as fast as a geometric series. Such convergence is called linear convergence in the context of iterative numerical methods [143]. The number of iterations required for $\psi(\rho)$ to converge depends on the gap between $\psi(\rho(1))$ and $\psi(\rho^*)$, ϵ, and z. Given the gap and the value of ϵ, a smaller z enables faster convergence. By properly selecting ς and ξ, we can reduce the value of z and hence reduce the number of iterations required for the convergence. However, the optimization of ς and ξ is beyond the scope of this study.

7.3.3.2 vGALA Optimality

As the vBSs' traffic load vector converges to ρ^*, we shall next show that the corresponding user association minimizes $\psi(\rho)$.

Theorem 7.3.2 *Suppose that F is not empty, and that the traffic load vector converges to ρ^*, then the user association corresponding to ρ^* minimizes $\psi(\rho)$.*

Proof: Define $\eta^* = \{\eta_j^*(x)|\eta_j^*(x) = \{0,1\}, \forall j \in B, \forall x \in A\}$ and $\eta = \{\eta_j(x)|\eta_j(x) = \{0,1\}, \forall j \in B, \forall x \in A\}$ as the user association corresponding to ρ^* and any other traffic load vector $\rho \in F$, respectively.

Let $\triangle \rho^* = \rho - \rho^*$. Since $\psi(\rho)$ is a convex function over ρ, proving the theorem is equivalent to proving:

$$\langle \nabla \psi(\rho)|_{\rho=\rho^*}, \triangle \rho^* \rangle \geq 0. \tag{7.49}$$

$$\langle \nabla \psi(\rho)|_{\rho=\rho^*}, \triangle \rho^* \rangle \tag{7.50}$$

$$= \sum_{j \in B} (\rho_j - \rho_j^*) \phi_j(\rho_j^*)$$

$$= \sum_{j \in B} \frac{\int_{x \in A} \lambda(x)v(x)(\eta_j(x) - \eta_j^*(x))dx}{r_j(x)\phi_j^{-1}(\rho_j^*)}$$

$$= \int_{x \in A} \lambda(x)v(x) \sum_{j \in B} \frac{\eta_j(x) - \eta_j^*(x)}{r_j(x)\phi_j^{-1}(\rho_j^*)}dx.$$

According to the user side algorithm,

$$\eta_j^*(x) = \begin{cases} 1, & j = \arg max_{i \in B} \frac{r_i(x)}{\phi_i(\rho_i^*)} \\ 0, & otherwise. \end{cases} \tag{7.51}$$

Therefore,

$$\sum_{j \in B} \frac{\eta_j^*(x)}{r_j(x)\phi_j^{-1}(\rho_j^*)} \leq \sum_{j \in B} \frac{\eta_j(x)}{r_j(x)\phi_j^{-1}(\rho_j^*)}. \tag{7.52}$$

Hence, $\langle \nabla \psi(\rho)|_{\rho=\rho^*}, \triangle \rho^* \rangle \geq 0.$ □

In conclusion, we have proved the convergence and optimality of the proposed vGALA traffic load balancing scheme. The analysis further shows that the vGALA scheme converges to the optimal solution within a small number of iterations.

7.3.3.3 vGALA Generalization

In determining user association, the vGALA scheme strives for a balance between green energy utilization and network performance. In the problem formulation, $w_j(\rho_j)$ and $L(\rho_j)$ model green energy utilization and network performance, respectively. Since $w_j(\rho_j)$ and $L(\rho_j)$ are functions of the traffic load ρ_j, they are coupled by ρ_j. $L(\rho_j)$ is a general latency indicator derived under the M/G/1 processor sharing queue model. In practical networks, traffic arrival may follow an arbitrary distribution rather than a Poisson distribution. In addition, network operators may aim to represent the network performance with other metrics than average traffic delivery latency. It is desirable that the vGALA framework can be applied to a collection of network performance models. Define $f(\rho_j)$ as a function of the traffic load ρ_j that models the jth BS's performance. Define the user association problem with a generalized network performance model, $f(\rho_j)$, as the general user association problem (GUA) problem, which can be expressed as:

$$\min_{\rho} \sum_{j\in B} w_j(\rho_j)f(\rho_j) \tag{7.53}$$

subject to: $0 \le \rho_j \le 1 - \epsilon.$ (7.54)

Lemma 7.3.5 *If $f(\rho_j)$ is positive, convex, and non-decreasing over ρ_j, $\forall j \in B$, $\tilde{\psi}(\rho) = \sum_{j\in B} w_j(\rho_j)f(\rho_j)$ is convex over $\rho \in \tilde{F}$.*

Proof: Since $f(\rho_j)$ is positive, convex, and non-decreasing, $f(\rho_j) > 0$, $f''(\rho_j) \ge 0$, and $f'(\rho_j) \ge 0$. Because $w_j''(\rho_j) > 0$, $w_j'(\rho_j) > 0$, and $w_j(\rho_j) > 0$,

$$\frac{\partial^2 \sum_{j\in B} w_j(\rho_j)f(\rho_j)}{\partial \rho_j^2}$$

$$= w_j''(\rho_j)f(\rho_j) + 2w_j'(\rho_j)f'(\rho_j) + w_j(\rho_j)f''(\rho_j)$$

$$\ge q. \tag{7.55}$$

Here, q is a positive number. Let I be an identity matrix. Since

$$\frac{\partial^2 \sum_{j\in B} w_j(\rho_j)f(\rho_j)}{\partial \rho_j \partial \rho_i} = 0, \ \forall i \ne j, \tag{7.56}$$

$\nabla^2 \tilde{\psi}(\rho) \ge qI$. Therefore, $\tilde{\psi}(\rho)$ is a strong convex function over ρ, $\rho \in \tilde{F}$. □

Theorem 7.3.3 *If the jth BS's network performance metric, $f(\rho_j)$, is positive, convex, and non-decreasing over ρ_j, $\forall j \in B$, then the GUA problem can be solved by the vGALA scheme.*

Proof: In order to guarantee the convergence and the optimality of the vGALA scheme, $\tilde{\psi}(\rho)$ has to be strongly convex over $\rho \in \tilde{F}$. According to the above lemma, if $f(\rho_j)$ is positive, convex, and non-decreasing, $\tilde{\psi}(\rho)$ is a strong convex function. Thus, the vGALA framework can be utilized to solve the GUA problem in which $f(\rho_j)$ is the jth BS's network performance metric. □

Based on the above analysis, we have shown that the vGALA scheme can be applied to optimize the traffic loads among BSs under different traffic models. Next, we will discuss how the vGALA algorithm can be implemented in practical networks.

7.3.4 The Practicality of the vGALA Scheme

In this section, we first present how to put the vGALA framework into practice and evaluate the assumptions made in developing the scheme. Then, we discuss two related issues on applying the vGALA scheme: the energy–latency trade-off and the admission control mechanism.

Figure 7.9 The practical implementation of vGALA. *Source:* Han 2016 [157]. Reproduced with permission of IEEE.

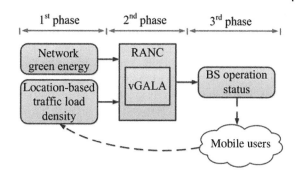

7.3.4.1 Practical Implementation

In practical cellular networks, traffic load balancing among BSs is usually triggered by network level events (for example, some BSs are congested while others are lightly loaded) rather than by user-level events (for example, users' movement and data rate changes). Since a BS's traffic load is determined by the average traffic load density of its coverage area, without considering green energy, it is reasonable to reduce a BS's coverage area to avoid traffic congestion if the traffic load density of its coverage area is increasing. Therefore, a BS's traffic load can be derived based on location-based traffic load density. Thus, for practical implementation, the vGALA scheme collects location-based traffic load density and network green energy information in the first phase as shown in Figure 7.9. Given a specific location, it is realistic to assume that BSs' downlink data rates to users at the location are not changing during a traffic load balancing period. Note that when modeling the traffic load in the UA problem, we differentiated users by their locations. Therefore, the vGALA scheme is compatible with input of location-based traffic load density and location-based downlink data rates.

In the second phase, the vGALA scheme implemented in the RANC optimizes user association and derives the optimal BS operation status based on the network information collected in the first phase. The optimization can be triggered either periodically or by some predefined events, for example, if a BS's traffic load exceeds a threshold or a BS's green energy utilization falls below a threshold. Deciding on the best strategies for triggering traffic load balancing can be determined by network operators and is beyond the scope of this study. The output of the second phase is BS operation status, based on which the user association is determined in the third phase. In this phase, a user's BS association can be determined in either a centralized or distributed fashion. In the first case, users send their data rate measurements to the RANC, and the RANC determines the users' BS associations based on the BS operation status and the users' date rates. In the second case, the RANC may simply let BSs broadcast their operation statuses, based on which individual users decide their own BS associations. The users' BS selections may change the location-based traffic load density. An individual BS translates the users' BS selections into location-based traffic load density and reports it to the RANC.

In the vGALA scheme, user association is optimized in consideration of both average traffic delivery latency and green energy usage. Users, who may not care about green energy usage, may seek to maximize their performance and violate the BS selection rule in the vGALA scheme. However, users, in fact, do not have the ability to maximize their own QoS. According to Eq. (7.21), a user's BS selection is based on both $r_j(x)$ and $\phi_j(\rho_j)$.

Here, $\phi_j(\rho_j)$ is determined by both the jth BS's traffic load and its available green energy. A user's average traffic delivery latency is determined by both the downlink data rate and the traffic load of the associated BS. Since users do not know the traffic loads of BSs, they cannot know which BS can provide them the best QoS. Simply selecting a BS with the largest $r_j(x)$ may lead them to a highly congested BS and degrade their QoS. Thus, users do not have obvious incentives to counterfeit their measurement reports.

7.3.4.2 Energy–Latency Trade-Off Adaptation

The vGALA scheme provides two parameters for adapting the trade-off between on-grid power consumption and average traffic delivery latency. The parameters are θ and κ. θ is the energy-latency coefficient of a BS. It reflects individual BSs' operation strategies. A BS with a large θ ($\theta \rightarrow 1$) indicates that the BS is energy-sensitive. When a BS chooses a small θ ($\theta \rightarrow 0$), the BS is latency-sensitive. Therefore, by choosing the value of θ, a BS adapts its sensitivity to on-grid power consumption and average traffic delivery latency. Hence, θ is chosen by individual BSs based on their operational strategies.

Here, κ is chosen by the RANC based on the global view of green energy status and mobile traffic demands. Given θ and the available green energy, $w_j(\rho_j)$ grows exponentially as traffic demand increases. For a large κ, $w_j(\rho_j)$ grows faster than it does with a small κ. This indicates that the vGALA scheme is more energy-sensitive when κ is assigned a larger value. When κ keeps increasing, the vGALA scheme will perform similarly as a solely energy aware user association scheme. On the other hand, when $\kappa = 0$, the vGALA scheme is a solely latency aware user association scheme. In addition, since $0 \leq \theta_j \leq 1, 0 \leq \kappa\theta_j \leq \kappa$. Thus, the value of κ restricts individual BSs' capabilities in adapting their energy-latency trade-off. The adaptation of κ can be triggered by either green energy changes or mobile traffic demand changes. For example, when the network experiences heavy traffic loads, the RANC will focus on balancing traffic loads to reduce network congestion. In this case, the RANC may choose a small κ to give a high priority to latency awareness in balancing traffic loads. On the other hand, if the network experiences light traffic loads, the RANC may increase κ to emphasize green energy usage.

Traffic load balancing parameters do impact the convergence of traffic load balancing algorithms such as the one reported in [175]. For the vGALA scheme, κ and θ determine the energy-latency trade-off of the network. As a result, they determine the optimal traffic loads of individual BSs. According to Lemma 7.3.4, the number of iterations required for the vGALA scheme to converge depends on $\psi(\rho(1)) - \psi(\rho^*)$. Here, $\rho(1)$ and ρ^* are the initial and the optimal traffic load vectors, respectively. The optimal traffic load vector is related to the energy-latency trade-off of the network. Therefore, κ and θ affect the convergence of the vGALA scheme: when κ and θ enable the vGALA scheme to achieve a higher network performance enhancement in terms of minimizing $\psi(\rho)$, the difference between $\psi(\rho(1))$ and $\psi(\rho^*)$ will be larger, and as a result, the vGALA scheme requires more iterations to converge.

7.3.5 The Admission Control Mechanism

A necessary condition for the convergence and optimality of the vGALA scheme is that the UA problem is feasible. In other words, the BSs' traffic loads should be within the feasible set defined in Eq. (7.26). When traffic loads are beyond the network capacity, the UA problem is no longer feasible. As a result, the properties of the vGALA scheme will not hold. Therefore, the admission control mechanism is necessary for the vGALA scheme to ensure the feasibility of the UA problem. Thus, the purpose of proposing a simple admission control mechanism is to ensure that the vGALA scheme works even under very heavy traffic load conditions (when the UA problem is not feasible) rather than to reduce either the energy consumption or average traffic delivery latency of the network.

Define $\mu(x)$ as the admission control coefficient for a user located at x. $0 \le \mu(x) \le 1$ indicates the probability that a user at location x is admitted to the network. The RANC assigns $\mu(x)$ to a user at location x. $\mu(x)$ does not depend on the user's BS selection. In other words, no matter which BS is selected by a user, the user's admission control coefficient does not change. Thus, integrating the admission control mechanism does not change the BS selection rule of the users. The coverage area of a BS, for example, $\tilde{A}_j(k)$, is still calculated by Eq. (7.22). Owing to the admission control, the traffic load measurement in the jth vBS is revised as

$$M_j(\rho(k)) = \min \left(\int_{x \in \tilde{A}_j(k)} \mu(x)\varrho_j(x)dx, 1 - \epsilon \right). \tag{7.57}$$

The vBS updates its traffic load based on Eq. (7.24).

With the admission control, the RANC is able to restrict the traffic loads in the network to ensure the UA problem being feasible. The relaxed feasible set for the UA problem with admission control is

$$\hat{\mathcal{F}} = \{ \rho | \rho_j = \int_{x \in \mathcal{A}} \mu(x)\varrho_j(x)dx,$$

$$0 \le \rho_j \le 1 - \epsilon, \sum_{j \in \mathcal{B}} \eta_j(x) = 1,$$

$$0 \le \eta_j(x) \le 1, \ \forall j \in \mathcal{B}, \ \forall x \in \mathcal{A} \}. \tag{7.58}$$

Since $0 \le \mu(x) \le 1$ is a constant, Lemma 7.3.1 still holds, implying that $\hat{\mathcal{F}}$ is a convex set. Integrating admission control does not change the objective function of the UA problem. Thus, Lemma 7.3.2 also holds. By applying a similar analysis, we can prove that the vGALA scheme still enables the convergence of the traffic loads and obtains the optimal solution to the UA problem with admission control.

7.3.6 Performance Evaluation

We set up system level simulations to investigate the performance of the vGALA scheme for downlink traffic load balancing in HetNet. In the simulation, three MBSs and seven SCBSs are randomly deployed in a $2000\,m \times 2000\,m$ area. The traffic arrival in the area follows a Poisson point process with the average arrival rate equal to 200. The traffic size per arrival is 250 *kbits*. The area is divided into 40 000 locations with each location representing a $10\,m \times 10\,m$ area. The location-based traffic load density is calculated based on the traffic model. The static power consumption of the MBS and the SCBS are 750 W and 37 W, respectively [3]. The load-power coefficients of the MBS and SCBS are 500 and 4, respectively [3]. The solar cell power efficiency is 17.4% [176]. We assume that the weather condition is the standard condition which specifies a temperature of 25 °C, an irradiance of 1000 $W\,m^{-2}$, and an air mass of 1.5 spectrum. Thus, the green energy generation rate is 174 $W\,m^{-2}$. The solar panel sizes are randomly selected but ensure the

Table 7.1 Channel model and parameters.

Parameters	Value
PL_{MBS} (dB)	$PL_{MBS} = 128.1 + 37.6 \log_{10}(d)$
PL_{SCBS} (dB)	$PL_{SCBS} = 38 + 10 \log_{10}(d)$
Rayleigh fading	$9\ dB$
Shadowing fading	$5\ dB$
Antenna gain	$15\ dB$
Noise power level	$-174\ dBm$
Receiver sensitivity	$-123\ dBm$

green power generation capacity of MBSs from 750 W to 1300 W and that of SCBS from 37 W to 48 W. BSs' energy-latency coefficients are set to be the same. In other words, we let $\theta_i = \theta_j = \theta$, $\forall i, j \in B$ in the simulation. The value of θ varies for different simulation scenarios. The total bandwidth is 20 MHz of which 10 MHz is exclusively used by MBSs and the other 10 MHz is allocated to SCBSs. The frequency reuse factor for each system (MBSs and SCBSs) is one. The channel propagation model is based on COST 231 Walfisch-Ikegami [144]. The model and parameters are summarized in Table 7.1. Here, PL_{MBS} and PL_{SCBS} are the path loss between the users and MBSs and that between the users and SCBSs, respectively. d is the distance between users and BSs. The values of various parameters used to implement the vGALA scheme are summarized in Table 7.2. While the values of θ and κ vary for different simulation scenarios, ϵ, ς, and ξ are fixed.

7.3.6.1 Performance Comparison

We compare the vGALA scheme with a green energy aware (GA) user association scheme and a latency aware (LA) user association scheme. The GA scheme solves the green energy aware problem (GAP) formulated as

$$\min_{\rho} \sum_{j \in B} \max(\rho_j - e_j, 0) \tag{7.59}$$

$$subject\ to: 0 \le \rho_j \le 1 - \epsilon. \tag{7.60}$$

The LA scheme solves the LAP as

$$\min_{\rho} \sum_{j \in B} L(\rho_j) \tag{7.61}$$

$$subject\ to: 0 \le \rho_j \le 1 - \epsilon. \tag{7.62}$$

Table 7.2 vGALA implementation parameters.

Parameters	θ	κ	ϵ	ς	ξ
Value	$0 < \theta < 1$	$1 \le \kappa \le 20$	0.001	0.4	0.7

Figure 7.10 Green energy aware (GA). *Source:* Han 2016 [157]. Reproduced with permission of IEEE.

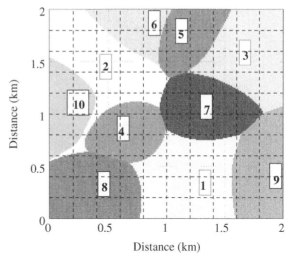

As shown in Figs. 7.10, 7.11, and 7.12, different user association schemes result in different traffic load distribution among BSs. In the figure, the coverage areas of different BSs are shaded differently.[1] A larger coverage area indicates that the BS serves a heavier traffic load. The first, second, and third BSs are MBSs and the other BSs are SCBSs. Taking the coverage area of the 5th BS as an example, as compared with the GA scheme (Figure 7.10), the LA scheme significantly reduces the BS's coverage area as shown in Figure 7.11. The 5th BS has sufficient green energy. Therefore, the GA scheme will redirect more traffic load to the BS to minimize the on-grid power consumption. The LA scheme, which does not consider the energy usage, balances the traffic loads among BSs to minimize the average traffic delivery latency. As a result, the LA scheme

Figure 7.11 Latency aware (LA). *Source:* Han 2016 [157]. Reproduced with permission of IEEE.

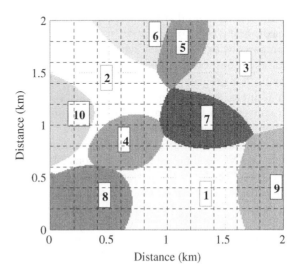

[1] The white areas indicate the coverage area of the second BS.

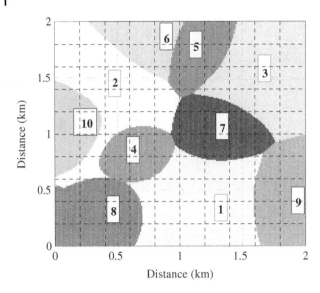

Figure 7.12 vGALA ($\theta = 0.8$, $\kappa = 4$). *Source:* Han 2016 [157]. Reproduced with permission of IEEE.

limits the traffic load in the BS. In terms of power consumption and average traffic delivery latency, as compared with the GA scheme, the vGALA scheme slightly reduces the BS's coverage area as shown in Figure 7.12 to obtain a trade-off between on-grid power consumption and average traffic delivery latency.

Figure 7.13 shows the trade-off achieved by the vGALA scheme between on-grid energy consumption and average traffic delivery latency. Figure 7.13(a) shows the on-grid power consumption of the LA, vGALA, and GA schemes, respectively. As compared with the LA scheme, the vGALA scheme consumes 30% less on-grid power. Figure 7.13(b) shows that the average traffic delivery latency of the vGALA scheme is only 8% more than that of the LA scheme. While the GA scheme significantly reduces the on-grid power consumption, it increases the traffic delivery latency by about 48% as compared with the vGALA scheme. Here, the latency indicator is equal to $\sum_{j \in B} L(\rho_j)$. The above observation indicates that the vGALA scheme achieves a preferable trade-off: saving 30% on-grid power at the cost of 8% increase in average traffic delivery latency. In addition, as shown in Figure 7.13, the vGALA scheme requires about 60 iterations to converge to the optimal solution. On the one hand, this proves that the vGALA scheme converges fast. On the other hand, it indicates that the vGALA scheme avoids the communications overhead of an over the air interface by virtualizing users and BSs in the RANC to simulate the interactions between users and BSs.

Figure 7.14 shows the performance of the LA, vGALA and GA schemes for different average traffic arrival rates. As shown in the figure, as the average traffic arrival rate increases, the on-grid power consumption and the average traffic delivery latency of these schemes also increase. When the average traffic arrival rate is very small, for example 100, these schemes exhibit similar performance because the traffic loads in individual BSs, in spite of different traffic load balancing schemes, are less than their green traffic capacity. As the average traffic arrival rate increases, the performance gap between the LA and GA schemes increases both in terms of on-grid power consumption and average traffic delivery latency. The performance of the vGALA scheme, which is determined by

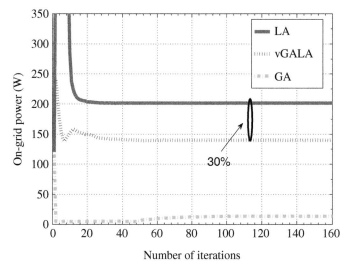

(a) The on-grid power consumption.

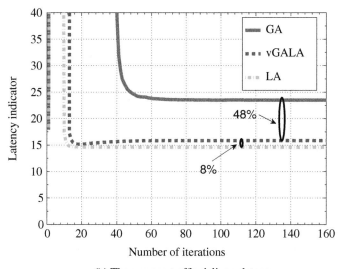

(b) The average traffic delivery latency.

Figure 7.13 The comparison of different user association schemes ($\theta = 0.8$, $\kappa = 4$).
Source: Han 2016 [157]. Reproduced with permission of IEEE.

θ and κ, is between that of the LA and the GA scheme. When the traffic arrival rate increases further, the traffic load balancing problem becomes infeasible.

7.3.6.2   Performance Adaptation

The trade-off between the on-grid power consumption and the average traffic delivery latency in the vGALA scheme can be adapted by adjusting κ and θ. Figure 7.15 shows the performance of the vGALA scheme with different κ. By varying κ, the vGALA scheme

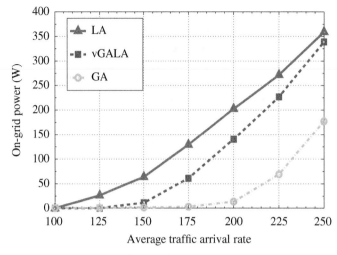

(a) The on-grid power consumption.

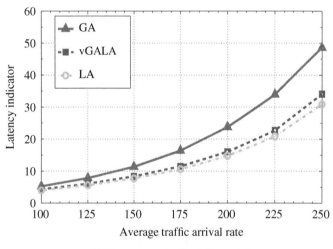

(b) The average traffic delivery latency.

Figure 7.14 The performance of user association schemes with different average traffic arrival rates ($\theta = 0.8$, $\kappa = 4$). *Source:* Han 2016 [157]. Reproduced with permission of IEEE.

may act like an LA scheme when $\kappa \to 0$ and like a GA scheme when $\kappa \to \infty$. As shown in Figure 7.16, given κ, adjusting θ has a limited performance adaptation. In other words, κ defines a performance adaptation range, and adjusting θ can only adapt the performance within the range. As discussed in the previous section, the selection of θ is determined by the operation strategies of BSs while the value of κ is chosen based on network conditions, for example, the traffic load intensity and the available green energy. In addition, the figures show how the values of κ and θ affect the number of iterations required for convergence. However, the optimization of these values is beyond the scope of this study.

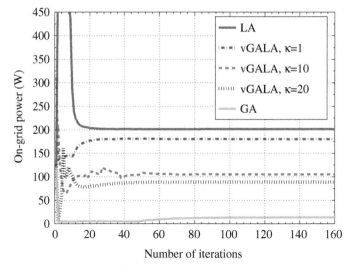

(a) The on-grid power consumption.

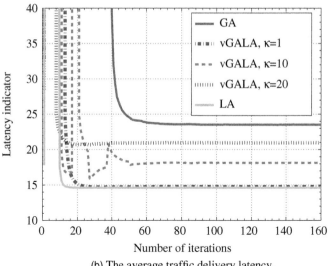

(b) The average traffic delivery latency.

Figure 7.15 The performance of vGALA with various κ ($\theta = 1$). *Source:* Han 2016 [157]. Reproduced with permission of IEEE.

7.3.6.3 Green Energy Generation Rate Evaluation

The amount of green energy in BSs impacts the performance of the vGALA scheme. In Figure 7.17, the x-axis is the solar cell power efficiency. As the efficiency is enhanced, the amount of green energy in BSs will increase. As shown in Figure 7.17(a), the on-grid power consumption of BSs decreases as solar cell power efficiency increases. This is because more green energy is available in BSs. Considering the increase in solar cell power efficiency, the performance of the average traffic delivery latency can be divided

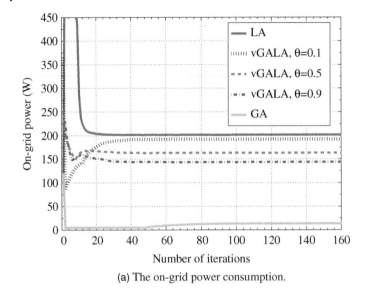

(a) The on-grid power consumption.

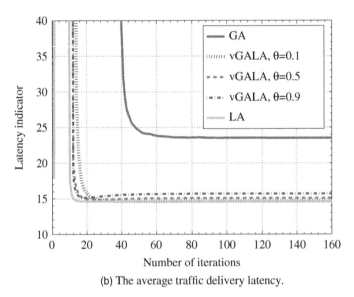

(b) The average traffic delivery latency.

Figure 7.16 The performance of vGALA with various θ ($\kappa = 4$). *Source:* Han 2016 [157]. Reproduced with permission of IEEE.

into four regions as shown in Figure 7.17(b). In the first region (R1), no BSs have sufficient green energy to offset their static power consumption. As a result, BSs' green traffic capacities are zero. In this condition, the vGALA scheme performs an an LA scheme. In the second region (R2), the green energy capacities of BSs start to impact the traffic load balancing. The traffic loads will be directed to BSs that have sufficient green energy. Meanwhile, the vGALA scheme avoids excessively increasing the average traffic delivery latency. In this region, green energy is not sufficient in the network. Thus,

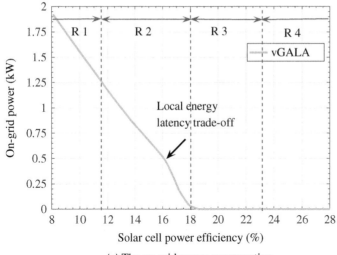

(a) The on-grid power consumption.

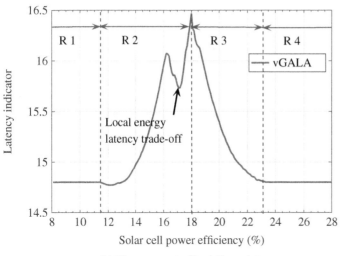

(b) The average traffic delivery latency.

Figure 7.17 The performance of vGALA versus solar cell power efficiency ($\theta = 0.8$, $\kappa = 4$). *Source:* Han 2016 [157]. Reproduced with permission of IEEE.

the major strategy is to trade the average traffic delivery latency for savings in on-grid power. However, as the solar power efficiency increases, some BSs may have sufficient green energy and they start trading their green energy for reductions in the average traffic delivery latency in the network (the solar power efficiency for this falls between 16% and 17%). This event reflects a local energy-latency trade-off among several BSs. In the third region (R3), as the solar cell power efficiency further increases, the traffic load balancing becomes more flexible with respect to the green energy constraint, which enables the vGALA scheme to further reduce average traffic delivery latency. In both region R2 and R3, the vGALA scheme determines the trade-off between on-grid power consumption and average traffic delivery latency. In the fourth region (R4), all BSs have sufficient green energy to operate with full traffic loads. In other words, the green traffic capacities of all the BSs are equal to one. Thus, green energy is no longer a concern in balancing the traffic load and the vGALA scheme acts as an LA scheme.

7.3.6.4 Practicality Evaluation

The CRE approach is one of the most practical traffic load balancing approaches and has been proven to have similar performance as optimal traffic balancing schemes in terms of maximizing network utilities [36, 161]. This simulation evaluates the traffic balancing performance of the vGALA scheme and the CRE approach. For the vGALA scheme, the simulation obtains the optimal BS operation status based on location-based traffic load density of the coverage area and the available green energy. We adopt a two-tier data rate bias approach as the CRE approach and assume that BSs in the same tier have the same data rate bias. In the simulation, MBSs are in the first tier while SCBSs are in the second tier. In the data rate bias approach, a user selects the BS to maximize the biased data rate

$$b(x) = \arg\max_{j \in B} Z_j r_j(x). \tag{7.63}$$

Here, $b(x)$ and Z_j are the index of the selected BS and the data rate bias of the jth BS, respectively. The data rate bias of a MBS is one. The data rate biases are selected for SCBSs to minimize (1) the average traffic delivery latency, (2) the overall on-grid power consumption, and (3) $\psi(\rho)$. Denote these data rate biases as (1) CRE_LA (latency aware), (2) CRE_GA (green energy aware), and (3) CRE_LG (latency and green energy aware), respectively.

In the simulation, the BS operation status and the data rate biases are calculated based on the location-based traffic load density generated in previous simulations. We randomly generate users' locations using a Poisson point process[2] with the average rate equal to 200 in the area. The average traffic size per user is 250 *kbits*. We run the simulation 10 000 times to evaluate the performance of different approaches in terms of the average traffic delivery latency and the average on-grid power consumption. As shown in Figure 7.18, CRE_GA achieves the lowest on-grid energy consumption among all the schemes. However, the average traffic delivery latency of CRE_GA is significantly larger than other schemes. As compared with CRE_LA and CRE_LG, the vGALA scheme not only saves on on-grid energy consumption but also reduces the average traffic delivery latency. For the vGALA scheme, when κ increases, the scheme is to gradually prioritize saving on-grid energy in balancing the traffic loads, as shown in Figure 7.18(a), at the cost of a small increase of the average traffic delivery latency as shown in Figure 7.18(b). For the CRE_LG scheme, increasing κ does not adjust the energy-latency trade-off as effectively as the vGALA scheme does. This indicates that the tier-based data rate bias approach may not perform well on jointly optimizing the utilization of green energy and network utilities.

---

[2] The Poisson point process used is the same as the Poisson point process used to generate the location-based traffic load density.

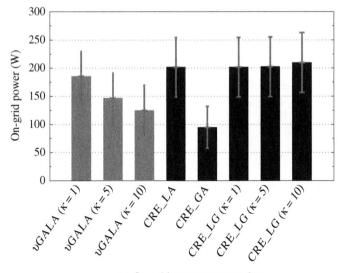

(a) On-grid power consumption.

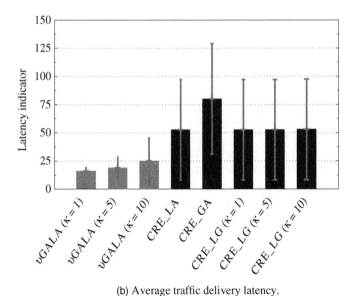

(b) Average traffic delivery latency.

Figure 7.18 The performance of vGALA versus CRE ($\theta = 0.8$). *Source:* Han 2016 [157]. Reproduced with permission of IEEE.

7.4 Energy Efficient Traffic Load Balancing in Backhaul Constrained Small Cell Networks

Although SCBSs can effectively enhance network capacity and offload traffic load from MBSs, cost-effective backhaul may not be readily available for SCBSs, leading to backhaul constraints in SCNs. Enabling caching in BSs may mitigate the backhaul

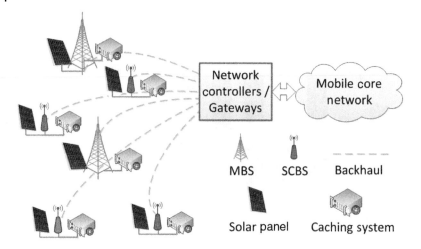

Figure 7.19 The small cell network. *Source:* Han 2017 [158]. Reproduced with permission of IEEE.

constraints in SCNs. In this section, we investigate traffic load balancing in backhaul constrained cache-enabled small cell networks with hybrid power supplies. The network architecture is shown in Figure 7.19. Traffic load balancing in such a network requires consideration of green power utilization, BS capacity, backhaul constraints, and the performance of cache systems. Therefore, we introduce four network utilities for balancing traffic loads among BSs: (1) green power utilization, (2) the traffic delivery latency in BSs, (3) the traffic delivery latency in backhaul, and (4) the cache hit ratio. The last three network utilities jointly determine the traffic delivery latency of the network. Thus, awareness of these network utilities helps reduce the traffic delivery latency in the network. Since green power utilization and traffic delivery latency are not optimized simultaneously in most scenarios, the trade-off between the two should be determined based on network conditions. We propose a network utility aware (NUA) traffic load balancing scheme to adapt user associations according to the dynamics of these network utilities and strike an adjustable trade-off between brown power consumption and traffic delivery latency.

7.4.1 System Model and Problem Formulation

In this subsection, we present the system model and the problem formulation. The system model includes traffic and QoS models. The energy model is the same as the model presented in Section 7.3.1.2.

7.4.1.1 Traffic and QoS Model

Define B as the set of BSs, and A as the coverage area of all BSs. Here, a BS refers to either an MBS or a SCBS. Since BSs are equipped with caches, users' data requests can be fulfilled by the cache system if the requested content is cached; otherwise, the requested content is retrieved from the Internet. Retrieving content from the Internet generates traffic in a BS's backhaul. Therefore, we model the traffic delivery process as a queuing system as shown in Figure 7.20.

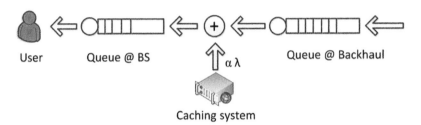

User   Queue @ BS $\alpha\lambda$ Queue @ Backhaul

Caching system

Figure 7.20 The traffic delivery process as a queuing system. *Source:* Han 2017 [158]. Reproduced with permission of IEEE.

The performance of a cache system is commonly evaluated based on the cache hit ratio, which is defined as the ratio between the number of cache hits and the total requests observed over a period of time [177]. Many analytical models have been proposed to estimate the hit ratio of a cache system and various content caching strategies have been designed to optimize the performance of cache systems [177–181]. Thus, in this section, we assume that the hit ratio of a cache system in a BS can be estimated for the time duration of one user association process and define $0 \leq \alpha_j \leq 1$ as the hit ratio of the cache system in BS j. Note that estimation and optimization of the hit ratio is outside the scope of this study.

Let a mobile user at location x be associated with BS j. We assume that traffic arrives at BS j's backhaul toward the user according to a Poisson process with arrival rate equal to $\tilde{\lambda}(x)$, and that the traffic load per arrival (packet sizes per arrival) has an exponential distribution with average traffic load $v(x)$. We assume that the users associated with BS j are uniformly distributed in its coverage area and the traffic arrival processes are independent. For simplicity of presentation, we further assume that no users share the same locations, that is, there is only one user at location x. Since the traffic arrival at a location is a Poisson process, the traffic arrival in BS j's backhaul, which is the sum of the traffic arrivals from its coverage area, is also a Poisson process. Although BSs may adopt different access technologies for their backhaul, it is reasonable to assume the expected data rates of the backhaul are constant in the time duration of one user association process [165]. Since the traffic load per arrival follows an exponential distribution, the traffic delivery time (service time) of the backhaul is also an exponential distribution. Therefore, the traffic delivery in backhaul simply realizes an M/M/1 queuing system.

Define R_j as the average data rate of BS j's backhaul. To fulfill the traffic demand of the user at location x, the required service time in BS j's backhaul is

$$\tilde{\gamma}(x) = \frac{v(x)}{R_j}. \tag{7.64}$$

The average traffic load density generated by a user at location x in BS j's backhaul is

$$\tilde{\varrho}_j(x) = \frac{\tilde{\lambda}(x)v(x)\eta_j(x)}{R_j}. \tag{7.65}$$

Here, $\eta_j(x) = \{0, 1\}$ is an indicator function. If $\eta_j(x) = 1$, the user at location x is associated with BS j; otherwise, the user is not associated with BS j. Since mobile users are uniformly distributed in the area, the traffic load in BS j's backhaul can be expressed as

$$\tilde{\rho}_j = \int_{x \in A} \tilde{\varrho}_j(x) dx. \tag{7.66}$$

According to the properties of the M/M/1 queue [152], the average waiting time for traffic load $v(x)$ in BS j's backhaul is

$$\tilde{W}_j(x) = \frac{\tilde{\rho}_j v(x)}{R_j(1 - \tilde{\rho}_j)}. \tag{7.67}$$

Define $\tilde{\mu}_j(x)$ as the latency ratio that measures how much time a user at location x must be sacrificed in waiting for per unit service time in BS j's backhaul.

$$\tilde{\mu}_j(x) = \frac{\tilde{W}_j(x)}{\tilde{\gamma}(x)} = \frac{\tilde{\rho}_j}{1 - \tilde{\rho}_j}. \tag{7.68}$$

Since $\tilde{\mu}_j(x)$ only depends on the traffic load in BS j's backhaul, all the users associated with BS j have the same latency ratio. Thus, we define

$$\tilde{\mu}_j(\tilde{\rho}_j) = \frac{\tilde{\rho}_j}{1 - \tilde{\rho}_j} \tag{7.69}$$

as the latency ratio of BS j's backhaul. A smaller $\tilde{\mu}_j(\tilde{\rho}_j)$ indicates that BS j's backhaul introduces less latency to its associated users.

According to Burke's Theorem [152], the traffic departure process at a BS's backhaul is a Poisson process with average departure rate equal to the average traffic arrival rate. Therefore, the average traffic arrival rate in BS j toward a user at location x is equal to $\tilde{\lambda}(x)$. Since the hit ratio of BS j's cache system is α_j, the data traffic from the backhaul accounts for $(1 - \alpha_j)$ of the total traffic load toward the user at location x. Therefore, the average traffic arrival rate in BS j toward the user at location x is

$$\lambda(x) = \frac{\tilde{\lambda}(x)}{(1 - \alpha_j)}. \tag{7.70}$$

Since α_j is assumed to be a constant during one user association process, the traffic arrival process toward the user at location x is a Poisson process. In BS j, users at different locations may have different data rates depending on channel conditions. When associating with BS j, the user's data rate, $r_j(x)$, can be generally expressed as a logarithmic function of the perceived SINR, $SINR_j(x)$, according to the Shannon–Hartley theorem [19],

$$r_j(x) = log_2(1 + SINR_j(x)), \tag{7.71}$$

where,

$$SINR_j(x) = \frac{P_j g_j(x)}{\sigma^2 + \sum_{k \in B, k \neq j} P_k g_k(x)}. \tag{7.72}$$

Here, P_j is the transmission power of BS j, and σ^2 denotes the noise power level. Since the users' data rate is generally distributed, the service time in BS j follows a general distribution. Therefore, a BS's downlink transmission process realizes a M/G/1 processor sharing (PS) queue, in which multiple users share the BS's downlink radio resource [152].

In mobile networks, various downlink scheduling algorithms have been proposed to enable proper sharing of the limited radio resource in a BS. According to the scheduling algorithm, users may be assigned different priorities on sharing the radio resource. For analytical simplicity, we assume that mobile users are served in a round robin (RR) fashion. Then, the average traffic load density at location x in BS j is calculated as

$$\varrho_j(x) = \frac{\lambda(x)v(x)\eta_j(x)}{r_j(x)}. \tag{7.73}$$

The traffic load in BS j can be expressed as

$$\rho_j = \int_{x \in A} \varrho_j(x)dx. \tag{7.74}$$

This value of ρ_j indicates the fraction of time during which BS j is busy. To fulfill the traffic demand of a user located at x, the required service time in BS j is

$$\gamma(x) = \frac{v(x)}{r_j(x)}. \tag{7.75}$$

Since the traffic delivery process in a BS realizes a M/G/1-RR queue, the average traffic delivery time for the user in BS j [152] is

$$T_j(x) = \frac{v(x)}{r_j(x)(1 - \rho_j)}. \tag{7.76}$$

The average waiting time for traffic load $v(x)$ in BS j is

$$W_j(x) = T_j(x) - \gamma(x) = \frac{\rho_j v(x)}{r_j(x)(1 - \rho_j)}. \tag{7.77}$$

Define $\mu_j(x)$ as the latency ratio of BS j for a user at location x.

$$\mu_j(x) = \frac{W_j(x)}{\gamma(x)} = \frac{\rho_j}{1 - \rho_j}. \tag{7.78}$$

$\mu_j(x)$ only depends on the traffic load in BS j. Therefore, all the users associated with BS j have the same latency ratio. Thus, we define

$$\mu_j(\rho_j) = \frac{\rho_j}{1 - \rho_j} \tag{7.79}$$

as the latency ratio of BS j. A smaller $\mu_j(\rho_j)$ indicates that BS j introduces less latency to its associated users. In this study, we use μ_j and $\tilde{\mu}_j$ to reflect the traffic delivery latency in BS j and its backhaul, respectively; we adopt $\mu_j(\rho_j) + \tilde{\mu}_j(\tilde{\rho}_j)$ as the QoS model that indicates the latency of delivering traffic through BS j.

7.4.1.2 Problem Formulation

In determining the user association, the network aims to not only enhance the network QoS by reducing the traffic delivery latency but also to reduce brown power consumption by improving green power utilization. Owing to the dynamics of data traffic and green power, the user association that minimizes the traffic delivery latency does not necessarily maximize green power utilization. Thus, the traffic load balancing problem strives for a trade-off between the traffic delivery latency and brown power consumption.

According to E.q. (7.12), brown power is consumed only when green power is not sufficient ($e_j < p_j$). Given e_j, the maximum traffic load can be supported by green power in BS j is

$$\hat{\rho}_j = max(0, min(\frac{e_j - p_j^s}{\beta_j}, 1 - \epsilon)). \tag{7.80}$$

Here, ϵ is an arbitrary small positive constant to guarantee $0 \le \hat{\rho}_j < 1$. Define $\hat{\rho}_j$ as BS j's green traffic capacity. When BS j's traffic load is larger than $\hat{\rho}_j$, BS j consumes brown power. In this case, it is desirable to offload data traffic from BS j to alleviate its brown power consumption. Let

$$\bar{\rho}_j = \rho_j - \hat{\rho}_j. \tag{7.81}$$

When $\bar{\rho}_j > 0$, BS j's traffic load is larger than its green power capacity and thus it is desirable to offload traffic from BS j to save brown power; when $\bar{\rho}_j < 0$, it is desirable to let BS j carry additional traffic load to enhance the usage of green power and thus reduce other BSs' brown power consumption. However, balancing traffic purely based on energy consumption may lead to heavy traffic congestion that increases the traffic delivery latency in BSs. In order to strike a balance, we introduce latency weights for individual BSs. Denote BS j's latency weight as

$$w_j(\rho_j) = e^{\kappa \bar{\rho}_j}, \tag{7.82}$$

where $\kappa \ge 0$ is a system parameter that adjusts the value of the latency weight.

Aiming to save brown power as well as to reduce the traffic delivery latency of the network, the traffic load balancing problem is formulated as:

$$\min_{\eta} \sum_{j \in B} w_j(\rho_j)(\mu_j(\rho_j) + \tilde{\mu}_j(\tilde{\rho}_j)) \tag{7.83}$$

$$subject\ to: \ \rho_j = \int_{x \in A} \frac{\lambda(x)v(x)\eta_j(x)}{r_j(x)} dx,$$

$$\tilde{\rho}_j = \int_{x \in A} \frac{\tilde{\lambda}(x)v(x)\eta_j(x)}{R_j} dx,$$

$$0 \le \rho_j \le 1 - \epsilon,$$

$$0 \le \tilde{\rho}_j \le 1 - \epsilon,$$

$$\eta_j(x) = \{0, 1\}, \forall j \in B, x \in A. \tag{7.84}$$

Here, $\eta = \{\eta_j | j \in B\}$ and $\eta_j = \{\eta_j(x) | x \in A\}$. Based on the formulation, if BS j has sufficient green power ($\tilde{\rho}_j \ge \rho_j$), $0 < w_j(\rho_j) \le 1$; otherwise, $w_j(\rho_j) > 1$. A large latency weight grants a BS a high priority in minimizing Eq. (7.83) as compared with those of the BSs having a small latency weight. In other words, a large latency weight grants a BS a high priority in offloading traffic. As compared with $w_j(\rho_j) \le 1$, $w_j(\rho_j) > 1$ enables

BS j to achieve a smaller latency ratio. Since $\frac{d\mu_j(\rho_j)}{d\rho_j} > 0$ and $\frac{d\tilde{\mu}_j(\tilde{\rho}_j)}{d\tilde{\rho}_j} > 0$, a small latency ratio indicates that BS j carries a lighter traffic load, which is desirable for a BS which is consuming brown power $(w_j(\rho_j) \leq 1)$. κ is a system parameter that enables the network to dynamically control the trade-off between brown power consumption and the traffic delivery latency.

7.4.2 Network Utility Aware Traffic Load Balancing

In this section, we propose a NUA traffic load balancing scheme and prove its properties. The network utilities considered in the traffic load balancing scheme consist of (1) green power utilization (brown power consumption), (2) the traffic delivery latency in BSs' backhaul, (3) the traffic delivery latency in BSs, and (4) the hit ratio of BSs' cache systems. The proposed network utility aware traffic load balancing scheme is able to adapt the traffic loads among BSs and their backhauls according to the dynamics of these network utilities.

7.4.2.1 Traffic Load Balancing Procedures

The traffic load balancing procedures can be implemented in either a distributed or a centralized fashion [157]. For distributed traffic load balancing, users select their serving BSs based on operating parameters, for example, traffic loads and data rates received from BSs. This will incur several interactions between BSs and users for updating the operating parameters and BS selections, respectively. For a centralized traffic load balancing scheme, the network collects the operating status information from both BSs and users, and determines the user association for individual users. A distributed traffic load balancing scheme can be implemented in a centralized fashion by leveraging virtualization techniques [157]. Thus, we propose a distributed traffic offloading procedure that includes four phases. The first phase is the initial user association and measurement of network utilities. When entering the network, a user simply attaches to any BS to retrieve network utility information. According to the initial user association, BSs measure their traffic loads and estimate the traffic loads in their backhaul. Based on the measurements, in the second phase, BSs advertise their network utility information. Define $\psi(\eta) = \sum_{j \in B} w_j(\rho_j)(\mu_j(\rho_j) + \tilde{\mu}_j(\tilde{\rho}_j))$. In the third phase, the users select their serving BSs according to the advertised network utility information and the downlink data rates, to minimize $\psi(\eta)$. In the fourth phase, the BSs and users iteratively update their network utilities (the second phase) and BS selections (the third phase), respectively, until the user association converges.

7.4.2.2 Network Utility Aware User Association

The network utility aware user association scheme consists of a user side algorithm and a BS side algorithm. The user side algorithm based on network utility advertisement selects the optimal serving BS for individual users, while the BS side algorithm updates individual BSs' network utility advertisements based on their user associations. In designing the network utility aware user association scheme, we make the following assumptions:

1) We assume that the timescale of the traffic arrival and departure process is faster than BSs in advertising their network utility information. That is to say, BSs advertise their network utility information after the system exhibits stationary performance.

2) We assume that the green power generation rate is consistent during the time period of establishing a stable user association [55].
3) We assume that all the BSs are synchronized and advertise their network utility simultaneously, and the system parameter κ does not change during one user association process.
4) We assume that a BS's cache hit ratio is constant for the duration of one user association.

The feasible set for the traffic load balancing problem in Eq. (7.83) is

$$
\mathcal{F} = \{\eta | 0 \leq \tilde{\rho}_j \leq 1 - \epsilon,
$$
$$
0 \leq \rho_j \leq 1 - \epsilon, \ \sum_{j \in B} \eta_j(x) = 1,
$$
$$
\eta_j(x) = \{0, 1\}, \ \forall j \in B, \ \forall x \in \mathcal{A}\}. \tag{7.85}
$$

Since $\eta_j(x) = \{0, 1\}$, $\psi(\eta)$ is not continuous differentiable. In order to derive the user side algorithm and the BS side algorithm for the NUA traffic load balancing scheme, we relax the feasible set by letting $0 \leq \eta_j(x) \leq 1$. Then, the relaxed feasible set is

$$
\tilde{\mathcal{F}} = \{\eta | 0 \leq \tilde{\rho}_j \leq 1 - \epsilon,
$$
$$
0 \leq \rho_j \leq 1 - \epsilon, \ \sum_{j \in B} \eta_j(x) = 1,
$$
$$
0 \leq \eta_j(x) \leq 1, \ \forall j \in B, \ \forall x \in \mathcal{A}\}. \tag{7.86}
$$

After presenting the NUA traffic load balancing scheme, we will prove that the proposed scheme achieves an optimal user association in the feasible set of the traffic load balancing problem.

We define the time interval between two consecutive network utility advertisements as a time slot. Let $\eta_j^k(x)$ denote whether a user at location x associates with BS j in the kth time slot or not. Define $\rho_j(k)$ and $\tilde{\rho}_j(k)$ as the traffic load in BS j and its backhaul in the kth time slot, respectively. Let

$$
\phi_j(k) = \frac{d\psi(\eta)}{d\eta_j(x)}
$$
$$
= \frac{\lambda(x)v(x)}{r_j(x)} e^{\kappa(\rho_j(k) - \hat{\rho}_j)} \left(\frac{\kappa \rho_j(k)}{1 - \rho_j(k)} + \frac{\kappa \tilde{\rho}_j(k)}{1 - \tilde{\rho}_j(k)} + \frac{1}{(1 - \rho_j(k))^2} \right)
$$
$$
+ \lambda(x)v(x)e^{\kappa(\rho_j(k) - \hat{\rho}_j)} \frac{1 - \alpha_j}{R_j(1 - \tilde{\rho}_j(x))^2}. \tag{7.87}
$$

Define

$$
\theta_j^a(k) = e^{\kappa(\rho_j(k) - \hat{\rho}_j)} \left(\frac{\kappa \rho_j(k)}{1 - \rho_j(k)} + \frac{\kappa \tilde{\rho}_j(k)}{1 - \tilde{\rho}_j(k)} + \frac{1}{(1 - \rho_j(k))^2} \right) \tag{7.88}
$$

and

$$\theta_j^b(k) = e^{\kappa(\rho_j(k) - \hat{\rho}_j)} \frac{1 - \alpha_j}{R_j(1 - \tilde{\rho}_j(x))^2}. \tag{7.89}$$

The User Side Algorithm

At the beginning of the kth time slot, BSs broadcast their network utility advertisements, for example, $\theta_j^a(k)$ and $\theta_j^b(k)$, to users. The BS selection rule for a user at location x is

$$b^k(x) = \arg\max_{j \in B} \frac{r_j(x)}{\theta_j^a(k) + r_j(x)\theta_j^b(k)}. \tag{7.90}$$

Here, $b^k(x)$ is the index of the BS selected by the user. Therefore,

$$\eta_j^k(x) = \begin{cases} 1, & j = b^k(x), \forall x \in \mathcal{A} \\ 0, & j \neq b^k(x), \forall x \in \mathcal{A}, \end{cases} \tag{7.91}$$

The BS Side Algorithm

After mobile users select their associating BSs, the user association in BS j, η_j^k, is updated. Given the user association, BS j updates its network utility advertisement. BS j calculates an intermediate user association $\bar{\eta}_j^k = \{\bar{\eta}_j^k | j \in B\}$ as

$$\bar{\eta}_j^k = (1 - \delta^k)\eta_j^k + \delta^k \bar{\eta}_j^{k-1}. \tag{7.92}$$

Here, $0 < \delta^k < 1$ is an exponential averaging parameter. With the intermediate user association, BS j calculates the intermediate traffic load in the BS and its backhaul. The intermediate traffic load in BS j's backhaul is

$$\tilde{\rho}_j(k+1) = \int_{x \in \mathcal{A}} \frac{\lambda(x)\nu(x)\bar{\eta}_j^k(x)}{R_j} dx, \tag{7.93}$$

and intermediate traffic load in BS j is

$$\rho_j(k+1) = \int_{x \in \mathcal{A}} \frac{\lambda(x)\nu(x)\bar{\eta}_j^k(x)}{r_j(x)} dx. \tag{7.94}$$

Based on the intermediate traffic load in both the BS and its backhaul, BS j calculates its network utility advertisements, $\theta_j^a(k+1)$ and $\theta_j^b(k+1)$, using Eqs. (7.88) and (7.89).

7.4.2.3 The Properties of the Network Utility Aware User Association Algorithm

In this subsection, we prove the convergence and the optimality of the proposed NUA traffic load balancing scheme. Since users select serving BSs based on BSs' network utility advertisements, the user association converges when BSs' network utility advertisements are stable. A BS's network utility advertisements are determined by its intermediate traffic loads in the BS and its backhaul. On calculating the intermediate traffic loads, the intermediate user association is the only variable. Therefore, when the intermediate user association converges, the intermediate traffic load is stabilized and so are the network utility advertisements. Therefore, we first prove any BS's network utility advertisements are stable by proving that its intermediate user association converges, and then show that the user association based on the stabilized network utility advertisements minimizes $\psi(\eta)$.

Lemma 7.4.1   *The relaxed feasible set $\tilde{F}$ is a convex set.*

Proof: The lemma is proved by showing that the set $\tilde{F}$ contains any convex combination of the user association vector η.   □

Lemma 7.4.2   *$\psi(\eta)$ is a convex function of η when η is defined in $\tilde{F}$.*

Proof: The lemma can be proved by showing $\nabla^2\psi(\eta) > 0$ when η is defined in $\tilde{F}$.   □

Let $\bar{\eta}^k = \{\bar{\eta}_j^k | j \in B\}$ and $\triangle\bar{\eta}^k = \bar{\eta}^k - \bar{\eta}^{k-1}$.

Lemma 7.4.3   *When $\triangle\bar{\eta}^k \neq 0$, $\triangle\bar{\eta}^k$ provides a descent direction of $\psi(\bar{\eta})$ at $\bar{\eta}^k$.*

Proof: Since $0 \leq \bar{\eta}_j^k \leq 1, \forall k, \forall j \in B$, $\bar{\eta}$ is defined in $\tilde{F}$. According Lemmas 7.4.1 and 7.4.2, $\psi(\bar{\eta})$ is a convex function of $\bar{\eta}$. Hence, the lemma can be proved by showing $\langle \nabla\psi(\bar{\eta})|_{\bar{\eta}=\bar{\eta}^k}, \triangle\bar{\eta}^k\rangle < 0$.

$$\langle \nabla\psi(\bar{\eta})_{\bar{\eta}=\bar{\eta}^k}, \triangle\bar{\eta}^k\rangle \tag{7.95}$$

$$= \int_{x\in A} \sum_{j\in B} \lambda(x)v(x)(\bar{\eta}_j^k(x) - \bar{\eta}_j^{k-1}(x)) \left(\frac{\theta_j^a(k)}{r_j(x)} + \theta_j^b(k) \right)$$

$$= (1 - \delta^k) \int_{x\in A} \lambda(x)v(x)$$

$$\sum_{j\in B} (\eta_j^k(x) - \bar{\eta}_j^{k-1}(x)) \frac{\theta_j^a(k) + r_j(x)\theta_j^b(k)}{r_j(x)} dx$$

Since

$$\eta_j^k(x) = \begin{cases} 1, & j = b^k(x) \\ 0, & j \neq b^k(x), \end{cases} \tag{7.96}$$

$$\sum_{j\in B} (\eta_j^k(x) - \bar{\eta}_j^{k-1}(x)) \frac{\theta_j^a(k) + r_j(x)\theta_j^b(k)}{r_j(x)} \leq 0. \tag{7.97}$$

Because $0 < \delta^k < 1$ and $\triangle\bar{\eta}_j^k \neq 0$,

$$\sum_{j\in B} (\eta_j^k(x) - \bar{\eta}_j^{k-1}(x)) \frac{\theta_j^a(k) + r_j(x)\theta_j^b(k)}{r_j(x)} < 0. \tag{7.98}$$

Thus, $\langle \nabla\psi(\bar{\eta})|_{\bar{\eta}=\bar{\eta}^k}, \triangle\bar{\eta}^k\rangle < 0$.   □

$\bar{\eta}^*$ denotes the optimal intermediate user association.

Lemma 7.4.4   When $\bar{\eta}^k \neq \bar{\eta}^*$, $\bar{\eta}^k \in \tilde{F}$, there exists $0 < \delta^k < 1$ such that $\psi(\bar{\eta}^k) < \psi(\bar{\eta}^{k-1})$.

Proof: Since

$$\triangle \bar{\eta}^k = \bar{\eta}^k - \bar{\eta}^{k-1} \tag{7.99}$$
$$= (1 - \delta^k)(\eta^k - \bar{\eta}^{k-1}),$$

$(\eta^k - \bar{\eta}^{k-1})$, according to Lemma 7.4.3, provides the descent direction for searching the optimal value in the iterations while $(1 - \delta^k)$ indicates the search step in the kth iteration. Since $\bar{\eta}^k \neq \bar{\eta}^*$, there exists $0 < \delta^k < 1$ that enables $\psi(\bar{\eta}^k) < \psi(\bar{\eta}^{k-1})$ □

Theorem 7.4.1   *If the traffic load balancing problem is feasible[3] and δ^k is properly selected, $\bar{\eta}^k = (1 - \delta^k)\eta^k + \delta^k \bar{\eta}^{k-1}$ converges to $\bar{\eta}^*$.*

Proof: Since 1) $\bar{\eta}_j^k - \bar{\eta}_j^{k-1}$ is a descent direction of $\psi(\bar{\eta})$ at $\bar{\eta}^k$ and 2) δ^k is properly selected such that $\psi(\bar{\eta}^k) < \psi(\bar{\eta}^{k-1})$, the mapping from η^k and $\bar{\eta}^{k-1}$ to $\bar{\eta}^k$ ($\bar{\eta}^k = (1 - \delta^k)\eta^k + \delta^k \bar{\eta}^{k-1}$) keeps decreasing $\psi(\bar{\eta})$. Since $\psi(\bar{\eta}) \geq 0$, $\bar{\eta}^k$ will eventually converge. According to Lemma 7.4.4, $\bar{\eta}^k$ converges to $\bar{\eta}^*$. Otherwise, $\psi(\bar{\eta})$ can be further reduced.   □

Corollary 7.4.1   *Any BS's network utility advertisements, $\theta_j^a(k)$ and $\theta_j^a(k)$, $j \in B$, are stabilized.*

Proof: $\theta_j^a(k)$ and $\theta_j^a(k)$, $j \in B$, are calculated by the traffic load in BS j and its backhaul, respectively. When the intermediate user association converges, the traffic loads are determined. As a result, individual BS's network utility advertisements are stabilized.   □

Theorem 7.4.2   *Given that the traffic load balancing problem is feasible, the user association, $\eta_j^* = \{\eta_j^*(x)|\eta_j^*(x) = \{0, 1\}, \ x \in A\}, j \in B$, based on the stabilized network utility advertisements is an optimal solution to the traffic load balancing problem.*

Proof: Define $\theta_j^a(*)$ and $\theta_j^b(*)$ as BS j's stabilized network utility advertisements. Let $\eta^* = \{\eta_j^*|j \in B\}$ and $\eta = \{\eta_j|j \in B\}$. Here, $\eta_j = \{\eta_j(x)|\eta_j(x) = \{0, 1\}, \ x \in A\}$. Suppose that η is an arbitrary user association in the feasible set F that is not equal to η^*.

$$\langle \nabla \psi(\eta)|_{\eta = \eta^*}, \eta - \eta^* \rangle \tag{7.100}$$

$$= \int_{x \in A} \sum_{j \in B} \lambda(x) v(x) (\eta_j(x) - \eta_j^*(x)) \left(\frac{\theta_j^a(*)}{r_j(x)} + \theta_j^b(*) \right)$$

$$= \int_{x \in A} \lambda(x) v(x) \sum_{j \in B} (\eta_j(x) - \eta_j^*(x)) \left(\frac{\theta_j^a(*)}{r_j(x)} + \theta_j^b(*) \right).$$

---

[3] The problem is feasible when the feasible set of the problem is not empty.

Since

$$b^*(x) = \arg\max_{j \in B} \frac{r_j(x)}{\theta_j^a(*) + r_j(x)\theta_j^b(*)} \tag{7.101}$$

and

$$\eta_j^*(x) = \begin{cases} 1, & j = b^*(x) \\ 0, & j \neq b^*(x), \end{cases} \tag{7.102}$$

$$\sum_{j \in B} \eta_j(x) \left(\frac{\theta_j^a(*)}{r_j(x)} + \theta_j^b(*) \right)$$

$$\geq \sum_{j \in B} \eta_j^*(x) \left(\frac{\theta_j^a(*)}{r_j(x)} + \theta_j^b(*) \right). \tag{7.103}$$

Hence, $\langle \nabla \psi(\eta)|_{\eta = \eta^*}, \eta - \eta^* \rangle \geq 0$. Therefore, η^* is an optimal solution to the UA problem.

$\square$

7.4.2.4 The Adaptation of the Energy–Latency Trade-off

The system parameter, κ, controls the trade-off between green power utilization and the traffic delivery latency. When $\kappa = 0$, $w_j(\rho_j) = 1$. In this case, the green power utilization is not modeled in the objective function. Thus, the NUA traffic load balancing scheme determines the user association based only on the traffic delivery latency. As κ increases, the awareness of green power utilization in determining the user association is enhanced. In other words, with a larger κ, green power utilization plays a more important role in determining the user association. If κ is large enough, the NUA traffic load balancing scheme achieves a user association that approximates the user association that only cares about green power utilization.

7.4.3 Performance Evaluation

7.4.3.1 Simulation Setup

We set up system level simulations to investigate the performance of the NUA traffic load balancing scheme for downlink traffic load balancing in backhaul constrained SCNs. In the simulation, three MBSs and seven SCBSs are randomly deployed in a $2000\,m \times 2000\,m$ area. The total bandwidth is $10\,MHz$ and the frequency reuse factor is one. The channel propagation model is based on COST 231 Walfisch-Ikegami [144]. The channel model and parameters are summarized in Table 7.1. Here, PL_{MBS} and PL_{SCBS} are the path loss between the users and MBSs and that between the users and SCBSs, respectively. d is the distance between users and BSs. The transmit power of an MBS and an SCBS are $43\,dBm$ and $33\,dBm$, respectively.

The static power consumptions of an MBS and an SCBS are $750\,W$ and $37\,W$, respectively [3]. The load-power coefficients of the MBS and the SCBS are 500 and 4, respectively [3]. The solar cell power efficiency is 17.4% [176]. We assume that the weather conditions are the standard conditions which specify a temperature of $25\,°C$, an irradiance of $1000\,W\,m^{-2}$, and an air mass of 1.5 spectrum [182]. Thus, the green power generation rate is $174\,W\,m^{-2}$. The solar panel sizes are randomly selected but

Table 7.3 Average cache hit ratio.

MBS 1	MBS 2	MBS 3	SCBS 4	SCBS 5
0.27	0.12	0.28	0.12	0.17
SCBS 6	SCBS 7	SCBS 8	SCBS 9	SCBS 10
0.22	0.22	0.24	0.24	0.19

ensure the green power generation capacity of MBSs from 750 W to 1300 W while that of SCBSs from 37 W to 4 W. BSs' energy-latency coefficients are set to be the same. In the simulation, the average data rate of SCBSs' backhaul is 5 $Mbps$. The hit ratio[4] of BSs' cache systems is shown in Table 7.3.

7.4.3.2   Traffic Load Balancing Algorithms and Network Utility Awareness

In the simulations, we investigate the performance of traffic load balancing schemes with different levels of network utility awareness. The network utilities considered are green power, the traffic delivery latency in BSs, the traffic delivery latency in backhaul, and the cache hit ratio. We implement three traffic load balancing schemes in the simulations. The first scheme is vGALA [157]. We realize the three traffic load balancing schemes by adapting the parameters of vGALA (κ and θ) to be: (1) BS latency aware, (2) green power aware, and (3) BS latency and green power aware, respectively. The second scheme is the NUA traffic load balancing scheme that simulates four traffic load balancing schemes: (1) BS latency and backhaul latency aware, (2) BS latency, backhaul latency, and cache hit ratio aware, (3) BS latency, backhaul latency, and green power aware, and (4) aware of all network utilities, respectively. For the NUA scheme, when $\alpha_j = 0$, $\forall j \in \mathcal{B}$, a BS estimates the traffic delivery latency in backhaul based purely on the traffic arrival rates in the BS. In fact, if the cache system is considered, the traffic load in backhaul should be less than that in the BS. Therefore, the NUA scheme with $\alpha_j = 0$, $\forall j \in \mathcal{B}$, simulates the cache unaware traffic load balancing scheme. The third scheme is the DRB scheme [163]. In the implementation, we assume that BSs in the same tier have the data rate bias. MBSs are in the first tier while SCBSs are in the second tier. In the data rate bias scheme, a user selects the serving BS to maximize the biased data rate. The data rate bias of an MBS is set to one. We vary the data rate bias of an SCBS to investigate the performance of the scheme. We set $\psi(\eta)$ as the performance metric for selecting the optimal data rate bias. Thus, the implemented DRB scheme is aware of all network utilities and is referred to as DRB-NU (the data rate bias with network utility awareness). The network utility awareness of the schemes with different settings are shown in Table 7.4.

7.4.3.3   Simulation Results

Figs. 7.21, 7.22, and 7.23 show the convergence of the NUA scheme and its energy-latency trade-off with different κ. Figure 7.21 shows that the value of $\psi(\eta)$ converges with less than 100 iterations, and so do the traffic delivery latency (Figure 7.22) and brown

---

[4] The cache hit ratio is randomly selected from 0.1 to 0.3. For analytical simplicity, we fix the hit ratio of BSs in the simulations.

Table 7.4 Network utility aware user association schemes.

UA Scheme	Green power	BS Latency	Backhaul Latency	Cache
vGALA ($\kappa = 0$)		x		
vGALA ($\kappa = 6, \theta = 1$)	x			
vGALA ($\kappa = 4, \theta = 0.5$)	x	x		
NUA ($\kappa = 0, \alpha_j = 0$)		x	x	
NUA ($\kappa = 0, \alpha_j > 0$)		x	x	x
NUA ($\kappa = 2, \alpha_j = 0$)	x	x	x	
NUA ($\kappa = 2, \alpha_j > 0$)	x	x	x	x
DRB-NU	x	x	x	x

power consumption (Figure 7.23). Figs. 7.22 and 7.23 show the energy-latency trade-off. As κ increases, the network emphasizes green power utilization in determining the user association. As a result, with a large κ, for example $\kappa = 7$, the network consumes less brown power at the cost of introducing additional traffic delivery latency. vGALA with a large κ and θ, for example $\kappa = 6$ and $\theta = 1$, realizes the user association that is only aware of green power utilization [157]. Figure 7.23 shows that as κ increases, the performance of the NUA scheme in terms of brown power consumption approaches that of the traffic load balancing scheme that optimizes green power utilization (i.e., is only aware of green power utilization).

Figs. 7.24, 7.25, and 7.26 show the performance of the NUA scheme versus different solar panel efficiencies. Figure 7.26 shows that brown power consumption is reduced as solar panel efficiency increases. This is because a higher solar panel efficiency enables solar panels to generate a larger amount of electricity and thus lessen brown power consumption. As shown in Fig 7.25, the performance of the traffic delivery latency can be divided into four regions. In the first region (R1), the traffic delivery

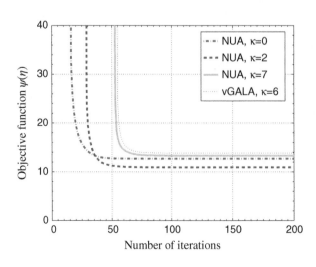

Figure 7.21 The value of $\psi(\eta)$ with different κ. *Source:* Han 2017 [158]. Reproduced with permission of IEEE.

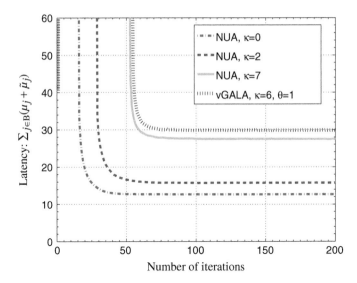

Figure 7.22 The traffic delivery latency with different κ. *Source:* Han 2017 [158]. Reproduced with permission of IEEE.

latency does not change. This is because the green power generated in individual BSs is less than their static power consumption when the solar panel efficiency is in R1. In other words, the green capacity of all BSs is zero when the solar panel efficiency is within R1. As a result, increasing the solar panel efficiency in R1 does not impact the traffic delivery latency, and neither does the value of $\psi(\eta)$. In the second region, as shown in Figure 7.25, the traffic delivery latency increases as the solar panel efficiency increases. When

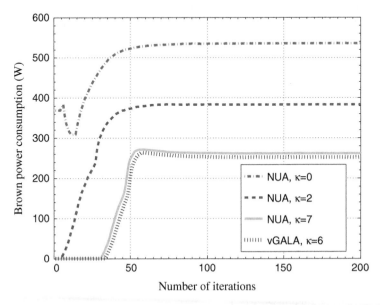

Figure 7.23 The brown power consumption with different κ. *Source:* Han 2017 [158]. Reproduced with permission of IEEE.

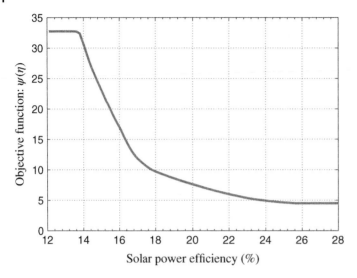

Figure 7.24 The value of $\psi(\eta)$ versus the solar power efficiency. *Source:* Han 2017 [158]. Reproduced with permission of IEEE.

the solar panel efficiency is within this region, the network trades the traffic delivery latency for reducing brown power consumption. In the third region (R3), the network trades power consumption for reducing the traffic delivery latency. This can be seen from Figure 7.26. The rate of brown power consumption reduction decreases when the solar panel efficiency is about 17% (the start point of R3) as shown in Figure 7.25. This indicates that the network emphasizes reduction of the traffic delivery latency when the solar panel efficiency is within R3. In both R2 and R3, the energy-latency trade-off reduces the value of $\psi(\eta)$ as shown in Figure 7.24. When the solar panel efficiency is

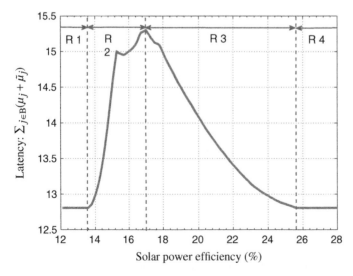

Figure 7.25 Traffic delivery latency versus solar power efficiency. *Source:* Han 2017 [158]. Reproduced with permission of IEEE.

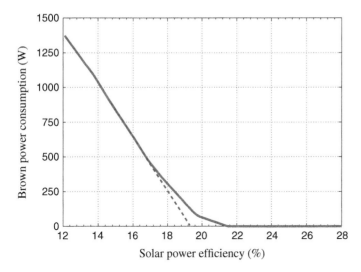

Figure 7.26 Brown power consumption versus solar power efficiency. *Source:* Han 2017 [158]. Reproduced with permission of IEEE.

within the fourth region (R4), the solar panel efficiency is high enough to enable zero brown power consumption in all BSs while minimizing the traffic delivery latency.

Figure 7.27 shows the value of $\psi(\eta)$ versus the small cell data rate biases under the DRB-NU scheme. The value of $\psi(\eta)$ is minimized when the small cell data rate bias is equal to 2.74. Given the data rate bias, the network's traffic delivery latency is 17.24 and the brown power consumption is 480.9 W. Under the NUA scheme with $\kappa = 2$, the network's traffic delivery latency and brown power consumption are 15.74 and

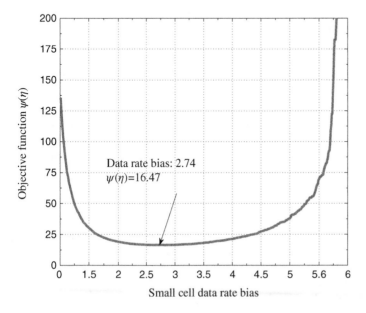

Figure 7.27 The value of $\psi(\eta)$ versus data rate bias. *Source:* Han 2017 [158]. Reproduced with permission of IEEE.

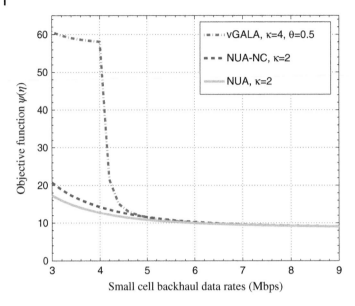

Figure 7.28 The value of $\psi(\eta)$ versus backhaul data rates. *Source:* Han 2017 [158]. Reproduced with permission of IEEE.

383.35 W, respectively. Therefore, as compared with the DRB-NU scheme, the NUA scheme reduces the traffic delivery latency and brown power consumption by 8.7% and 20.28%, respectively. The NUA scheme has achieved enhanced performance because it allows individual BSs to adapt their network utility advertisements, while the DRB-NU scheme only allows changing the data rate bias for an entire tier rather than for individual BSs. Another drawback of the DRB-NU scheme is that it does not dynamically respond to network utility changes. The small cell data rate bias is optimized based on previous rather than current network conditions, for example, traffic intensities, backhaul constraints, and green power availabilities.

We now compare the performance of three traffic load balancing schemes with varying network conditions. The first scheme is a green power and BS latency aware traffic load balancing scheme realized by vGALA with $\kappa = 4$ and $\theta = 0.5$ [157]. The second is a NUA scheme that is aware of all network utilities. The third one, referred to as NUA-NC (no cache), is NUA without awareness of the cache hit ratio. This scheme is realized by the NUA scheme with $\alpha_j = 0$, $\forall j \in B$. Figs. 7.28, 7.29, and 7.30 show the performance of these traffic load balancing schemes versus different backhaul data rates. When the backhaul data rate is very low, for example, less than 5 *Mbps* in the simulation, the value of $\psi(\eta)$ and the traffic delivery latency under vGALA are very large. This indicates that, without an awareness of backhaul data rates, the traffic load balancing under vGALA congests some BSs in the network. The brown power consumption under vGALA does not change versus the backhaul data rates because vGALA does not consider the traffic delivery latency in backhaul as a performance metric in determining user associations. As the backhaul data rates increase, the value of $\psi(\eta)$ under these schemes converges because the backhaul constraint is gradually mitigated. However, the NUA scheme achieves smaller traffic delivery latency compared to the vGALA scheme because of its awareness of traffic delivery latency in backhaul.

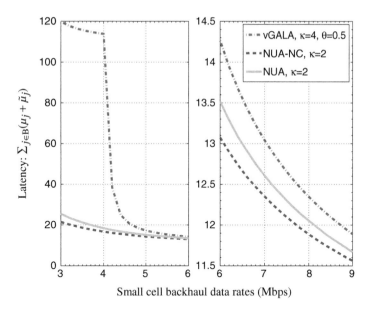

Figure 7.29 The traffic delivery latency versus backhaul data rates. *Source:* Han 2017 [158]. Reproduced with permission of IEEE.

In the simulation, the value of $\psi(\eta)$ is minimized by the NUA scheme. However, the NUA-NC scheme achieves the lowest traffic delivery latency, as shown in Figure 7.29. This is because the NUA scheme aims to minimize the value of $\psi(\eta)$ and thus strikes a trade-off between traffic delivery latency and brown power consumption. As a result, as compared with the NUA-NC scheme, the NUA scheme consumes less brown power at the cost of an increase in traffic delivery latency.

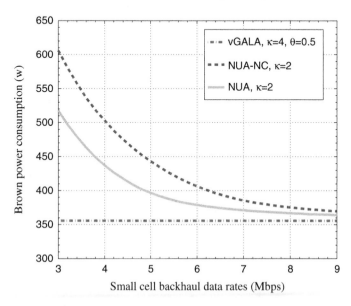

Figure 7.30 Brown power consumption versus backhaul data rates. *Source:* Han 2017 [158]. Reproduced with permission of IEEE.

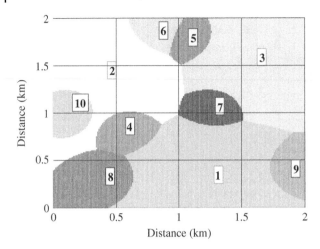

Figure 7.31 The coverage area of vGALA. *Source:* Han 2017 [158]. Reproduced with permission of IEEE.

The BSs' coverage areas under these schemes are shown in Figs. 7.31, 7.32, and 7.33. The NUA-NC scheme is aware of the backhaul limitation and thus reduces the coverage area of the backhaul constrained SCBS, for example, SCBS 8. However, owing to its being unaware of the cache hit ratio, the NUA-NC scheme overestimates the traffic load in backhaul and constrains the coverage area of SCBSs, for example SCBS 8. The NUA scheme, being aware of the cache hit ratio of individual BSs, accurately estimates the traffic load in backhaul and thus achieves optimal coverage areas for BSs, for example increasing the coverage area of SCBS 8 to minimize the value of $\psi(\eta)$.

As shown in Figs. 7.34, 7.35, and 7.36, when the backhaul data rate of a SCBS changes, for example, R_5 is reduced from 5 *Mbps* to 1 *Mbps*, the NUA and NUA-NC schemes are able to adapt their traffic load balancing according to the backhaul data rate changes. However, the vGALA scheme, without awareness of backhaul data rate, incurs excessive traffic delivery latency, which is 667% of the traffic delivery latency of the NUA scheme as shown in Figure 7.35. As shown in Figs. 7.37, 7.38, and 7.39, both the NUA and NUA-NC schemes are able to reduce the coverage area of SCBS 5. The NUA-NC scheme,

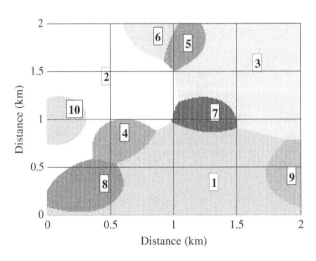

Figure 7.32 The coverage area of NUA-NC. *Source:* Han 2017 [158]. Reproduced with permission of IEEE.

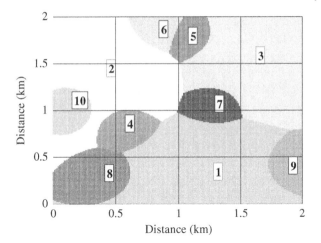

Figure 7.33 The coverage area of NUA. *Source:* Han 2017 [158]. Reproduced with permission of IEEE.

because it is not aware of the cache hit ratio, shrinks the coverage area more than the NUA scheme does.

Figure 7.40 shows the impact of cache awareness on the traffic delivery latency. In the simulation, we set $\kappa = 0$ for both the NUA scheme and the NUA-NC scheme to focus on the performance of the traffic delivery latency. Thus, both schemes are unaware of the green power utilization. As shown in Figure 7.40, when the backhaul data rate is small, cache awareness helps reduce the traffic delivery latency.

7.5 Traffic Load Balancing in Smart Grid Enabled Mobile Networks

As smart grid advances, power trading among distributed power generators and energy consumers will be enabled. In this section, we have investigated the optimization of

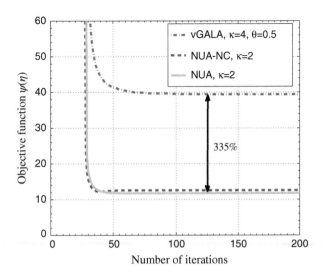

Figure 7.34 The value of $\psi(\eta)$ with R_5 reduced from 5 *Mbps* to 1 *Mbps*. *Source:* Han 2017 [158]. Reproduced with permission of IEEE.

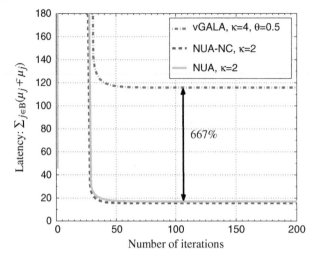

Figure 7.35 The traffic delivery latency with R_5 reduced from 5 *Mbps* to 1 *Mbps*. *Source:* Han 2017 [158]. Reproduced with permission of IEEE.

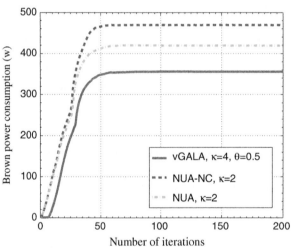

Figure 7.36 Brown power consumption with R_5 reduced from 5 *Mbps* to 1 *Mbps*. *Source:* Han 2017 [158]. Reproduced with permission of IEEE.

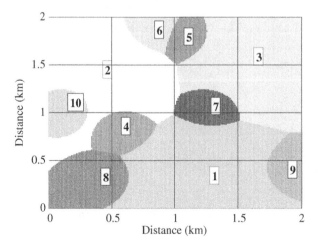

Figure 7.37 vGALA. *Source:* Han 2017 [158]. Reproduced with permission of IEEE.

Figure 7.38 NUA-NC. *Source:* Han 2017 [158]. Reproduced with permission of IEEE.

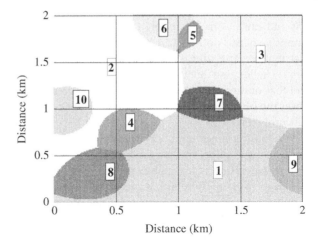

Figure 7.39 NUA. *Source:* Han 2017 [158]. Reproduced with permission of IEEE.

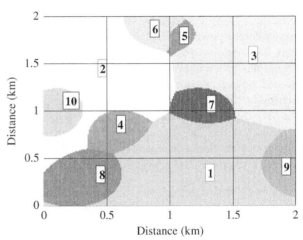

Figure 7.40 The impact of cache awareness on traffic delivery latency. *Source:* Han 2017 [158]. Reproduced with permission of IEEE.

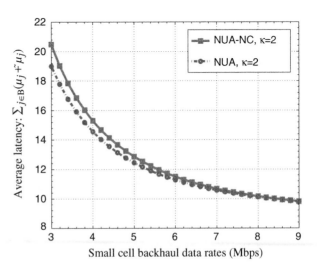

smart grid enabled mobile networks in which green energy is generated in individual BSs and can be shared among the BSs. In order to minimize the on-grid power consumption of this network, we have proposed to jointly optimize BS operation and power distribution. The joint BS operation and Power distribution Optimization (BPO) problem is challenging due to the complex coupling of the optimization of mobile networks and that of the power grid. We have proposed an approximation solution that decomposes the BPO problem into two sub-problems and solves the BPO by addressing these sub-problems.

By adopting green energy powered BSs, mobile service providers may save on-grid power consumption and thus reduce their CO_2 emissions [109]. However, since green energy generation is not stable, green energy may not be a reliable energy source for mobile networks. Therefore, future mobile networks are likely to adopt hybrid energy supplies: on-grid power and green energy. Green energy is utilized to reduce the on-grid power consumption and thus reduce CO_2 emissions while on-grid power is utilized as a backup power source.

In smart grid, electricity can be traded among distributed power generators and consumers [184]. Powered by smart grid, mobile networks are able to buy electricity from different power generators according to energy market information and optimize their BS operation to minimize their operating expenses [17, 185]. Via smart grid, the surplus green energy generated by a BS can be shared with other BSs to reduce on-grid power consumptions. The power supplies of future mobile networks will incorporate both green energy and smart grid as shown in Figure 7.41. In the network, BSs have three power supplies: the power generated from the green energy generators in individual BSs, the power shared from other BSs, and the power generated from remote power plants. The power generated from a remote power plant is the least efficient because of the energy loss during power transmission and distribution. We

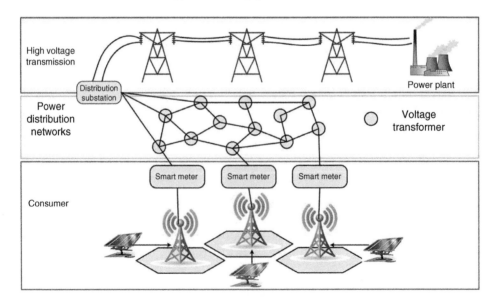

Figure 7.41 The smart grid enabled mobile network. *Source:* Han 2014 [183]. Reproduced with permission of IEEE.

refer to the power generated from remote power plants as on-grid power and we aim to minimize the on-grid power consumption of mobile networks.

We propose to jointly optimize the BS operation and the power distribution to minimize mobile networks' on-grid power consumption. Although the optimization of the BS operation [19], the green energy utilization [109], and the power distribution [186] have been individually well studied, the joint BPO problem is not well investigated. Solving the BPO problem is challenging due to the complex coupling of the BS operation, the green energy utilization, and the power distribution. We propose an approximation solution which decomposes the BPO problem into two sub-problems: the weighted user–BS association (WUA) problem and the BS energy sharing (BES) problem. By addressing the WUA problem, we realize energy efficiency aware BS operation. By solving the BES problem, we optimize the power sharing among BSs and thus minimize the BSs' on-grid power consumption.

7.5.1   System Model and Problem Formulation

In this section, we consider a mobile network with multiple BSs. The BSs are equipped with a green energy system that generates electricity from green energy sources. The BSs are also connected with smart grid via smart meters. Thus, in the network, BSs are powered by both on-grid power and green energy. We consider solar energy as the green energy source. We focus on the downlink data transmission scenario.

7.5.1.1   Traffic Model

Assume N BSs are deployed to provide data service in an area. Define $\mathcal{B}$ and $\mathcal{A}$ as the set of BSs and the coverage area of the BSs, respectively. We assume that the traffic arrives according to a Poisson process with the arrival rate per unit area at location x equal to $\lambda(x)$, and that the traffic loads have a general distribution with an average traffic load of $v(x)$. Assuming a mobile user at location x is associated with the jth BS, then the user's data rate $r_j(x)$ can be generally expressed as Eq. (7.71),

Then, the average traffic load density at location x in the jth BS is

$$\varrho_j(x) = \frac{\lambda(x)v(x)\eta_j(x)}{r_j(x)}. \tag{7.104}$$

Here, $\eta_j(x)$ is an indicator function. If $\eta_j(x) = 1$, then the user at location x is associated with the jth BS; otherwise, the user is not associated with the jth BS. Assuming mobile users are uniformly distributed in the area, the traffic load in the jth BS can be expressed as

$$\rho_j = \int_{x \in \mathcal{A}} \varrho_j(x)dx. \tag{7.105}$$

This value of ρ_j indicates the fraction of time during which the jth BS is busy.

7.5.1.2   Energy Model

A BS's power consumption consists of two parts: static power consumption and dynamic power consumption [3]. The static power consumption is the power consumption of the BS without any traffic load. The dynamic power consumption refers to the additional power consumption caused by the traffic load in the BS, which can be well approximated by a linear function of the traffic load [3]. Define p_j^s as the static power consumption of the jth BS. Then, the jth BS's power consumption can be expressed as Eq. (7.11).

In the network, since all BSs have their own solar panels for generating green energy, energy consumed in a BS can be drawn from the green energy system. We assume that green energy can be shared among BSs via power distribution networks. When green energy in a BS is not sufficient to satisfy its energy demand, the BS can draw power from other BSs which have surplus green energy. If the BS's energy demand is still not satisfied, the BS consumes on-grid power which is pulled from a remote power plant through the high voltage transmission line, the distribution substations and the power distribution network. Owing to the disadvantages of "banking" green energy [48], we do not assume that green energy can be stored. In other words, if green energy is not consumed when it is generated, it is wasted.

Define $0 < \theta_{0,j} < 1$ as the power transmission efficiency reflecting the power loss in transferring power from the remote power plant to the jth BS. Define $0 < \theta_{i,j} < 1$, $i \neq j$ as the power transmission efficiency of transferring power from the ith BS to the jth BS. We assume $\theta_{j,j} = 1$. Define $\delta_{j,i} \geq 0$ as the green energy drawn from the ith BS to the jth BS. If the ith BS does not share energy with the jth BS, $\delta_{j,i} = 0$. We define $\delta_{j,j} = 0$. Define e_j as the green energy generation rate in the jth BS.

The on-grid power consumption in the jth BS is

$$p_j^o = \frac{1}{\theta_{0,j}} \max((\max(p_j - e_j, 0) - \sum_{i \in B \setminus \{j\}} \delta_{j,i}\theta_{i,j}), 0). \tag{7.106}$$

7.5.1.3 Problem Formulation

In jointly optimizing BS operation and power distribution, the network aims to minimize on-grid power consumption while satisfying mobile users' QoS requirements. In this section, we use the average traffic delivery latency to represent mobile users' QoS.

Define A_j as the jth BS's coverage area. We assume that the users associated with the jth BS are uniformly distributed in its coverage area, and the traffic arrival processes are independent. Since the traffic arrival at a location is a Poisson process, the traffic arrival in the jth BS, which is the sum of the traffic arrivals from its coverage area, is also a Poisson process. Since the service rate follows a general distribution, the BS realizes an $M/G/1$ queuing system. Assuming mobile users are served in a round robin fashion, the traffic delivery in the BS can be modeled as an $M/G/1 - PS$ queue [152].

A mobile user who is located at x and associated with the jth BS is assumed to have the traffic load $v(x)$. To fulfill the user's traffic demand, the required service time is

$$\gamma(x) = \frac{v(x)}{r_j(x)}. \tag{7.107}$$

According to [152], the average traffic delivery time for the user in the jth BS is

$$T_j(x) = \frac{v(x)}{r_j(x)(1 - \rho_j)}. \tag{7.108}$$

In the jth BS, the average waiting time for traffic load $v(x)$ is

$$W_j(x) = \frac{\rho_j v(x)}{r_j(x)(1 - \rho_j)}.$$ (7.109)

Denote the latency ratio, which measures how much time a user at location x must be sacrificed in waiting for per unit service time, as $\mu_j(x)$.

$$\mu_j(x) = \frac{W_j(x)}{T_j(x)} = \frac{\rho_j}{1 - \rho_j}.$$ (7.110)

According to Eq. (7.110), $\mu_j(x)$ only depends on the traffic load in the jth BS. Therefore, all the users associated with BS j have the same latency ratio. Thus, we define

$$\mu_j = \frac{\rho_j}{1 - \rho_j}$$ (7.111)

as the latency ratio of the jth BS. A smaller μ_j indicates that the jth BS introduces less latency to its associated users. Denote the maximum allowable latency ratio for satisfying users' QoS requirement in the network as τ.

The network aims to satisfy users' QoS requirement with the minimum on-grid power consumption. Therefore, the joint BPO problem can be formulated as

$$\min_{(\rho,\delta)} \sum_{j \in B} p_j^o$$ (7.112)

$$subject\ to: 0 \le \mu_j \le \tau.$$ (7.113)

Here, $\rho = (\rho_1, \rho_2, \ldots, \rho_N)$, and

$$\delta = \begin{pmatrix} 0, & \delta_{1,2}, & \cdots, & \delta_{1,N} \\ \delta_{2,1}, & 0, & \cdots, & \delta_{2,N} \\ \vdots, & \vdots, & \ddots, & \vdots \\ \delta_{N,1}, & \delta_{N,2}, & \cdots, & 0 \end{pmatrix}.$$ (7.114)

To solve the BPO problem, both the traffic load in the BSs and power sharing among the BSs should be optimized. A BS's traffic load is determined by the user–BS association scheme. The power transmission efficiency of the power grid is one of the factors in determining the optimal user–BS association. Optimal power sharing depends on the BSs' traffic load, the green energy generation rate, and the power transmission efficiency. Owing to the complex coupling of these variables and parameters, solving the BPO problem is challenging.

7.5.2 An Approximation Solution

Solving the BPO problem involves optimizing the user–BS association and the power distribution in the power grid. These two items are highly coupled, which complicates the problem. In order to solve the BPO problem, we make two assumptions about the power transmission efficiency of the power grid. The first is that the power flow within the power grid is optimized in terms of the power transmission efficiency.

The power transmission efficiency of transferring power from the ith BS to the jth BS is determined by the efficiency of the power transformer and the power transmission line in the power distribution networks. Define $\theta_{i,j}^T$ and $\theta_{i,j}^L$ as the power transformer efficiency and the power transmission line efficiency. Then, $\theta_{i,j} = \theta_{i,j}^T \theta_{i,j}^L$. The second assumption is that all BSs have the same power transformer efficiency and $\theta_{0,j} \ll \theta_{i,j}^L$, $\forall i \in B \setminus \{j\}$, which indicates that the power transmission efficiency of transferring power from the remote power plant to a BS is much lower than the power transmission line efficiency between BSs. This assumption is reasonable because the power transferred from the remote power plant experiences power line loss during high voltage transmission, power loss in the distribution substations, and power loss in the power distribution networks (including both power transformer loss and power transmission line loss).

7.5.2.1 Problem Decomposition

Based on these assumptions, we are able to decompose the BPO problem. Since all BSs have the same power transformer efficiency, we define $\theta^T = \theta_{i,j}^T$. If the jth BS consumes on-grid power,

$$p_j^o = \frac{(p_j - e_j)}{\theta_{0,j}} - \theta^T \sum_{i \in B \setminus \{j\}} \frac{\theta_{i,j}^L}{\theta_{0,j}} \delta_{j,i}. \tag{7.115}$$

The jth BS's on-grid power consumption is determined by two components: $\frac{(p_j - e_j)}{\theta_{0,j}}$ and

$$\theta^T \sum_{i \in B \setminus \{j\}} \frac{\theta_{i,j}^L}{\theta_{0,j}} \delta_{j,i}. \tag{7.116}$$

Given a BS's green energy generation rate and the power transmission efficiency of the power grid, the first component is determined by the BS's power consumption while the second component is determined by power sharing among the BSs. Since $\theta_{0,j} \ll \theta_{i,j}^L$, $\forall i \in B$, we assume

$$\frac{\theta_{i,j}^L}{\theta_{0,j}} \simeq \frac{\theta_{k,j}^L}{\theta_{0,j}}, \ \forall i, k \in B, \ i \neq k. \tag{7.117}$$

In this case, the power loss of transferring power between any BSs is considered the same. Thus, optimizing $\frac{(p_j - e_j)}{\theta_{0,j}}$ does not have to consider the power transmission loss caused by power sharing among BSs. The jth BS's power consumption depends on its traffic load which is determined by the user–BS association scheme. Hence, we decompose the BPO problem into two sub-problems: the WUA problem and the BES problem.

The WUA problem can be expressed as

$$\min_{\rho} \sum_{j \in B} \frac{\max(p_j - e_j, 0)}{\theta_{0,j}} \tag{7.118}$$

$$subject\ to: \ 0 \leq \mu_j \leq \tau;$$

$$r(x) \geq \omega r^{\max}(x). \tag{7.119}$$

Here, $r(x)$ and $r^{\max}(x)$ denote the data rate of a user located at x and the maximum data rate achieved by a user located at x while the network's QoS requirement is satisfied, respectively. $0 < \omega < 1$ is a linear coefficient that constrains the minimum data rate of a user at location x. When minimizing

$$\sum_{j \in B} \frac{\max (p_j - e_j, 0)}{\theta_{0,j}}, \tag{7.120}$$

more traffic load will be offloaded to the BSs with a higher green energy generation rate. During traffic offloading, the mobile network's energy efficiency in terms of bits per Joule may be reduced. As a result, the network consumes more power in serving the same traffic demands. The additional power consumption can be recognized as the power loss caused by traffic offloading. We define the power efficiency of traffic offloading as the ratio of the network's minimum power consumption when the QoS requirements are satisfied to the network's power consumption after traffic offloading. If the power efficiency of traffic offloading is less than the power efficiency of sharing energy among BSs, a BS's surplus green energy should be shared with other BSs rather than being utilized to absorb more traffic to the BS. In other words, traffic offloading is preferred only when the power efficiency of traffic offloading is higher than that of sharing energy among BSs. In the WUA problem, we restrict the problem to this case by constraining a user's minimum data rate at location x.

Define ρ^* as the solution to the WUA problem and p_j^*, $j \in B$ as the corresponding power consumption of the jth BS. Define B^o and B^g as the set of BSs which consume on-grid power and the set of BSs which have surplus green energy, respectively. $B^o = \{j | p_j^* > e_j, j \in B\}$ and $B^g = \{j | p_j^* < e_j, j \in B\}$. Then, the BES problem can be expressed as

$$\min_{\delta} \sum_{j \in B^o} \frac{\max(p_j^* - e_j - \sum_{i \in B^g} \theta_{i,j} \delta_{i,j}, 0)}{\theta_{0,j}} \tag{7.121}$$

$$subject\ to: \sum_{j \in B^o} \delta_{i,j} \le e_i - p_i^* \ \forall i \in B^g. \tag{7.122}$$

7.5.2.2 Solving the WUA Problem

In this section, we present the energy loss and latency aware (ELLA) user association scheme that solves the WUA problem. We map the constraint, $0 \le \mu_j \le \tau$, to $0 \le \rho_j \le \frac{\tau}{1+\tau}$. In the ELLA scheme, we approximate

$$\frac{\max(p_j - e_j, 0)}{\theta_{0,j}} \tag{7.123}$$

as

$$\log(e^{\frac{p_j - e_j}{\theta_{0,j}}} + 1), \tag{7.124}$$

and approximately reformulate the WUA problem using the logarithmic barrier function [143]. Then, the approximate WUA (AWUA) problem is

$$\min_{\rho} \sum_{j\in B} \log\left(e^{\frac{p_j - e_j}{\theta_{0,j}}} + 1\right)$$
$$- \frac{1}{t} \sum_{j\in B} \log\left(\frac{\tau}{\tau+1} - \rho_j\right)$$
$$subject\ to: r(x) \geq \omega r^{\max}(x). \tag{7.125}$$

Here, $t > 0$ is a parameter that sets the accuracy of the approximation. The quality of the approximation increases as t grows.

The ELLA scheme solves the AWUA problem in a distributed fashion and consists of a user side algorithm and a BS side algorithm. The BS side algorithm measures the traffic load in the BS and updates its advertising traffic load. The BS broadcasts the traffic load information, its green energy generation rate, and its operation parameters including β_j, p_j^s, and $\theta_{0,j}$ to users. Based on the broadcasting messages, the user side algorithm selects the optimal BS to minimize the objective function of the AWUA problem. In order to guarantee convergence of the distributed user–BS association scheme, we assume that the time scale of the traffic arrival and departure process is fast relative to that of BSs advertising their traffic loads. In other words, BSs broadcast their traffic loads after the system exhibits stationary performance. We assume that all the BSs are synchronized and advertise their traffic loads simultaneously. We define the time interval between two consecutive traffic load advertisements as a time slot. We assume that the green energy generation rate is consistent during the time period of establishing a stable user– BS association.[5]

The user side algorithm
At the beginning of the kth time slot, the mobile users receive BSs' broadcasting messages including $\rho_j(k)$, e_j, β_j, p_j^s, and $\theta_{0,j}$. Let

$$f(\rho_j(k)) = e^{\frac{\beta_j \rho_j(k) + p_j^s - e_j}{\theta_{0,j}}}, \tag{7.126}$$

and

$$\phi_j(k) = \frac{\theta_{0,j}t(\tau - \rho_j(k)\tau - \rho_j(k))(f(\rho_j(k)) + 1)}{f(\rho_j(k))(\beta_j t(\tau - \rho_j(k)\tau - \rho_j(k)) + \theta_{0,j}) + \theta_{0,j}}. \tag{7.127}$$

The BS selection rule for a user at location x can be expressed in Algorithm 14. Here, $b^k(x)$ is the index of the BS selected by the user.

---

[5] The timescale of the traffic arrival and departure process is typically a few minutes; the user–BS association process is at a timescale of several minutes; solar power generation is usually modeled at a timescale of an hour. Thus, this assumption is reasonable.

Algorithm 14: The user side algorithm

---

1 Calculate $b_{tmp} = \arg\max_{j \in B} r_j(x)\phi_j(k)$;
2 **if** $r_{b_{tmp}}(x) \geq \omega r^{max}(x)$ **then**
3 $\quad\lfloor$ Assign the $b^k(x) = b_{tmp}$;
4 Return $b^k(x)$.

The BS side algorithm

Upon receiving BSs' broadcasting messages, mobile users select BSs according to the user side BS selection algorithm. Then, the coverage area of BS j is updated as

$$\hat{A}_j(k) = \{x|j = b^k(x), \forall x \in A\}. \tag{7.128}$$

Then, given $\rho(k) = (\rho_1(k), \rho_2(k), \dots, \rho_N(k))$, BS j's perceived traffic load in the kth time slot is

$$M_j(\rho(k)) = \min(\int_{x \in \hat{A}_j(k)} \varrho_j(x)dx, \frac{\tau}{\tau + 1} - \epsilon). \tag{7.129}$$

Here, ϵ is an arbitrary small positive constant to guarantee $\frac{\tau}{\tau+1} - \rho_j > 0$. After having derived the perceived traffic load, BSs update their next advertising traffic load.

$$\rho(k + 1) = \xi\rho(k) + (1 - \xi)M(\rho(k)). \tag{7.130}$$

Here, $M(\rho(k)) = (M_1(\rho(k)), M_2(\rho(k)), \dots, M_{|B|}(\rho(k)))$, and $0 < \xi < 1$ is an exponential averaging parameter.

7.5.2.3   Convergence and Optimality

Since both $M(\rho(k))$ and $\rho(k)$ are defined on $[0, \frac{\tau}{\tau+1} - \epsilon]$, $M(\rho(k))$ is a continuous mapping to itself. According to Brouwer's fixed-point theorem [187], there exists a solution $\rho^* = M(\rho^*)$. Let

$$h(\rho) = \sum_{j \in B} \log\left(e^{\frac{p_j - e_j}{\theta_{0,j}}} + 1\right) \tag{7.131}$$

$$-\frac{1}{t}\sum_{j \in B} \log(\tau - \mu_j), \tag{7.132}$$

and

$$\psi(\rho) = \sum_{j \in B} \log\left(e^{\frac{p_j - e_j}{\theta_{0,j}}} + 1\right). \tag{7.133}$$

We shall first demonstrate the following lemma in order to prove convergence and optimality.

Lemma 7.5.1 $h(\rho)$ *is a convex function of ρ when ρ_j is defined in $[0, \frac{\tau}{\tau+1} - \epsilon], \forall j \in B$.*

Proof: The lemma can be proved by showing $\nabla^2 h(\rho) > 0$ when $\rho_j \in [0, \frac{\tau}{\tau+1} - \epsilon], \forall j \in B$.
□

Lemma 7.5.2   *When $\rho(k) \neq \rho^*$, $M(\rho(k))$ provides a descent direction of $h(\rho)$ at $\rho(k)$.*

Proof: The lemma is proved by showing $\langle \nabla h(\rho)|_{\rho=\rho(k)}, M(\rho(k)) - \rho(k) \rangle < 0$. Let $\eta_j^m(x)$ and $\eta_j(x)$ be the user association indication of BS j that result in the traffic load $M_j(\rho(k))$ and $\rho_j(k)$, respectively.

$$\langle \nabla h(\rho)|_{\rho=\rho(k)}, M(\rho(k)) - \rho(k) \rangle \tag{7.134}$$

$$= \sum_{j \in B} \frac{(M_j(\rho(k)) - \rho_j(k))}{\phi_j(k)}$$

$$= \int_{x \in A} \lambda(x) v(x) \sum_{j \in B} \frac{\eta_j^m(x) - \eta_j(x)}{r_j(x)\phi_j(k)} dx.$$

Since $\rho_j(k) \neq \rho_j^*$ and

$$\eta_j^m(x) = \begin{cases} 1, & j = b^k(x); \\ 0, & otherwise. \end{cases} \tag{7.135}$$

$$\sum_{j \in B} \frac{\eta_j^m(x) - \eta_j(x)}{r_j(x)\phi_j(k)} < 0. \tag{7.136}$$

Thus,

$$\langle \nabla h(\rho)|_{\rho=\rho(k)}, M(\rho(k)) - \rho(k) \rangle < 0. \tag{7.137}$$
□

Theorem 7.5.1   *The traffic load vector ρ converges to the optimal traffic load vector ρ^* that minimizes $h(\rho)$.*

Proof:

$$\rho(k+1) - \rho(k)$$
$$= \delta\rho(k) + (1 - \delta)M(\rho(k)) - \rho(k)$$
$$= (1 - \delta)(M(\rho(k)) - \rho(k)). \tag{7.138}$$

Since $M(\rho(k))$ gives a descent direction of $h(\rho)$ at $\rho(k)$ and $0 < \delta < 1$, $\rho(k+1)$ also provides a descent direction of $h(\rho)$ at $\rho(k)$. Then, $h(\rho(k+1)) < h(\rho(k))$ until $\rho(k+1) = \rho(k)$. Therefore, the traffic load vector ρ converges to the optimal traffic load vector ρ^* that minimizes $h(\rho)$.
□

7.5.2.4 Solving the BES Problem

To obtain optimal energy sharing among BSs, we transform the BES problem into a linear programming problem. We eliminate the max function in the objective function of the BES problem by introducing an additional constraint, $\sum_{i \in B^g} \theta_{i,j} \delta_{i,j} \leq p_j^* - e_j \; \forall j \in B^o$. By introducing this constraint, the BES problem can be mapped into:

$$\min_{\delta} \sum_{j \in B^o} \frac{p_j^* - e_j}{\theta_{0,j}} - \sum_{j \in B^o} \sum_{i \in B^g} \frac{\theta_{i,j} \delta_{i,j}}{\theta_{0,j}} \tag{7.139}$$

$$subject \; to: \; \sum_{j \in B^o} \delta_{i,j} \leq e_i - p_i^* \; \forall i \in B^g,$$

$$\sum_{i \in B^g} \theta_{i,j} \delta_{i,j} \leq p_j^* - e_j \; \forall j \in B^o. \tag{7.140}$$

Since $\sum_{j \in B^o} \frac{p_j^* - e_j}{\theta_{0,j}}$ is known, the BES problem is equivalent to the following linear programming problem, which can be efficiently solved using optimization software such as CVX [188]:

$$\max_{\delta} \sum_{j \in B^o} \sum_{i \in B^g} \frac{\theta_{i,j} \delta_{i,j}}{\theta_{0,j}} \tag{7.141}$$

$$subject \; to: \; \sum_{j \in B^o} \delta_{i,j} \leq e_i - p_i^* \; \forall i \in B^g,$$

$$\sum_{i \in B^g} \theta_{i,j} \delta_{i,j} \leq p_j^* - e_j \; \forall j \in B^o. \tag{7.142}$$

7.5.3 Performance Evaluation

Simulations are set up to evaluate the performance of the proposed approximation solution in a mobile network with 12 BSs deployed in a $6\,km \times 6\,km$ area. The BS's transmit power is $20\,W$. The BS's static power consumption is $700\,W$ and $\beta_j = 500, \; \forall j \in B$ [3]. We adopt COST 231 Walfisch-Ikegami [144] as the propagation model with $9\,dB$ Rayleigh fading and $5\,dB$ shadowing fading. The carrier frequency is $2110\,MHz$, the bandwidth is $5\,MHz$, the antenna feeder loss is $3\,dB$, the transmitter gain is $1\,dB$, the noise power level is $-104\,dBm$, and the receiver sensitivity is $-97\,dBm$. The green energy generation rate is consistent during the simulation. We compare the proposed algorithm with the GALA scheme that optimizes user–BS association by considering both green energy utilization and average traffic delivery latency [55].

The convergence of the ELLA algorithm is shown in Figure 7.42. At the beginning of the ELLA algorithm (the first 35 time slots), $h(\rho)$ tends to infinity. This is because at the beginning of the ELLA algorithm, traffic demands are directed to the BSs with higher green energy generation rates, and thus these BSs' QoS constraints are violated. Since traffic loads concentrate on the BSs with higher green energy rates, the on-grid power consumption is low. Therefore, $\psi(\rho)$ is small at the beginning of the ELLA algorithm. As the ELLA algorithm evolves, traffic demands are gradually offloaded from the congested BSs to other BSs until the BSs' QoS requirements are satisfied. The ELLA algorithm converges after around 40 time slots.

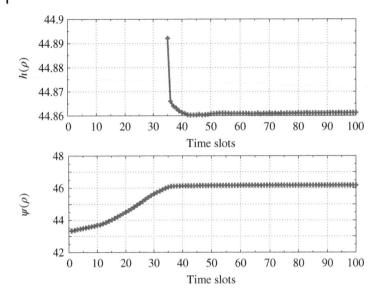

Figure 7.42 ELLA convergence. *Source:* Han 2014 [183]. Reproduced with permission of IEEE.

The coverage area of the ELLA algorithm is shown in Figure 7.43. In associating users with BSs, the ELLA algorithm considers the energy efficiency of traffic offloading, green energy generation rates, and power transmission efficiency. By considering the energy efficiency of traffic offloading, the maximum coverage area of a BS is constrained. By accommodating the green energy generation rate and power transmission efficiency, more traffic is directed to BSs with higher green energy generation rates and power transmission efficiency. In this simulation, as compared with its neighboring BSs, the fifth BS has higher green energy generation rate and power transmission efficiency. Thus, the fifth BS covers a relatively larger area.

Figs. 7.44 and 7.45 compare the on-grid power consumption and the traffic delivery latency of the ELLA and GALA algorithms, respectively. As shown in Figure 7.44, as

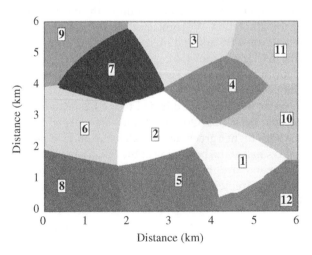

Figure 7.43 Coverage area ($\tau = 8$). *Source:* Han 2014 [183]. Reproduced with permission of IEEE.

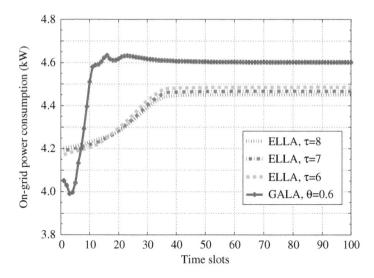

Figure 7.44 On-grid power consumption. *Source:* Han 2014 [183]. Reproduced with permission of IEEE.

compared with the GALA algorithm, the ELLA algorithm achieves additional on-grid power savings. The power savings reduce as τ decreases. When $\tau = 8$, the ELLA algorithm achieves about 150 W power savings as compared with the GALA algorithm. The power saving is not very significant because (1) we constrain the maximum coverage area of each BS to restrict the energy efficiency of traffic offloading to being higher than that of sharing energy among BSs, and (2) the GALA algorithm is also green energy aware and is designed to optimize green energy utilization. In Figure 7.45, the traffic delivery latency of the ELLA algorithm is larger than that of the GALA algorithm

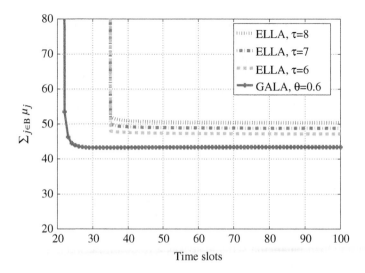

Figure 7.45 Traffic delivery latency. *Source:* Han 2014 [183]. Reproduced with permission of IEEE.

Table 7.5 BSs' on-grid power consumption (kW).

GALA	BPO ($\tau = 6$)	BPO ($\tau = 7$)	BPO ($\tau = 8$)
4.6	3.80	3.77	3.76

because the ELLA algorithm aims to minimize on-grid power consumption with a targeted traffic delivery latency while the GALA algorithm tries to optimize the trade-off between on-grid power consumption and traffic delivery latency.

Solving the BES problem yields the optimal power sharing among BSs. With BS energy sharing, the network's on-grid power consumption is further reduced as shown in Table 7.5. As compared with the GALA algorithm which only optimizes BS operation, the proposed approximation solution to solve the BPO problem saves around 18% of on-grid power.

7.6 Summary

In this chapter, we have presented four traffic load balancing schemes for optimizing energy efficiency in mobile networks under different networking scenarios. First, we have designed the ICE traffic load balancing algorithm to optimize the utilization of green energy in future cellular networks, and therefore to minimize the energy consumption from the main grid. We have derived and demonstrated the low computational complexity of ICE. Through simulations, we show that ICE balances the energy consumptions among LBSs, enables more users to be served with green energy, and reduces user outage.

Second, we have proposed a traffic load balancing framework referred to as vGALA. During the procedure of establishing user association, the vGALA scheme not only considers network performance, for example, the average traffic delivery latency, but also adapts to the availability of green energy. Various properties, in particular, convergence of vGALA, have been proved. The vGALA scheme reduces on-grid power consumption with a little sacrifice in average traffic delivery latency. The trade-off between network performance and on-grid power consumption is adjustable in individual BSs and controllable by the radio access network controller. The vGALA scheme includes both the user side algorithm and the BS side algorithm. To avoid extra communication overheads, the vGALA scheme, leveraging the SoftRAN architecture, introduces virtual users and vBSs to simulate the interactions between users and BSs, thus significantly reducing the information exchanges required over the air interface. The extensive simulation results have validated the performance and practicality of the vGALA scheme.

Third, we have designed a NUA traffic load balancing scheme for backhaul constrained cache-enabled SCNs with hybrid power supplies. During the procedure of establishing user associations, the NUA traffic load balancing scheme considers four network utilities: green power utilization, the traffic delivery latency in BSs, the traffic delivery latency in backhaul, and the cache hit ratio. By optimizing the user association, the NUA traffic load balancing scheme strikes a trade-off between green power utilization and traffic delivery latency in the network. The NUA traffic load balancing scheme adapts the user association according to the dynamics of green power, BS capacity, backhaul data rates, and the cache hit ratio. It significantly reduces the traffic delivery latency when the network is constrained by the backhaul data rate. Moreover, by adjusting system parameters, for example κ, the NUA scheme is able to adjust the trade-off between brown power consumption and traffic delivery latency.

Fourth, we have jointly optimized the BS traffic load and power distribution for mobile networks powered by smart grid. The joint BS operation and power distribution optimization (BPO) problem is difficult to solve because of the strong coupling of BS operation and power distribution. We have proposed an approximation solution that solves the BPO problem, which saves about 18% on-grid power as compared with solutions that only optimize BS operation.

7.7   Questions

7.1   What is traffic load balancing in mobile networks?

7.2   How does a mobile network balance traffic loads and what are the procedures?

7.3   What network performance metrics should be considered in determining user–BS association?

7.4   How difficult is the traffic load balancing problem? Prove whether it is NP-hard.

7.5   Recap the solutions that have been proposed for traffic load balancing in mobile networks.

8

Enhancing Energy Efficiency via Device-to-Device Proximity Services

This chapter presents three D2D proximity services-based energy efficient communications schemes, namely, energy efficient cooperative wireless multicasting, green relay assisted D2D communications and green content brokerage. In the energy efficient cooperative wireless multicasting scheme, BS and mobile users cooperate to reduce the total power consumption in wireless multicasting. Minimizing power consumption under this scenario is, however, an NP-hard problem. The gradient guided approximation algorithm is thus proposed to achieve low computational complexity. The proposed energy efficient wireless multicasting scheme consumes less energy than other multicasting schemes without cooperation. The energy savings are validated through well designed simulations.

Green relay assisted D2D communications is a novel D2D communications architecture as shown in Figure 8.1. Here, we define the UE that delivers the content as the source node and the UE that retrieves the content as the destination node. In green relay assisted D2D communications, we introduce green relay nodes to increase the link data rate of D2D communications by leveraging cooperative communications [190]. Green relay assisted D2D communications exhibits several advantages: (1) green relay nodes can forward the source nodes' information and increase the transmission data rate between source and destination nodes; (2) performing as relays, green relay nodes do not require backhaul connections; (3) green relay nodes are powered by renewable energy, for example, solar energy and wind energy, thus do not require on-grid power supplies; (4) green relay nodes can cache popular content and enable localized content delivery [191]; and (5) a green relay node can facilitate service and peer discovery and D2D communications session management [192]. We investigate the utilization of green relay nodes to enhance the data rates of D2D communications. In order to enhance data rates between the source and destination nodes, the green relay nodes should be properly tailored for the SD pairs. Distinct from traditional relay assignment problems, which assume a relay node is assigned to at most one SD pair, a green relay node can be assigned to multiple SD pairs to increase their data rates. In practical implementations, a green relay node can be a powerful SCBS but configured as a relay node. Therefore, a green relay node is able to handle multiple transmissions simultaneously. Hence, a green relay node may be assigned to multiple SD pairs. In addition, since green relay nodes are powered by green energy, green relay node assignments should also consider the availability of green energy in the relay nodes. We propose a heuristic GRA algorithm to

Green Mobile Networks: A Networking Perspective, First Edition. Nirwan Ansari and Tao Han.
© 2017 John Wiley & Sons Ltd. Published 2017 by John Wiley & Sons Ltd.

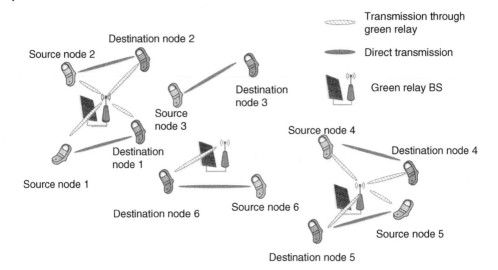

Figure 8.1 Green relay assisted D2D communications. *Source:* Han 2013 [189]. Reproduced with permission of IEEE.

optimize the relay assignments in green relay assisted D2D communications. The GRA algorithm maximizes the minimum data rates of the SD pairs under the green energy constraint.

Green content brokerage is, in fact, a novel mobile traffic offloading scheme, which combined the merits of both infrastructure based and ad-hoc mobile traffic offloading. The content brokerage scheme introduces a new network node called the green content broker (GCB), which arranges content delivery between content owners (the servers), for example, UEs or BSs, and content requesters (the clients). The content broker discovers the available content among the UEs under its coverage, collects the content requests, and forwards the content from content owners to the content requester.

The proposed content brokerage scheme has several advantages:

- As compared with mobile traffic offloading by deploying SCBSs, the content broker-age scheme does not require backhaul connections. This not only reduces the operation costs of offloading mobile traffic, but also increases the flexibility of deploying the GCB and increases the efficiency of traffic offloading.
- As compared with ad-hoc based mobile traffic offloading, the GCB can facilitate service and content discovery and D2D communications session management [192]. In addition, the GCB may anonymize mobile traffic to protect user privacy, and may maintain content copyrights for the content providers.
- The content broker is powered by green energy, for example, solar energy, to avoid on-grid energy consumption, thus reducing the carbon footprint. Powering by green energy further increases the flexibility of deploying the GCB.

In the content brokerage scheme, the GCB handles multiple content requests from UEs. Given the available number of radio channels and amount of green energy, the GCB may not be able to serve all content requests. To maximize traffic offloading, and thus minimize the energy consumption of the MBS, the GCB should selectively serve a portion of content requests. In selecting the content to serve, the GCB should consider the radio channel constraints in both the uplink and downlink transmissions. Here, the uplink transmissions refer to radio channels used by the GCB to retrieve content from UEs or the MBS, and the downlink transmissions refer to the radio channels the GCB utilizes to deliver content to clients. Since the GCB is powered by green energy, it also experiences green energy constraints, which limit the amount of mobile traffic that can be offloaded to the GCB. In addition, the requested content may be owned by multiple UEs. Selecting different content owners (UEs) leads to a different uplink radio channel requirements. The traffic offloading maximization (TOM) problem including consideration of these constraints is an NP-hard problem. To approximate the optimal solution with low computational complexity, the TOM problem is decomposed into two sub-problems: the serving content selection (SCS) problem and the content owner selection (COS) problem. A heuristic traffic offloading (HTO) algorithm is designed to iteratively solve these sub-problems, and hence solve the TOM problem.

8.1 Energy Efficient Cooperative Wireless Multicasting

Wireless multicasting is the central feature of next generation cellular networks. Therefore, minimizing energy consumption in wireless multicasting is important for green communications. Multicast beamforming is a promising technique for wireless multicasting [193, 194]. Cooperative networking is another promising technique that potentially empowers wireless networks to reduce energy consumption [195]. We propose a novel wireless multicasting scheme, which integrates multicast beamforming and cooperative networking. It contains two phases: in phase 1, the BS transmits the signal to the subscribers using antenna arrays with multicast beamforming; in Phase 2, users who successfully received the signal in phase 1 forward the signal to other users. Unsatisfied users combine the signals from the BS and from relays via maximal ratio combining (MRC). In addition, we design a low complexity gradient guided algorithm that minimizes the transmit power in the multicasting, thus reducing BS energy consumption.

8.1.1 System Model and Problem Formulation

Consider a two phase amplify-and-forward cooperative communications strategy. In phase 1, a BS transmits its information to the subscribers, and the received signal at user i is y_i^b. In phase 2, user j, who successfully received the signal, forwards the signal to other users, for example, user i, who did not, and the received signal at user i is $y_{j,i}^r$. Here, we assume that the signal is successfully received if its SNR is larger than the user's minimum SNR requirement, and that the signal relay in phase 2 is well scheduled to avoid transmission collisions. Consider a BS with N transmit antennas, and K multicasting service subscribers with a single receive antenna. Define h_i^b as the $N \times 1$ complex vector that models the transmit channels between the antenna array of the BS and user i, and define w^H as the beamforming weight vector applied to the transmitting antenna array, where $(\cdot)^H$ denotes the hermitian transpose. Define $h_{j,i}^r$ as the relay channel between user j and user i. Let x be the transmit signal, and then the received signal at user i in phase 1 is

$$y_i^b = w^H h_i^b x + n_i^b. \tag{8.1}$$

Let $P_i^b = |w^H h_i^b|^2$, and the relay users forward the signal with the same transmit power P^r. Assume subscriber j, who successfully received the signal in phase 1, forwards the signal to user i in phase 2. The received signal at user i from relay user j is

$$y_{j,i}^r = \sqrt{\frac{P^r}{P_j^b + \mathcal{N}_0}} h_{j,i}^r y_j^b + n_{j,i}^r. \tag{8.2}$$

In Eqs. (8.1) and (8.2), the noise items n_i^b and $n_{j,i}^r$ are modeled as zero-mean, complex Gaussian variables with variance $\mathcal{N}_0$. The transmit signal is assumed to have an average energy of 1. According to [195], the instantaneous SNR of MRC output at receiver i is

$$\gamma_i = \gamma_i^b + \gamma_{j,i}^{br}, \tag{8.3}$$

where $\gamma_i^b = P_i^b / \mathcal{N}_0$, and

$$\gamma_{j,i}^{br} = \frac{\frac{P_i^b P^r |h_{j,i}^r|^2}{P_i^b + \mathcal{N}_0}}{\left(\frac{P^r |h_{j,i}^r|^2}{P_i^b + \mathcal{N}_0} + 1\right) \mathcal{N}_0}. \tag{8.4}$$

To satisfy the users' minimum instantaneous SNR requirement with the minimum transmit power, the multicast design problem can be cast as follows:

$$\min_{w \in \mathbb{C}^N} \|w\|^2 \tag{8.5}$$

$$s.t. : \gamma_i^b + \beta(\gamma_i^b - \rho_i^{\min})(1 - \beta(\gamma_j^b - \rho_j^{\min}))\gamma_{j,i}^{br} \geq \rho_i^{\min},$$
$$i, j \in 1, 2, \cdots, K. \tag{8.6}$$

Here,

$$\beta(x) = \begin{cases} 0, & x \geq 0, \\ 1, & x < 0. \end{cases} \tag{8.7}$$

The above problem is NP-hard since when $\gamma_i^b \geq \rho_i^{\min}$, $\beta(\gamma_i^b - \rho_i^{\min})$ is 0. It is also equivalent to the transmit beamforming problem [193], which has been shown to be NP-hard.

8.1.2 Gradient Guided Algorithm

Let $\gamma_{j,i}^r = P^r |h_{j,i}^r|^2 / \mathcal{N}_0$; then, $\gamma_{j,i}^{br}$ in the constraint Eq. (8.6) can be expressed as

$$\gamma_{j,i}^{br} = \frac{\gamma_i^b \gamma_{j,i}^r}{\gamma_i^b + \gamma_{j,i}^r + 1}. \tag{8.8}$$

When $\gamma_i^b < \rho_i^{\min}$ and $\gamma_j^b \geq \rho_j^{\min}$, the constraint Eq. (8.6) becomes

$$\gamma_i^b + \frac{\gamma_j^b \gamma_{j,i}^r}{\gamma_j^b + \gamma_{j,i}^r + 1} \geq \rho_i^{\min}; \tag{8.9}$$

otherwise, the constraint Eq. (8.6) is

$$\gamma_i^b \geq \rho_i^{\min}. \tag{8.10}$$

In Eq. (8.9), γ_j^b is only related to the transmitting power in user j's direction and the transmit channel between BS and user j, and $\gamma_{j,i}^r$ is only related to the channel between user i and user j since all relay users use the same transmit power. Thus, increasing the power allocation in user i's direction will not reduce the relay SNR, $\gamma_{j,i}^{br}$, but will reduce the gap between γ_i and $\rho_i^{\min}$. Therefore, the idea behind the algorithm is to increase the transmit power toward the weakest user at every iteration. The signal power at receiver i from the BS is $P_i^b = w^H h_i h_i^H w$. The gradient of the received power with respect to w is

$$\nabla_w P_i^b = h_i h_i^H w. \tag{8.11}$$

Let $w^{\langle t \rangle}$ be the value at iteration t; to increase the power allocation toward user i, $w^{\langle t \rangle}$ is updated as

$$w^{\langle t+1 \rangle} = w^{\langle t \rangle} + \mu h_i h_i^H w^{\langle t \rangle}, \tag{8.12}$$

where μ is the step size of the update.

The proposed algorithm is to identify the user with the smallest MRC output SNR, and then to increase the power allocation toward him/her at every iteration. The algorithm continues until all the subscribers' minimal SNR requirements are satisfied. It starts by initializing $w^{\langle 0 \rangle}$, and the complete flowchart of the algorithm is shown in Figure 8.2.

Here, **S** and **U** are the sets of satisfied and unsatisfied users regarding their minimal SNR requirements, respectively.

8.1.3   Performance Evaluation

Simulations are set up with parameters as shown in Table 8.1. Here, we evaluate the performance in a single sector and assume the users are uniformly distributed in that sector. We further assume that the channel state information is perfectly estimated by both BS and users.

Figure 8.3 compares the minimal transmit power of different multicasting strategies. With transmit beamforming, BS saves more than 3 dBm transmit power. To identify the benefits from the cooperation, we set up three relay strategies: (1) relay with the full transmit power of user equipment, that is 23 dBm; (2) relay with half of the transmit power, that is 20 dBm; and (3) relay with WIFI transmit power, that is 15 dBm. We compare these with Lozano's algorithm [194], which does not consider cooperation. As the number of users increases, the performance of the proposed algorithm becomes better because there are more cooperative opportunities. When the number of users is larger than 40, the performance becomes steady, at which point it uses about 3.5 dBm, 2 dBm, and 1 dBm less transmit power than that of Lozano's algorithm, respectively. It becomes steady because when the number of users is large enough (40 in the simulation), the cooperation gain is not limited by the cooperative opportunities.

In Figure 8.4, the user SNRs under different multicasting strategies are compared. Both the proposed algorithm and Lozano's algorithm have almost the same

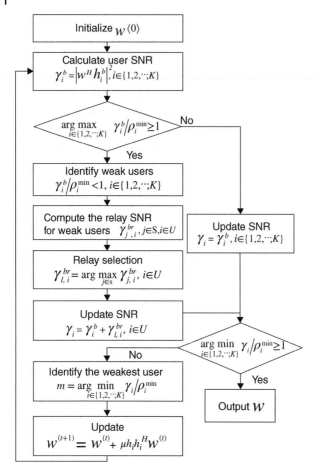

Figure 8.2 Algorithm flowchart. *Source:* Han 2011 [196]. Reproduced with permission of IEEE.

Table 8.1 Simulation parameters.

Parameters	Assumption
Cellular layout	Hexagonal grid, 3 sectors per site
Inter-site distance	1.8 km
Transmit antenna gain	12 dBi
Antenna array	8 elements ULA with $d/\lambda = 0.5$
Path loss between BS and UE	$128.1 + 37.6\log_{10}(d)$ (d in km)
Shadowing	log-norm 10 dB std.
Path loss between UE and UE	$41 + 22.7\log_{10}(d)$ (d in km)
Receiver antenna gain	0 dBi
Receiver sensitivity	−97 dB
Minimum SNR requirement	10 dB

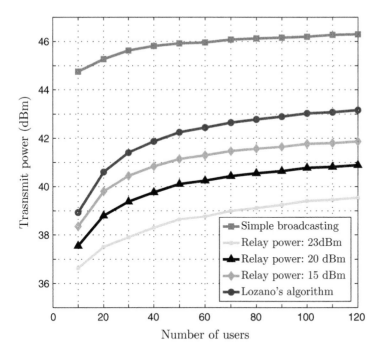

Figure 8.3 Transmit power versus number of users. *Source:* Han 2011 [196]. Reproduced with permission of IEEE.

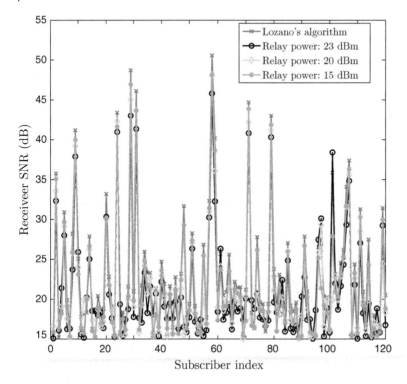

Figure 8.4 User SNR. 120 users are uniformly distributed in the sector. *Source:* Han 2011 [196]. Reproduced with permission of IEEE.

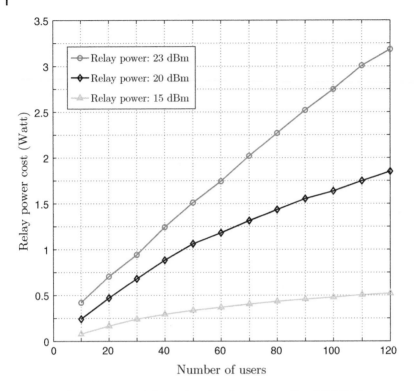

Figure 8.5 Power consumed during the relay phase. *Source:* Han 2011 [196]. Reproduced with permission of IEEE.

performance, and all the users' minimum SNR requirements are satisfied. In Figure 8.5, as the number of users increases, more power is consumed during the relay phase. When there are more users, more relay nodes are needed, and thus more relay power is consumed.

Figure 8.6 shows the number of iterations required to achieve the optimal transmit power of the proposed algorithm and Lozano's algorithm. When the number of users is small, our proposed algorithm is slightly better than Lozano's algorithm. However, as the number of users increases, the number of iterations of the proposed algorithm is much less than that of Lozano's algorithm. This is because there are more cooperation opportunities when the number of users is large, and thus the cooperation gain is large. With cooperation gain, the users' minimum SNR requirements are satisfied faster because some users' SNR requirements may be satisfied during the relay phase, thus reducing the number of iterations in the beamforming process. The performance regarding the number of iterations becomes steady when the number of users is larger than 40 because the cooperation gain becomes steady.

In Figure 8.7, we compare the BS power consumptions under different multicasting strategies. We apply the BS power consumption model in [197] to calculate BS power consumption regarding transmit power. The line with star markers indicates the power consumption of the standard LTE MBS [197], which can be considered as the power constraint of BS. Note that simply broadcasting without beamforming and

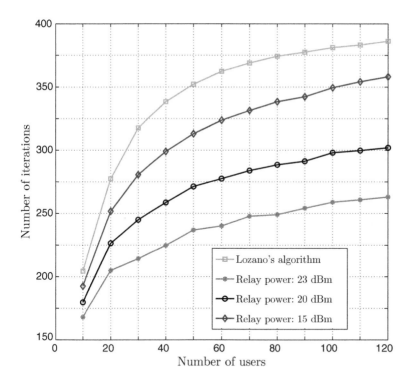

Figure 8.6 The number of iterations versus the number of users ($\mu = 0.01$). *Source:* Han 2011 [196]. Reproduced with permission of IEEE.

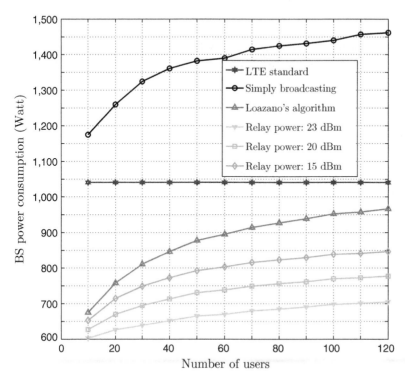

Figure 8.7 Power consumption versus number of users. *Source:* Han 2011 [196]. Reproduced with permission of IEEE.

cooperation cannot satisfy users' requirement under the constraint. As compared to Lozano's multicast beamforming algorithm, our proposed algorithm can save at least 100 Watts when the number of users is larger than 60. The power savings are benefits from the cooperation between the BS and users.

8.2 Green Relay Assisted D2D Communications

The idea of D2D communications is a promising concept to improve data rates and the energy efficiency of mobile networks. User devices (UEs) participating in D2D communications may increase their own data rates by retrieving content from their neighboring peers instead of from BSs. Meanwhile, UEs may drain their batteries while performing as content providers and transmitting content to their peers. Increasing the data rates of D2D communications is desirable to alleviate UEs' power consumption. In this chapter, we propose a novel green relay assisted D2D communications architecture, in which relay nodes powered by green energy are deployed to increase the data rates of D2D communications. However, achieving the optimal relay assignment for green relay assisted D2D communications is challenging. We propose a heuristic green relay assignment algorithm which maximizes the minimum data rates of D2D pairs while considering the green load capacity of the relay nodes. We show that the proposed algorithm approximates the optimal solution with low computational complexity, and validate its performance by using simulation results.

In designing green relay assisted D2D communications, we consider a wireless network consisting of $2N$ UEs and M green relay nodes, with the UEs forming N SD pairs. Define $\mathcal{R} = (r_1, r_2, \cdots, r_m)$, $\mathcal{S} = (s_1, s_2, \cdots, s_n)$, and $\mathcal{D} = (d_1, d_2, \cdots, d_n)$ as the sets of the green relay nodes, the source nodes and the destination nodes, respectively. We assume that orthogonal channels are available in the network, for example, OFDMA, to avoid interference [32]. We assume all the nodes can only either transmit or receive at one time.

8.2.1 System Model and Problem Formulation

8.2.1.1 Communications Model

Define p_i^s and p_j^r as the transmission power of the ith source node and the jth relay node, respectively. If a node μ is transmitting a signal to the node v with power p_μ, the perceived SNR, $\Upsilon_{\mu v}$, at the receiving node is

$$\Upsilon_{\mu v} = \frac{p_\mu}{N_0 ||\mu, v||^\alpha}. \tag{8.13}$$

Here, N_0 is white noise, $||\mu, v||$ is the Euclidean distance between node μ and v, and α is the path loss component [33].

In green relay assisted D2D communications, the SD pairs can either transmit data directly or via a green relay node. If the ith SD pair transmits data directly, the achievable data rate is

$$C_D(s_i, d_i) = W \log_2(1 + \Upsilon_{s_i, d_i}). \tag{8.14}$$

Figure 8.8 Illustration of cooperative communications. *Source:* Han 2013 [189]. Reproduced with permission of IEEE.

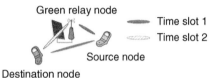

Here, W is the available bandwidth. The SD pair can also transmit via a green relay node with either the AF or DF cooperative communications mode. In this chapter, we adopt the DF cooperative communications mode. Cooperative communications is illustrated in Figure 8.8.

In the first time slot, the source node broadcasts the data. Both the destination node and the relay node receive the data. In the second time slot, the relay node forwards the received data to the destination node. For DF cooperative communications, the green relay node decodes the data received in the first time slot, and transmits the decoded data to the destination in the second time slot [198]. Based on DF cooperative communications, the achievable data rate of the ith SD pair via the jth green relay node is

$$
C_{DF}(s_i, r_j, d_i) =
$$
$$
\frac{W}{2} \min(\log_2(1 + \Upsilon_{s_i, r_j}), \log_2(1 + \Upsilon_{s_i, d_i} + \Upsilon_{r_j, d_i})). \tag{8.15}
$$

8.2.1.2 Traffic Model

We assume that the traffic generated from the ith source node follows a Poisson process with average rate equal to λ_i, and that the traffic load is distributed according to a general distribution with average traffic load l_i. Then the traffic load generated from the ith SD pair in the jth green relay node is

$$
\varrho_{i,j} = \frac{\lambda_i l_i \eta_{i,j}}{C_{DF}(s_i, r_j, d_i)}. \tag{8.16}
$$

Here, $\eta_{i,j}$ is an indicator function. If $\eta_{i,j} = 1$, the ith SD pair transmits via the jth green relay node. The average traffic load in the jth green relay node is

$$
\rho_j = \sum_{i \in S} \varrho_{i,j}. \tag{8.17}
$$

This value of ρ_j indicates the fraction of the time that BS j is busy. The green relay node is assumed to serve SD pairs in a round robin fashion. Then, the green relay node realizes a M/G/1-PS queue [152]. If the jth green relay node is assigned to multiple SD pairs, the effective data rate of the ith SD pair transmitting via the jth relay node is

$$
C_{eff}(s_i, r_j, d_i) = C_{DF}(s_i, r_j, d_i)(1 - \rho_j). \tag{8.18}
$$

8.2.1.3 Energy Model

In green relay assisted D2D communications, the relay nodes are powered by green energy generated by their solar panels. Owing to the disadvantages of "banking" green energy [48], we do not assume that green energy can be stored. The relay node's power consumption consists of two parts: static power consumption and dynamic power consumption [140]. Static power consumption is the power consumption of a relay node without any traffic load. Dynamic power consumption refers to the additional power consumption caused by traffic load on the relay node, which can be well approximated by a linear function of the traffic load [140]. Define c_j^s as the static power consumption of the jth relay node. Then, the jth relay node's power consumption can be expressed as

$$c_j = \beta_j \rho_j + c_j^s. \tag{8.19}$$

Here, β_j is a linear coefficient which reflects the relationship between the traffic load and the dynamic power consumption in the jth relay node. Define e_j as the energy generation rate in the jth relay node. Define the green load capacity as the maximum traffic load that the relay node can support with a given green energy generation rate. Define ρ_j^g as the green load capacity of the jth relay node. If $e_j \leq c_j^s$, which indicates the relay node does not have enough green energy to be turned on, then $\rho_j^g = 0$; otherwise,

$$\rho_j^g = \min\left(\frac{(e_j - c_j^s)}{\beta_j}, 1 - \epsilon\right). \tag{8.20}$$

Here, $0 < \epsilon < 1$ is a small positive real number to guarantee $\rho_j^g < 1$.

8.2.1.4 Problem Formulation

In green relay assisted D2D communications, we aim to maximize the minimum data rate of SD pairs by optimizing the relay assignment. Thus, the green relay assignment problem can be formulated as:

$$\max_{\eta} \min_{i \in S} C(s_i, d_i) \tag{8.21}$$

$$subject\ to: \sum_{j \in R} \eta_{i,j} = 1, \tag{8.22}$$

$$0 \leq \rho_j \leq \rho_j^g. \tag{8.23}$$

Here,

$$\eta = \begin{pmatrix} \eta_{1,1}, & \eta_{1,2}, & \cdots, & \eta_{1,M} \\ \eta_{2,1}, & \eta_{2,2}, & \cdots, & \eta_{2,M} \\ \vdots, & \vdots, & \ddots, & \vdots \\ \eta_{N,1}, & \eta_{N,2}, & \cdots, & \eta_{N,M} \end{pmatrix}. \tag{8.24}$$

$C(s_i, d_i)$ is the effective data rate of the ith SD pair. The constraint in Eq. (8.22) requires an SD pair to be assigned to at most one green relay node. The constraint in Eq. (8.23) is to guarantee that the total power consumption is not larger than the available green energy. The above optimization problem is a mixed integer quadratic constraint problem, which is challenging to solve. A brute-force search, in the worst case, requires $O(M^N)$ iterations to find the optimal green relay assignments. The computational complexity of the brute-force search increases exponentially with respect to the number of the SD pairs in the network, and is thus not practical for real time applications.

8.2.2 A Heuristic Green Relay Assignment Algorithm

In this section, we propose a heuristic GRA algorithm which approaches the optimal solution with low computational complexity. Define $\hat{\eta}$ as a relay assignment derived by the $(\hat{\eta}, flag) = new\_relay\_search(s_i, \eta)$ algorithm. *flag* is an indicator for whether $\hat{\eta}$ is a better relay assignment or not. If *flag* = 1, $\hat{\eta}$ is a better assignment. The GRA algorithm, as shown in Algorithm 15 below, starts with an initial green relay assignment in which each green relay is assigned to one source node. Then, GRA adjusts the green relay assignment to increase the minimum data rate of the SD pairs during each iteration. Specifically, in each iteration, GRA finds the SD pair, (s_i, d_i), with the minimum data rate and searches for a new green relay assignment to increase the data rate of the SD pair. If a new assignment is found, GRA starts another iteration; otherwise, GRA checks whether (s_i, d_i) shares a green relay, r_j, with other SD pairs. If true, GRA searches for an alternative relay assignment for the SD pairs assigned to r_j. If such a relay assignment is found, an SD pair that was originally assigned to r_j is assigned to another relay. As a result, ρ_j is reduced, and therefore $C(s_i, d_i)$ is increased. If alternative relay assignments are not found, GRA is terminated.

Algorithm 15: The Heuristic GRA Algorithm

Step 1: Initial relay assignment;
Step 2: Set *flag* = 0; find $i = \arg\min_{j \in S} C(s_j, d_j)$;
$(\hat{\eta}, flag) = new\_relay\_search(s_i, \eta)$;
if (*flag* == 1) **then**
  $\eta = \hat{\eta}$;
  Go to step 2;
else
  Step 3:
  if (s_i share a relay node with other SD pairs) **then**
    Find $S_k = \{j | \eta_{j,k} = 1, \eta_{i,k} = 1, j \neq i, j \in S\}$,
    for $j = 1 : |S_k|$ **do**
      Clear the markers;
      $(\hat{\eta}, flag) = new\_relay\_search(s_j, \eta)$;
      if (*flag* == 1) **then**
        $\eta = \hat{\eta}$, and break;
      end if
    end for
    if (*flag* == 1) **then**
      Go to step 2;
    end if
  end if
end if
Return η.

The key component of the GRA algorithm is a recursive algorithm, $(\hat{\eta}, flag) = new\_relay\_search(s_i, \eta)$, as shown in Algorithm 16, which searches for a new relay assignment for a specified source node. Since Algorithm 16 is a recursive algorithm, we define $\bar{\eta}$ as an intermediate relay assignment during the recursions. We use an example shown in Figure 8.9 to illustrate the recursive algorithm. Suppose s_1 is the source node with

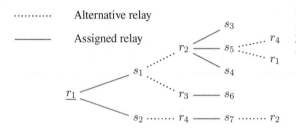

Figure 8.9 Illustrative example of the recursive GRA algorithm. *Source:* Han 2013 [189]. Reproduced with permission of IEEE.

the minimum data rate, and under the relay assignment, s_1 and s_2 share the green relay node r_1. GRA first tries to find a better relay assignment for s_1. The dotted lines in the figure indicate the alternative relay nodes for a source node. Given s_1 and η, the recursive algorithm finds that s_1 has two alternative relays, r_2 and r_3. Since s_1 is to be assigned to either r_2 or r_3, the traffic load on r_1 is reduced. Therefore, r_1 may be assigned to other source nodes. To increase the possibility that s_1 can successfully join r_2 or r_3, the recursive algorithm first tries to offload traffic from r_2 and r_3, and then offloads traffic from other relays. After that, r_1 is marked to avoid being searched again. If r_2 is not marked, the algorithm checks whether s_1 can be assigned to r_2 successfully.[1] If it is successful, a new assignment, $\hat{\eta}$, is found; otherwise, the algorithm checks whether s_3, s_4, or s_5 can be assigned to other relay nodes to enable s_1 to join r_2. In this example, s_3 and s_4 can only be assigned to r_2, $C_D(s_3, d_3) < C(s_1, d_1)$, and $C_D(s_4, d_4) < C(s_1, d_1)$. Therefore, s_3 and s_4 cannot be removed from r_2. s_5 has an alternative relay, r_4. Then, the recursive algorithm calls itself with s_5 as input, in order to check whether s_5 can be successfully assigned to r_4. Suppose it is not successful, then r_2 is marked, and the algorithm tries to offload load traffic from unmarked relay nodes to r_2. Then, the algorithm checks whether s_1 can be successfully assigned to r_3 following the same procedure. If it fails, then s_1 cannot find a better relay node. The GRA algorithm clears the marks on all the relay nodes and tries to remove s_2 from r_1 following the same procedure.

There are two key innovations in the proposed heuristic GRA algorithm. The first is the relay node marking mechanism, which reduces the computational complexity in solving the relay assignment problem. Suppose we want to search for a new relay assignment for s_1; without the relay node marking mechanism, it requires M^N iterations in the worst case. With the marking mechanism, the GRA algorithm only checks M relay nodes in the worst case. However, the marking mechanism may deteriorate the performance of the algorithm. For example, suppose in the optimal relay assignment, s_5 is assigned to r_1 and s_1 is assigned to r_2. If this is the case, with the marking mechanism, the GRA algorithm may fail to find the optimal solution. To reduce the probability of such failure, we introduce the traffic offloading mechanism, which tries to offload traffic from the unmarked relay nodes to the marked ones. Thus, when r_1 is marked, the GRA algorithm offloads traffic from the unmarked node, for example, r_2. As a result, s_5 may be offloaded to r_1, and s_1 can be assigned to r_2 successfully. Thus, the optimal assignment is obtained. In each iteration, the computational complexity of the offloading mechanism in the worst case is $O(N^M)$. Since an SD pair has a choice of up to M relay nodes, and on each relay node the SD pair can have N different data rates, the total data rate

---

[1] Here, a successful assignment means that the assignment increases the minimum data rates of SD pairs in the network.

improvements an individual SD pair can achieve is limited by *NM*. Therefore, the computational complexity of the heuristic GRA algorithm is $O(M^2 N^{(M+3)})$. If we fix the total number of green relay nodes deployed in an area, then the computational complexity of the heuristic GRA algorithm is polynomial with respect to the number of SD pairs.

Algorithm 16: $(\hat{\eta}, flag) = new\_relay\_search(s_i, \eta)$

Set $flag = 0, \hat{\eta} = \eta$;
Find $\mathcal{R}_i = \{k | C_{DF}(s_i, r_k, d_i) > \min_{j \in S} C(s_i, d_i), k \in \mathcal{R}\}$;
if $(C_{DF}(s_i, r_k, d_i) > \min_{j \in S} C(s_i, d_i), \exists k \in \mathcal{R})$ **then**
    if $(\eta_{i,j} = 1, \exists j \in \mathcal{R})$ **then**
        Add SD pairs to the *j*th relay, and update η;
    end if
    Mark the *j*th relay;
    for $k = 1 : |\mathcal{R}_i|$ **do**
        if (r_k is not marked) **then**
            Mark the relay, r_k;
            if (s_i can be assigned to r_k) **then**
                Update η, set $flag = 1$, break;
            else
                Find $S_k = \{j | \eta_{j,k} = 1, \eta_{i,k} = 1, j \neq i, j \in S\}$
                for $j = 1 : |S_k|$ **do**
                    $(\bar{\eta}, flag) = new\_relay\_search(s_j, \eta)$;
                    if ($flag == 1$) **then**
                        $\hat{\eta} = \bar{\eta}$, and break;
                    end if
                end for
            end if
        end if
        if ($flag == 0$) **then**
            Add SD pairs to the *j*th relay, and update η;
            Set $\hat{\eta} = \eta$;
        end if
    end for
end if

8.2.2.1 Performance Evaluation

Simulations are set up to evaluate the performance of the proposed GRA algorithm in green relay assisted D2D communications. We consider a wireless network in which SD pairs are randomly distributed in a $1000\,m \times 1000\,m$ square area. In the simulation, we adopt two green relay deployments as shown in Figure 8.10. The first deployment is utilized to compare the performance of the proposed GRA algorithm, the brute-force search, and the optimal relay assignment (ORA) algorithm [32]. In this relay deployment, we only consider four green rely nodes to reduce the runtime of the brute-force search. The second deployment is utilized to evaluate the performance of the GRA

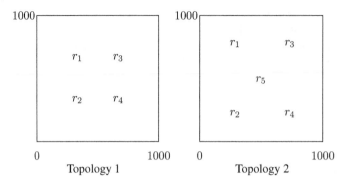

Topology 1   Topology 2

Figure 8.10 The relay locations for the simulations. *Source:* Han 2013 [189]. Reproduced with permission of IEEE.

algorithm versus different numbers of users and various green load capacities. In the simulations, the transmission power of the source node and the green relay node are $500\,mW$ and $4\,W$, respectively. The path loss exponent $\alpha = 4$, the available bandwidth for each channel $W = 10\,MHz$, and the white noise $N_0 = 10^{-10}$. We assume $\lambda_i l_i = 0.4\,Mbps$, $\forall i \in S$.

Figure 8.11 compares the minimum data rate of the SD pairs under the proposed GRA algorithm, the brute-force search, and the ORA algorithm [32]. We assume the green load capacity is 1 for all the green relay nodes. As shown in the figure, the proposed GRA algorithm approximates the optimal solution resulting from the brute-force search. One

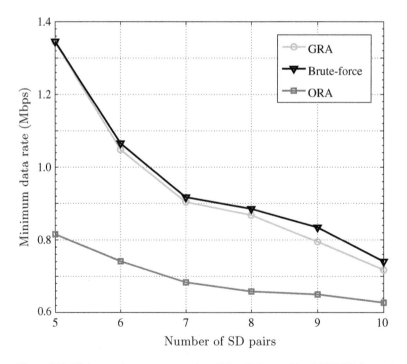

Figure 8.11 Minimum data rate comparison ($M = 4$). *Source:* Han 2013 [189]. Reproduced with permission of IEEE.

reason for the performance gap between the GRA algorithm and the brute-force search is that during the traffic offloading procedure, the GRA algorithm may not correctly offload traffic to enable a better relay assignment. For example, as shown in Figure 8.9, when GRA searches for a better relay assignment for s_1, r_1 may offload traffic from r_2 and r_3. Suppose either s_3 or s_6 can be offloaded to r_1 and offloading s_6 to r_1 will enable a better relay assignment for s_1. However, since GRA offloads the source node sequentially, s_3 is offloaded to r_1 which disables s_1 from obtaining a better relay assignment. The minimum data rate under the ORA algorithm is lower than that of under the GRA algorithm because the ORA algorithm only considers assigning one relay node to only one SD pair.

Figures 8.12 and 8.13 compare the minimum data rates of the SD pairs and the total data rates of the network under the GRA algorithm and the ORA algorithm, respectively. In the simulation, we assume the green load capacity is 1 for all green relay nodes. As shown in Figure 8.12, as the number of SD pairs increases, the minimum data rates under both algorithms converge to the same value. This is because as the number of SD pairs increases, the traffic loads on the relay nodes become heavier. Owing to the heavy traffic, transmitting via relay nodes may not increase the SD pairs' data rates. Therefore, the minimum data rate of the SD pairs is the direct transmission data rate of a certain SD pair. Thus, both GRA and ORA result in the same minimum data rate. However, owing to allowing a relay node to be assigned to multiple SD pairs, as shown in Figure 8.13, the total data rate of the network under GRA is much higher than that of the network under ORA. Figure 8.14 shows the minimum data rate of the SD pairs versus the green load capacity of the relay nodes. In the simulation, $M = 5$ and $N = 15$. We assume all the

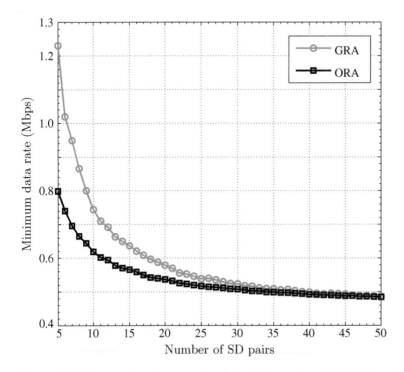

Figure 8.12 Minimum data rate comparison ($M = 5$). *Source:* Han 2013 [189]. Reproduced with permission of IEEE.

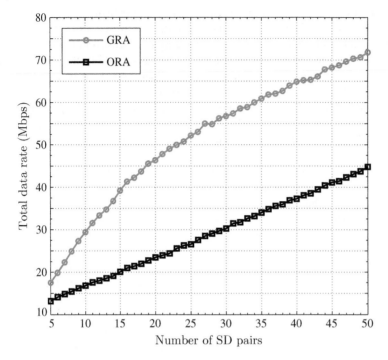

Figure 8.13 Total data rate comparison ($M = 5$). *Source:* Han 2013 [189]. Reproduced with permission of IEEE.

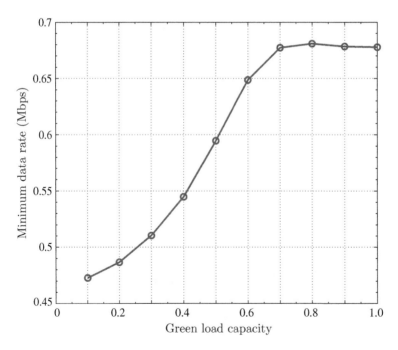

Figure 8.14 Minimum data rate vs. green load capacity. *Source:* Han 2013 [189]. Reproduced with permission of IEEE.

relay nodes have the same green load capacity. As shown in the figure, as the green load capacity increases, the minimum data rate of the SD pairs increases. However, when the green load capacity is larger than 0.7, the minimum data rate of the SD pair almost keeps constant. This indicates that the green load capacity of the green relay nodes no longer limits the performance of the network in terms of the minimum data rate.

8.3 Green Content Brokerage

By leveraging D2D communications, we now design a novel mobile traffic offloading scheme—content brokerage. In the content brokerage scheme, a new network node called the green content broker is introduced to arrange content delivery between the content requester and the content owner. The green content broker is powered by green energy, for example, solar energy, to reduce the CO_2 footprints of the network. In the scheme, maximizing traffic offloading with the constraints of the amount of green energy and bandwidth is an NP-hard problem. We propose a HTO algorithm to approximate the optimal solution with low computational complexity. Our simulation results validate the performance of the content brokerage scheme and the HTO algorithm.

In designing the green content brokerage scheme, we consider a cellular network consisting of one MBS, N UEs and one GCB. Define $\mathcal{U} = (u_1, u_2, \cdots, u_N)$ as the set of UEs. It is assumed that orthogonal channels are available in the network, for example, OFDMA, to avoid interference [32] among the MBS, the GCB and UEs. It is assumed that UEs and the GCB can transmit and receive simultaneously. The time horizon is divided into multiple content slots with duration of τ seconds. One content slot contains L transmission frames, which can be further divided into multiple uplink and downlink subframes. The total number of uplink and downlink subframes in one transmission frame are defined as N^u_{max} and N^d_{max}, respectively. The bandwidth allocated to each subframe is w. The content slot structure is shown in Figure 8.15.

At the beginning of the kth content slot, the GCB collects the UEs' content requests and registrations which are submitted in the $(k-1)$th content slot. Here, it is assumed that the UE's content requests are postponed at most one content slot for seeking traffic offloading opportunities. In other words, the UEs' content requests in the $(k-1)$th content slot either are fulfilled by the GCB in the kth content slot or are served by the MBS.

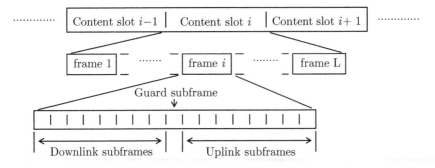

Figure 8.15 The content slot structure. *Source:* Han 2014 [199]. Reproduced with permission of IEEE.

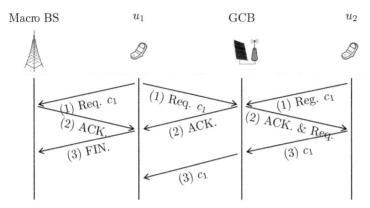

Figure 8.16 The communications procedure of the content brokerage scheme. *Source:* Han 2014 [199]. Reproduced with permission of IEEE.

The content request indicates which content a UE is retrieving while the content registration shows which content is available in the UE. Then, the GCB tries to match the content requests and registrations in offloading traffic from the MBS. If a UE's content request can be fulfilled locally, the GCB sends an acknowledgment message to inform the UE; otherwise, the GCB sends a FIN message to terminate the connection with the UE. Upon receiving the acknowledgment, the UE sends a FIN message to the MBS to cancel the content request. Meanwhile, the GCB retrieves the requested content from the UE which registers the content and delivers it to the requesting UE. Figure 8.16 shows an example to illustrate the communications procedures in the content brokerage scheme. At the content slot, u_1 requests the content c_1 from the MBS and the GCB. Meanwhile, u_2 registers the content c_1. Then, since c_1 is locally available, the GCB sends an acknowledgment message to u_1 to indicate that its content request can be fulfilled locally. Upon receiving the acknowledgment, u_1 cancels the content request from the MBS by sending a FIN message. The GCB acknowledges u_2's content registration and retrieves the content c_1 from it. Then, the GCB forwards the content c_1 to u_1.

The content brokerage scheme introduces additional signaling overhead. However, these additional signaling and control messages can be integrated with the control messages required for establishing D2D networks. Forming a D2D network usually involves processes such as device discovery, device association, and channel allocation. The content request and registration information can be integrated within the device discovery and association messages. For example, when a mobile device is associating with the D2D network, it informs the GCB which content is to be cached in the device and which content is to be requested by the device. In this way, the control and signaling overhead of the content brokerage scheme will be reduced and will not significantly impact the efficiency of D2D communications.

We assume that there are M items of content in the network, and define $C = (c_1, c_2, \cdots, c_M)$ and $S = (s_1, s_2, \cdots, s_M)$ as the sets of content and their sizes, respectively. It is assumed a UE can only request one content in a content slot, but can register multiple items of content in a content slot. Define $\eta_{i,j}(k)$ and $\varphi_{i,j}(k)$ as a content request indicator and a content registration indicator, respectively. If $\eta_{i,j}(k) = 1$, it indicates the ith UE requests the jth content item in the kth content slot. Since a UE can only request one item of content in a content slot, $\sum_{j \in C} \eta_{i,j}(k) \leq 1$. If the ith UE registers the jth item of content in the kth content slot, $\varphi_{i,j}(k) = 1$; otherwise, $\varphi_{i,j}(k) = 0$.

Define p^c as the transmission power of the GCB, and $h_i^c(k)$ as the channel gain between the ith UE and the GCB in the kth content slot. Here, it is assumed that UEs' channel gains do not change within a content slot. The downlink data rate of a subframe for the ith UE associated with the GCB is

$$r_i^c(k) = w \log_2 \left(1 + \frac{p^c h_i^c(k)}{w \sigma^2} \right). \tag{8.25}$$

Here, σ^2 is the thermal noise density. Supposing the ith UE requests the jth item of content in the kth content slot, the number of subframes required in the GCB to serve the content request is

$$n_i^c(k) = \frac{s_j}{r_i^c(k)}. \tag{8.26}$$

Define p^m as the transmission power of the MBS, and $h_i^m(k)$ as the channel gain between the ith UE and the MBS in the kth content slot. The downlink data rate of a subframe for the ith UE associated with the MBS is

$$r_i^m(k) = w \log_2 \left(1 + \frac{p^m h_i^m(k)}{w \sigma^2} \right). \tag{8.27}$$

The number of subframes in the MBS required to serve the ith UE's content request is

$$n_i^m(k) = \frac{s_j}{r_i^m(k)}. \tag{8.28}$$

The GCB's energy consumption consists of two parts: static power consumption and dynamic power consumption [140]. Static power consumption is the power consumption of the GCB without any traffic load. Dynamic power consumption refers to the additional power consumption caused by traffic loads in the GCB, which can be well approximated by a linear function of the traffic load [140]. Since a higher traffic load requires a large number of subframes, the energy consumption of the GCB is modeled as a linear function of the number of transmitting subframes [200]. The more subframes that the GCB is transmitting, the higher the energy consumption of the GCB. Then, the energy consumption of the GCB in the kth content slot can be expressed as

$$C^c(k) = \alpha^c \sum_{j \in C} (\gamma_j(k) \max_{i \in U} \{ n_i^c(k) \eta_{i,j}(k) \}) + C_s^c. \tag{8.29}$$

Here, α^c is a linear coefficient which reflects the relationship between the number of transmitting subframes and the GCB's dynamic energy consumption, and C_s^c is the GCB's static energy consumption. $\gamma_j(k)$ is a content serving indicator. If the jth content is served by the GCB in the kth content slot, $\gamma_j(k) = 1$; otherwise, $\gamma_j(k) = 0$.

The same energy consumption model is adopted for the MBS. Define α^m and C_s^m as the linear coefficient and the static energy consumption of the MBS, respectively. The MBS consumes energy in two cases. For the first case, the MBS consumes energy to deliver content to the UEs which are not served by the GCB. For the second case, the MBS consumes energy to deliver the requested content to the GCB. In this case, a UE is served by the GCB, but the requested content is not registered by the other UEs. Thus, the GCB retrieves the content from the MBS. Considering both cases, the MBS's energy consumption is

$$C^m(k) = \alpha^m \sum_{j \in C} \left[(1 - \gamma_j(k)) \max_{i \in \mathcal{U}} \{ n_i^m(k) \eta_{i,j}(k) \} \right.$$
$$\left. + \mu_j(k) \frac{s_j}{r_c^m} \right] + C_s^m. \tag{8.30}$$

Here, $\mu_j(k)$ indicates whether the GCB retrieves content j from the MBS. If the GCB retrieves the jth item of content from the MBS, $\mu_j(k) = 1$; otherwise, $\mu_j(k) = 0$. r_c^m is the per subframe data rate for the transmission from the MBS to the GCB, and

$$r_c^m = w \log_2 \left(1 + \frac{p^m h_c^m}{w\sigma^2} \right). \tag{8.31}$$

Here, h_c^m is the channel fading between the MBS and the GCB.

In this chapter, it is assumed that the GCB is powered by green energy generated by its solar panel. Define $e(k)$, $b^{\max}$, and $b^r(k)$ as the energy generation rate, the maximum battery capacity, and the residual energy storage in the kth content slot, respectively. The GCB's available green energy in the kth content slot is

$$b(k) = \min\{e(k)\tau + b^r(k), b^{\max}\}. \tag{8.32}$$

Here,

$$b^r(k) = \min\{e(k-1)\tau + b^r(k-1) - C^c(k-1), b^{\max}\}. \tag{8.33}$$

8.3.1 Problem Formulation and Analysis

The content brokerage scheme aims to maximize traffic offloading, and thus reduce the energy consumption of the MBS. Since the GCB is powered by green energy, the maximum amount of traffic offloading is constrained by the available green energy. It is assumed that the number of downlink subframes in each transmission frame within the same content slot is the same. Define the GCB's capacity as the number of downlink subframes that can be supported in a content slot by green energy. Define $N_{\max}^c(k)$ as the GCB's capacity in the kth content slot.

$$N_{\max}^c(k) = \min \left\{ \frac{C^c(k) - C_s^c}{\alpha^c}, N_{\max}^d L \right\}. \tag{8.34}$$

Define p^u as the transmission power of the UE, and $h_i^u(k)$ as the channel gain between the ith UE and the GCB. The data rate of the ith UE in the uplink during the kth content slot is

$$r_i^u(k) = w \log_2 \left(1 + \frac{p^u h_i^u(k)}{w\sigma^2} \right). \tag{8.35}$$

The number of subframes required for the ith UE to upload the jth item of content to the GCB is

$$n_{i,j}^u(k) = \frac{s_j}{r_i^u(k)}. \tag{8.36}$$

If the ith UE does not have the jth content, $n_{i,j}^u(k) = \infty$. According to the system model, maximizing the traffic offloading is equivalent to minimizing the energy consumption of the MBS. Thus, the TOM problem can be formulated as:

$$\min_{\gamma(k),\beta(k)} \; C^m(k) \tag{8.37}$$

$$s.t.: \; \sum_{j \in C} \beta_{i,j}(k) \le 1, \forall i \in \mathcal{U},$$

$$\sum_{i \in \mathcal{U}} \beta_{i,j}(k) + \mu_j(k) \le 1, \forall j \in C,$$

$$\sum_{j \in C} \gamma_j(k) \max_{i \in \mathcal{U}} \left\{ n_i^c(k) n_{i,j}(k) \right\} \le N_{\max}^c(k),$$

$$\sum_{j \in C} \gamma_j(k) \left(\sum_{i \in \mathcal{U}} \beta_{i,j}(k) n_i^u(k) + \mu_j(k) \frac{s_j}{r_c^m} \right)$$
$$\le N_{\max}^u L. \tag{8.38}$$

Here, $\gamma(k) = (\gamma_1(k), \cdots, \gamma_j(k), \cdots \gamma_M(k))$.

$$\beta(k) = \begin{pmatrix} \beta_{1,1}(k), & \cdots, & \beta_{1,M}(k) \\ \vdots, & \ddots, & \vdots \\ \beta_{N,1}(k), & \cdots, & \beta_{N,M}(k) \end{pmatrix}. \tag{8.39}$$

$\beta_{i,j}(k)$ is a content upload indicator. If the ith UE uploads content s_j to the GCB in the kth content slot, $\beta_{i,j}(k) = 1$; otherwise, $\beta_{i,j}(k) = 0$. The TOM problem consists of four constraints. The first constraint is that in one content slot, a UE can only upload one item of content. The second constraint is that one item of content can be uploaded by either a UE or the MBS. The third constraint is to guarantee the total number of required subframes to be less than the GCB's capacity. The fourth constraint is to make sure that the total number of required uplink subframes is less than the maximum number of the uplink subframes.

To solve the TOM problem, two parameters should be determined. The first parameter is $\gamma(k)$, which determines which content is served by the GCB. This problem is referred to as the SCS problem. Given its capacity and the number of available uplink subframes, the GCB may not be able to serve all the requested content. Different content may require different resources in terms of uplink and downlink subframes from the GCB. To minimize the MBS's energy consumption, the GCB should select and serve a subset of the requested content. The second parameter, $\beta(k)$, determines which UEs are selected to upload the requested content items. This problem is referred to as the COS problem. Different UEs may have different uplink data rates. Thus, for a given content item, selecting a different UE to upload the content results in a different requirement on the number of uplink subframes. Therefore, the solutions of the SCS problem and the SOC problem are interdependent.

Theorem 8.3.1   *The TOM problem is an NP-hard problem.*

Proof: The theorem is proved by reducing any instance of the two dimensional knapsack problem to a simple case of the TOM problem. The two dimensional knapsack problem is a well known NP-hard problem.

In the simple case of the TOM problem, it is assumed that each item of content has only one content owner in a content slot. In other words, in this simple case, if the serving content is selected, the corresponding UE which uploads the content to the GCB is determined. Thus, solving the simple case of the TOM problem is equivalent to determining the content to serve. The GCB consumes two resources in serving a content item: the uplink subframes and the downlink subframes. Since the total number of uplink and downlink subframes is constrained, the simple case of the TOM problem is actually to offload as many content items as possible to minimize the energy consumption of the MBS within the constraints of the number of uplink and downlink subframes. Therefore, the simple case of the TOM problem can be translated into a two dimensional knapsack problem. In other words, any instance of the two dimensional knapsack problem can be reduced to a simple case of the TOM problem. □

8.3.2   The Heuristic Traffic Offloading Algorithm

This section presents an HTO algorithm to approximate the optimal solution of the TOM problem. The HTO algorithm iteratively solves the COS problem and the SCS problem, and thus subsequently solves the TOM problem.

8.3.2.1   Solving the COS Problem

In the content brokerage scheme, a UE may register multiple content items to the GCB, and a content item may be registered by multiple UEs. However, in a content slot, a UE can only upload one item of content to the GCB, and a content item can only be retrieved from one content owner by the GCB. From the MBS's point of view, delivering content to different UEs may consume different amounts of energy because (1) the UEs may experience different channel fading, and (2) the size of the content requested by different UEs may be different. Therefore, serving different content in the GCB can result in different amounts of energy saving in the MBS. Define $C_j^s(k)$ as the MBS's energy savings by offloading the jth content item to the GCB. Then,

$$C_j^s(k) = \alpha^m \left(\max_{i \in \mathcal{U}} \left\{ n_i^m(k)\eta_{i,j}(k) \right\} - \mu_j(k)\frac{s_j}{r_c^m} \right). \tag{8.40}$$

UEs may experience different channel conditions when uploading content to the GCB. Thus, as the content owners, different UEs require different numbers of uplink subframes from the GCB. The number of uplink subframes required by the ith UE to upload the jth content item to the GCB is shown in Eq. (8.36). The per uplink subframe energy savings achieved by selecting the ith UE to upload the jth content is

$$\delta_{i,j}(k) = \frac{C_j^s(k)}{n_{i,j}^u(k)}. \tag{8.41}$$

Since the total number of uplink subframes in the GCB is constrained, in order to optimize the utilization of the uplink subframes in terms of maximizing the energy savings,

Figure 8.17 An illustration of the COS problem. *Source:* Han 2014 [199]. Reproduced with permission of IEEE.

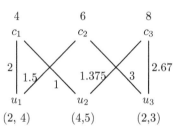

the GCB aims to maximize the summation of the per uplink subframe energy savings of all content served. Figure 8.17 shows an example of the COS problem. There are three content items in the network, and the MBS's energy savings by offloading these to the GCB are 4, 6, and 8 units, respectively. In the figure, if a UE registers a content item, then the UE is connected to that content with an edge. u_1 registers c_1 and c_2, and requires 2 and 4 uplink subframes to upload c_1 and c_2, respectively. u_2 may upload c_1 and c_3 at costs of 4 and 5 uplink subframes, respectively; and u_3 may upload c_2 and c_3 at costs of 2 and 3 uplink subframes, respectively. The weights on the edges are the per subframe energy savings for the content if it is uploaded by the connected UE. For instance, the weight of the edge between c_1 and u_1 is 2, which indicates that the per uplink subframe energy saving is 2 if u_1 uploads c_1 to the GCB. The optimal solution of the COS problem is that u_1 uploads c_1, u_2 uploads c_3, and u_3 uploads c_2. In this case, the total numbers of required uplink subframes, which is 9, is also minimized. Without solving the COS problem, it may require more uplink subframes to serve this content, which may violate the constraints of the total number of uplink subframes.

The COS algorithm translates the COS problem into a maximum weight bipartite matching problem, and solves the COS problem with the maximum weight bipartite matching (MWBM) algorithm [201]. Define $C^r = (j| \sum_{i \in \mathcal{U}} \eta_{i,j}(k) > 0, \forall j \in C)$ as the set of requested content in the kth content slot. The bipartite graph is built as shown in Figure 8.18. The requested content, the UES, and the MBS are the vertices of the graph; if u_i registers c_j, an edge is added between u_i and c_j. Since the MBS is able to deliver any content to the GCB, edges are added between the MBS and every content item. u_0 denotes the MBS during the graph construction. The weights of the edges are calculated according to Eq. (8.41). The pseudo code of the COS algorithm is shown in Algorithm 17.

8.3.2.2 The HTO Algorithm

Owing to the constraints of green energy and the number of uplink and downlink subframes, the GCB may not be able to serve all content requests. To maximize the MBS's energy savings, the GCB selects a subset of content requests to serve. As analyzed in Section 8.3.1, when the content owners are selected, the SCS problem can be transformed into a two dimensional knapsack problem. Therefore, the SCS problem is addressed

Figure 8.18 The formation of the bipartite graph. *Source:* Han 2014 [199]. Reproduced with permission of IEEE.

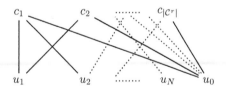

Algorithm 17: The COS Algorithm

Input : $\eta_{i,j}(k), \varphi_{i,j}(k), s_j, n_i^m(k), r_c^m, \forall j \in C, \forall i \in \mathcal{U}$;
Output: $\beta(k), \mu_j(k), \forall j \in C^r$;

1 Construct a set of $|C^r|$ vertices corresponding to C^r;
2 Construct a set of $N + 1$ vertices corresponding to $\mathcal{U} \cup \{u_0\}$;
3 Construct a set ε of edges, where $(u_i, c_j) \in \varepsilon$ if $\varphi_{i,j}(k) = 1$ or $i = 0$;
4 **for** $(u_i, c_j) \in \varepsilon, \forall u_i \in \mathcal{U} \cup \{u_0\}, \forall c_j \in C$ **do**
5   **if** $i = 0$ **then**
6          $\delta_{0,j}(k) = \frac{s_j}{r_c^m}$;
7   **else**
8   Calculate $\delta_{i,j}(k)$;
9 Apply an MWBM algorithm to find the maximum weight matching of the constructed graph;
10 Derive $\beta(k)$ and $\mu_j(k), \forall j \in C^r$.

---

based on Toyoda's primal effective gradient method in solving the multi-dimensional knapsack problem [202].

Given $\beta(k)$ and $\mu_j(k)$, the energy savings by offloading the jth content item is calculated based on Eq. (8.40), and the number of required uplink and downlink subframes for serving the jth content item is derived. The number of required uplink and downlink subframes for the jth content item in the kth content slot is normalized with respect to the total number of uplink and downlink subframes. Define $\psi_j^u(k)$ and $\psi_j^d(k)$ as the normalized number of required uplink and downlink subframes for the jth content items, respectively. Then,

$$\psi_j^u(k) = \frac{\mu_j(k) \sum_{i \in \mathcal{U}} \beta_{i,j}(k) n_{i,j}^u(k) + (1 - \mu_j(k)) \frac{s_j}{r_c^m}}{N_{\max}^u L} \tag{8.42}$$

and

$$\psi_j^d(k) = \frac{\max_{i \in \mathcal{U}} \{n_i^m(k) \eta_{i,j}(k)\}}{N_{\max}^d L}. \tag{8.43}$$

The SCS problem can be expressed as:

$$\max_{\gamma(k)} C_j^s(k) \gamma_j(k) \tag{8.44}$$

$$s.t. : \sum_{j \in C} \psi_j^u(k) \gamma_j(k) \leq 1;$$

$$\sum_{j \in C} \psi_j^d(k) \gamma_j(k) \leq 1. \tag{8.45}$$

Since serving a content item consumes both uplink and downlink subframes, it is diffi-cult to evaluate which content consumes less resources in terms of uplink and down-link subframes. In other words, it is difficult to determine which content to offload in order to yield the most profit in terms of energy savings of constrained resources (the uplink and downlink subframes). The evaluation of the resource consumption of content depends on the current status of resource consumption. For example, assume $\psi_j^u(k) = 0.4$, $\psi_j^d(k) = 0.1$, $\psi_i^u(k) = 0.1$, and $\psi_i^d(k) = 0.4$. In this case, the GCB requires more uplink subframes but less downlink subframes in serving the jth content item than in serving the ith content item. Without the current status of the usage of the uplink and downlink subframes, it is not possible to compare the resource consumption of the jth and ith content items. Define C^* as the set of content selected to be served by the GCB. Then, $\sum_{j \in C^*} \psi_j^u(k)$ and $\sum_{j \in C^*} \psi_j^d(k)$ represent the normalized usage of the uplink and downlink subframes, respectively. In order to compare the resource consumption of the different content items, $\sum_{j \in C^*} \psi_j^u(k)$ and $\sum_{j \in C^*} \psi_j^d(k))$ are defined as the penalty coefficients of the usage of the uplink and downlin subframes, respectively. Then, the aggregate resource requirement on serving the jth content item is defined as

$$\chi_j(k) = \frac{\psi_j^u(k) \sum_{i \in C^*} \psi_i^u(k) + \psi_j^d(k) \sum_{i \in C^*} \psi_i^d(k)}{[(\sum_{i \in C^*} \psi_i^u(k))^2 + (\sum_{i \in C^*} \psi_i^d(k))^2]^{\frac{1}{2}}}. \tag{8.46}$$

$\chi_j(k)$ reflects the additional burden introduced by serving the jth content item. For example, assuming $\sum_{j \in C^*} \psi_j^u(k) = 0.2$ and $\sum_{j \in C^*} \psi_j^d(k) = 0.6$, then $\chi_j(k) = 0.22$ and $\chi_i(k) = 0.41$. This indicates that serving the jth content item introduces less burden on the GCB than serving the ith content item. This result is reasonable because the current usage of the downlink subframes is greater than that of the uplink subframes. In other words, the GCB has fewer available downlink subframes than uplink subframes. The ith content item consumes many more downlink subframes than the jth content item does, and therefore, serving the ith content item introduces more burden on the GCB than serving the jth content item.

To further determine whether the jth content item should be served by the GCB, the per unit resource profit of the jth content item (in terms of reducing the MBS's energy consumption) is calculated as $\frac{C_j^s(k)}{\chi_j(k)}$. The HTO algorithm iteratively adds the content item with the largest per unit resource profit into C^*. Define

$$C^f = (j | \psi_j^u(k) \leq 1 - \sum_{i \in C^*} \psi_i^u(k),$$

$$\psi_j^d(k) \leq 1 - \sum_{i \in C^*} \psi_i^d(k), \forall j \in C) \tag{8.47}$$

as the set of feasible content requests which can be served by the GCB. The pseudo code of the HTO algorithm is shown in Algorithm 18.

In the HTO algorithm, *opt_flag* indicates the termination of the algorithm. When $C^f(k) = \emptyset$, *opt_flag* is set to 1, and the HTO algorithm is terminated. *loop_flag* indicates whether an optimal content owner selection has been found for a set of requested con-tent. When $C^r(k) = C^f(k)$, the HTO algorithm sets *loop_flag* = 0, and terminates the loop for computing $\beta(k)$ and $\mu_j(k)$, $\forall j \in C^r$. If $C^r(k) \neq C^f(k)$, it indicates that some of

---

Algorithm 18: The HTO Algorithm

Input : $\eta_{i,j}(k), \varphi_{i,j}(k), s_j, n_i^m(k), r_c^m, \forall j \in C, \forall i \in \mathcal{U}$;
Output: $C^*(k), \beta(k)$, and $\mu_j(k) \; \forall j \in C$;

1 Set $C^* = \emptyset$;
2 **while** *opt_flag equals to 0* **do**
3   Set *loop_flag* $= 1$;
4   **while** *loop_flag equals to 1* **do**
5   Derive $\beta(k)$ and $\mu_j(k)$ using the COS algorithm ;
6   Calculate $\psi_j^u(k), \psi_j^d(k), C^r(k)$ and $C^f(k)$;
7   **if** $C^r(k) \neq C^f(k)$ **then**
8   Set $\eta_{i,j}(k) = 0, \forall i \in \mathcal{U}, \; \forall j \in C^r(k) \setminus C^f(k)$;
9   **else**
10   Set *loop_flag* $= 0$;
11   **if** $C^f(k) \neq \emptyset$ **then**
12   **if** C^* *is empty* **then**
13   Calculate $\chi_j(k) = \dfrac{\psi_j^u(k) + \psi_j^d(k)}{2^{\frac{1}{2}}}$ and $\dfrac{C_j^s(k)}{\chi_j(k)}, \forall j \in C^f(k)$;
14   **else**
15   Calculate $\chi_j(k)$ based on Eq. (8.46) and $\dfrac{C_j^s(k)}{\chi_j(k)}, \forall j \in C^f(k)$;
16   Find $j = \arg\max_{j \in C^f(k)} \dfrac{C_j^s(k)}{\chi_j(k)}$;
17   Set $C^*(k) = C^*(k) \cup \{j\}$ and $\eta_{i,j}(k) = 0, \forall i \in \mathcal{U}$;
18   Find i that satisfies $\beta_{i,j}(k) = 1$ and set $\varphi_{i,j}(k) = 0, \; \forall j \in C$;
19   **else**
20   Set *opt_flag* $= 1$;
21 Calculate and Return $C^*(k), \beta(k)$, and $\mu_j(k) \; \forall j \in C$.

---

the requested content cannot be served by the GCB because of the lack of uplink and downlink subframes. In this case, the HTO algorithm ignores such content requests by setting $\eta_{i,j}(k) = 0, \forall j \in C^r(k) \setminus C^f(k), \forall i \in \mathcal{U}$. Then, the HTO algorithm calls the COS algorithm to recalculate $\beta(k)$ and $\mu_j(k), \; \forall j \in C^r$. This loop of the algorithm guarantees the maximization of the sum of the per uplink subframe energy saving for all the content being served. For instance, as shown in Figure 8.19, the energy savings of offloading c_1 and c_2 are 4 and 6, respectively. u_1 registers both content items at the costs of 2 and 4 uplink subframes, respectively. u_2 only registers c_1 at the cost of 3 uplink subframes. If both requests for c_1 and c_3 are feasible content requests, then the optimal content owner selection (in terms of maximizing the summation of the per uplink subframe energy savings) is that u_1 uploads c_2 while u_2 uploads c_1. However, if the content request for c_2 is not feasible because of a lack of uplink subframes, then the content request for c_1 is the

Figure 8.19 An illustration of the optimization. *Source:* Han 2014 [199]. Reproduced with permission of IEEE.

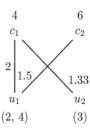

only feasible content request. In this case, u_1 rather than u_2 is the optimal content owner of c_1.

Theorem 8.3.2 *Ignoring infeasible content requests and redoing content owner selection do not decrease the sum of the per uplink subframe energy savings of feasible content items.*

Proof: Assume $C^r(k) \setminus C^f(k) = \{h\}$. Define $\bar{\eta}_{i,j}(k)$ and $\hat{\eta}_{i,j}(k)$ as the content requests for the content in $C^r(k)$ and $C^f(k)$, respectively.

$$\hat{\eta}_{i,j}(k) = \begin{cases} \eta_{i,j}(k), & j \in C^f; \\ 0, & j = h. \end{cases} \tag{8.48}$$

Define $\bar{\beta}(k)$ and $\bar{\mu}_j(k)$, $\forall j \in C$ as the optimization results of the COS algorithm with $\bar{\eta}_{i,j}(k)$ as the input, and define $\hat{\beta}(k)$ and $\hat{\mu}_j(k)$, $\forall j \in C$ as the optimization results of the COS algorithm with $\hat{\eta}_{i,j}(k)$ as the input. According to Eqs. (8.40) and (8.41), $\bar{C}^s_j(k)$, $\bar{\delta}_{i,j}(k)$, $\bar{n}^u_{i,j}(k)$, $\hat{C}^s_j(k)$, $\hat{\delta}_{i,j}(k)$, and $\hat{n}^u_{i,j}(k)$ are derived, respectively. Proving the theorem is equivalent to proving

$$\sum_{j \in C^f(k)} \sum_{i \in \mathcal{U}} \hat{\beta}_{i,j}(k) \hat{\delta}_{i,j}(k) \geq \sum_{j \in C^f(k)} \sum_{i \in \mathcal{U}} \bar{\beta}_{i,j}(k) \bar{\delta}_{i,j}(k).$$

1) If $\bar{\mu}_h(k) = 1$, the GCB retrieves the hth content from the MBS. Ignoring the content request for c_h does not change the content owner selections of the other content items, and thus

$$\sum_{j \in C^f(k)} \sum_{i \in \mathcal{U}} \hat{\beta}_{i,j}(k) \hat{\delta}_{i,j}(k) = \sum_{j \in C^f(k)} \sum_{i \in \mathcal{U}} \bar{\beta}_{i,j}(k) \bar{\delta}_{i,j}(k). \tag{8.49}$$

2) If $\bar{\mu}_h(k) = 0$ and $\bar{\beta}_{g,h}(k) = 1$, the GCB receives the content item from the gth UE. If

$$\bar{n}^u_{g,j}(k) < \sum_{i \in \mathcal{U}} \bar{\beta}_{i,j}(k) \bar{n}^u_{i,j}(k), \quad \exists j \in C^f, \tag{8.50}$$

then ignoring the request for the hth content item enables the gth UE to upload the jth content item, and thus

$$\sum_{j \in C^f(k)} \sum_{i \in \mathcal{U}} \hat{\beta}_{i,j}(k) \hat{\delta}_{i,j}(k) > \sum_{j \in C^f(k)} \sum_{i \in \mathcal{U}} \bar{\beta}_{i,j}(k) \bar{\delta}_{i,j}(k); \tag{8.51}$$

otherwise, ignoring the content request for c_h does not change the content owner selections of the other content items, and thus

$$\sum_{j\in C^f(k)}\sum_{i\in U}\hat{\beta}_{i,j}(k)\hat{\delta}_{i,j}(k) = \sum_{j\in C^f(k)}\sum_{i\in U}\bar{\beta}_{i,j}(k)\bar{\delta}_{i,j}(k). \tag{8.52}$$

Therefore,

$$\sum_{j\in C^f(k)}\sum_{i\in U}\hat{\beta}_{i,j}(k)\hat{\delta}_{i,j}(k) \geq \sum_{j\in C^f(k)}\sum_{i\in U}\bar{\beta}_{i,j}(k)\bar{\delta}_{i,j}(k). \tag{8.53}$$

The proof is complete. □

8.3.2.3 Computational Complexity

There are two loops in the HTO algorithm. The number of iterations of each loop in the worst case is M. The COS algorithm is called within the inner loop. The major computational complexity of the COS algorithm is from the MWBM algorithm which is called to find the maximum weighted match between the requested content and the content owner. The Kuhn-Munkres algorithm [201] is adopted to solve the maximum weight bipartite matching problem, that is, finding the maximum weight bipartite match for M content items and N UEs. The computational complexity of the Kuhn-Munkres algorithm for matching M content items and N UEs is $O((N+M)^3)$. Therefore, the overall computation complexity of the HTO algorithm is $O(M^2(N+M)^3)$.

8.3.3 Performance Evaluation

System level simulations are set up by using Matlab to evaluate the performance of the content brokerage scheme and the HTO algorithm. In the simulation, one MBS and one GCB are deployed, the distance between the MBS and the GCB is $700\,m$, and 100 UEs are randomly distributed around the GCB. The static power consumption of the MBS and the GCB are $100\,W$ and $20\,W$, respectively. For simplicity, the simulation sets $a^c = 2$ and $a^m = 20$. The SAM [148] and PVWatts model [149] are adopted to estimate the hourly solar energy generation, and b^{max} is set to $100\,W$. The total number of content items is 100, the content size is randomly distributed between $1\,MB$ and $10\,MB$, and the duration of a content slot is 5 seconds. COST 231 Walfisch-Ikegami [144] is adopted as the propagation model. The communications parameters in the simulation are summarized in Table 8.2.

The simulations compare the MBS's power consumption with and without the content brokerage scheme. Content delivery without content brokerage is defined as the traditional scheme. In order to evaluate the content brokerage scheme, two parameters, the content availability ratio and the content popularity ratio, are introduced. The content availability ratio is defined as the percentage of the requested content items that are stored in UEs while the content popularity ratio is defined as the percentage of the UEs that have the requested content items. In addition, to evaluate the HTO algorithm, its performance is compared with an energy saving greedy (ESG) algorithm. The ESG algorithm iteratively selects the requested content item which leads to the largest energy savings in the MBS, and assigns a content owner to that content. The ESG algorithm terminates when either the uplink or downlink subframes are exhausted.

Figure 8.20 shows the power consumption of the MBS over time. In this simulation, it is assumed that the total number of content requests in each content slot is 30. The UEs

Table 8.2 Simulation parameters.

Parameter	Value
MBS's transmit power	20 W
GCB's transmit power	2 W
UEs' transmit power	500 mW
Rayleigh fading	9 dB
Shadowing fading	5 dB
Antenna feeder loss	3 dB
Carrier frequency	2100 MHz
Transmitter gain	1 dB
Noise power level	−104 dBm
Receiver sensitivity	−97 dB

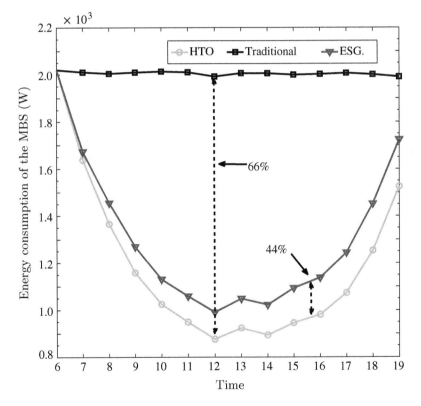

Figure 8.20 MBS energy consumption over time. *Source:* Han 2014 [199]. Reproduced with permission of IEEE.

are randomly selected to submit a content request. If selected, the UE randomly chooses and requests a content item. The total number of uplink and downlink subframes are both 50. The content availability ratio and popularity ratio are 0.9 and 0.5, respectively. In the figure, each marker represents the MBS's average power consumption for all the content slots within an hour. As shown in the figure, the content brokerage scheme achieves up to 66% energy savings in the MBS. The content brokerage scheme achieves the highest energy savings around noon. This is because the GCB is powered by solar energy, which peaks around noon. With a larger amount of green energy, the GCB is able to utilize more subframes, and thus can offload more traffic from the MBS. As compared with the ESG algorithm, the HTO algorithm achieves up to 14% energy savings in the MBS. This is because the HTO algorithm optimizes both the content owner selection and the selection of content served. As a result, with a given amount of resources (the uplink and downlink subframes), the HTO algorithm enables the GCB to offload more content.

Figures 8.21 and 8.22 show the performance of the content brokerage scheme at different content availability ratios. In this simulation, it is assumed that the total number of content requests in each content slot is 30. The total numbers of uplink and downlink subframes are both 40. The content popularity ratio is 0.5. To evaluate the impact of the content availability on the content brokerage scheme, it is assumed that the GCB has sufficient green energy to utilize all the subframes. Thus, the constraints of the GCB on offloading traffic are from the total numbers of subframes. Figure 8.21 shows

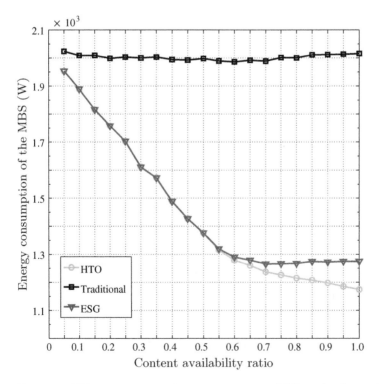

Figure 8.21 MBS energy consumption versus content availability ratios. *Source:* Han 2014 [199]. Reproduced with permission of IEEE.

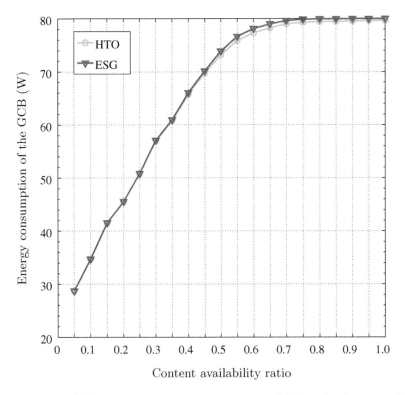

Figure 8.22 GCB energy consumption versus content availability ratios. *Source:* Han 2014 [199]. Reproduced with permission of IEEE.

the MBS's energy consumption with different content availability ratios. As shown in the figure, when the content availability ratio is small, the content brokerage scheme does not achieve much energy saving. However, as the content availability ratio increases, using the content brokerage scheme, the MBS's energy consumption reduces significantly. When the content availability ratio is larger than 0.5, the ESG algorithm is unable to further reduce MBS's energy consumption because of the limitations on total subframes. However, by optimizing content owner selection and serving content selection, the HTO algorithm is able to maximize the utilization of the uplink and downlink subframes, and thus achieves more energy savings than the ESG algorithm does. Figure 8.22 shows the GCB's energy consumption with the HTO algorithm and the ESG algorithm. The GCB's energy consumption under both algorithms is almost the same, implying that the number of subframes utilized by both algorithms is almost the same. Since the HTO algorithm achieves more energy savings, this indicates the HTO algorithm achieves higher resource utilization than the ESG algorithm does.

Figure 8.23 shows the MBS's energy consumption with different content popularity ratios. In this simulation, it is assumed that the total number of content requests in each content slot is 30. The total numbers of uplink and downlink subframes are both 30. The content availability ratio is 0.6. As shown in the figure, the content brokerage scheme significantly reduces the MBS's energy consumption. When the content popularity ratio is 0.01, the performance of the content brokerage scheme is limited by the number of content owners. Since the total number of UEs is 100, if the content popularity ratio

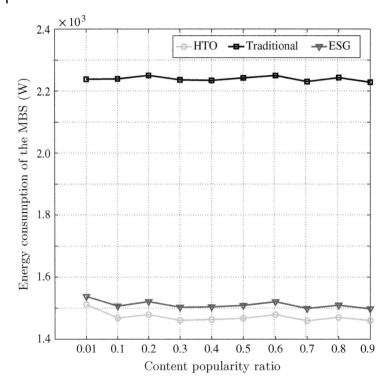

Figure 8.23 MBS energy consumption versus content popularity ratios. *Source:* Han 2014 [199]. Reproduced with permission of IEEE.

equals 0.01, this indicates that a requested content item is available in at most one UE. The UE may also hold other requested content; in this case content popularity to some extent impairs content availability because a UE can upload at most one content item to the GCB in a content slot. As the content popularity increases, such impairment decreases, and thus the performance of the brokerage scheme is almost stable.

8.4 Summary

This chapter presents three novel D2D proximity services-based energy efficient communications schemes: energy efficient cooperative wireless multicasting, green relay assisted D2D communications, and a green content brokerage scheme. A low complexity gradient guided algorithm is designed to implement the energy efficient cooperative wireless multicasting scheme. A heuristic green relay assignment algorithm is presented to maximize the minimum data rate of the source and destination pairs in green relay assisted D2D communications. The proposed GRA algorithm approximates the optimal solution with low computational complexity. In addition, GRA is green energy aware, and optimizes relay assignment under consideration of the green load capacity of the relay nodes. The green content brokerage scheme enables traffic offloading from a remote MBS to local UEs, and the HTO algorithm is proposed to maximize traffic offloading within the constraints of resources such as available green energy and the number of uplink and downlink subframes. As demonstrated via extensive simulations, the proposed content brokerage scheme together with the HTO algorithm significantly reduces MBS energy consumption.

8.5   Questions

8.1   How does D2D communications improve the energy efficiency of mobile networks?

8.2   What are the advantages and disadvantages of D2D communications in reducing energy consumption?

8.3   How does energy efficient cooperative multicasting work?

8.4   Is the relay assignment problem in the green relay assisted communications scheme NP-hard? If so, please prove this.

8.5   How does the green content brokerage scheme work?

9

Greening Mobile Networks via Optimizing the Efficiency of Content Delivery

The efficiency of content delivery in a mobile network has a significant impact on the energy consumption of the network. With the rapid development of radio access techniques and mobile devices, Internet applications are gradually migrating to mobile networks.

Mobile video applications generate the largest wireless data traffic volume. Different video applications behave differently in terms of bandwidth consumption. For example, applications such as Netflix and Hulu adopt adaptive HTTP streaming, which adjusts the bit rates according to the network conditions. Adaptive HTTP streaming usually consumes as much bandwidth as possible to maintain the best possible quality of the video. Video applications like YouTube, however, behave differently. They usually start at the lowest possible bit rate, and allow subscribers to choose the bit rate. Despite the difference in bandwidth usage, both categories of video application are bandwidth intensive, and their bandwidth consumption is growing rapidly owing to the availability of high definition video and the increased screen size of mobile devices. Therefore, accelerating mobile web and video content delivery is crucial to enhance the QoE of mobile network subscribers [14].

In this chapter, we present a comprehensive tutorial of mobile network optimization in terms of accelerating content delivery. We first investigate live network measurements and identify the network obstacles that dominate content delivery delays. Then, we classify existing content delivery acceleration solutions in mobile networks into three categories: mobile system evolution, content and network optimization, and mobile data offloading, and provide an overview of available solutions in each category. Finally, we survey content delivery acceleration solutions tailored for web content delivery and multimedia delivery. For web content delivery acceleration, we overview existing web content delivery systems and summarize their features. For multimedia delivery acceleration, we focus on accelerating HTTP-based adaptive streaming while briefly reviewing other multimedia delivery acceleration solutions. This chapter presents a timely survey on content delivery acceleration in mobile networks, and provides a comprehensive reference for further research in this field.

The rest of the chapter is organized as follows. Section 9.1 analyzes live network measurements and identifies the dominant obstacles that delay content delivery in mobile networks. Section 9.2 discusses the latency reduction achieved by mobile system evolution. Section 9.3 provides an overview of content and network optimization techniques.

Green Mobile Networks: A Networking Perspective, First Edition. Nirwan Ansari and Tao Han.
© 2017 John Wiley & Sons Ltd. Published 2017 by John Wiley & Sons Ltd.

Section 9.4 presents mobile data traffic offloading techniques. Section 9.5 discusses web content delivery acceleration systems for mobile networks. Section 9.6 presents multimedia delivery acceleration solutions. Section 9.7 summarizes the chapter and discusses several open issues related to content delivery acceleration in mobile networks.

9.1 Mobile Network Measurements

Understanding the performance of mobile networks and identifying performance bottlenecks are the first steps toward accelerating content delivery in mobile networks. In this section, we overview studies on mobile network measurement and generalize the dominant factors that deteriorate network performance with respect to content delivery delay.

9.1.1 Packet Retransmission

To alleviate the impact of wireless errors on network performance, mobile networks adopt two-layer retransmission mechanisms: MAC layer hybrid-ARQ (Automatic Repeat-reQuest) and RLC (Radio Link Control) layer ARQ [203]. Hybrid-ARQ located in NodeB targets fast retransmission and provides feedback on the decoding attempts to the transmitter after each transmission. Excessive feedback introduces additional cost to the mobile system. To keep a reasonable feedback overhead, hybrid-ARQ does not provide feedback on every decoding attempt that results in a hybrid-ARQ residual error. RLC layer ARQ located in the RNC (Radio Network Controller) requires relatively infrequent RLC status reports, and can achieve a low error rate with a small cost. However, RLC layer ARQ takes more packet recovery time than hybrid-ARQ. The two-layer retransmission architecture achieves fast retransmission attributed to hybrid-ARQ and reliable packet delivery facilitated by RLC layer ARQ.

These retransmission mechanisms, however, introduce delay spikes, and around 15% of the packets over the UMTS (Universal Mobile Telecommunications System) network are affected by delay spikes [204]. One reason for delay spikes is out of sequence delivery of transport blocks, PDUs (Protocol Data Units). Since the hybrid-ARQ processes operate independently, PDUs may be delivered out of sequence. In this case, the RLC layer ARQ, which assures in-sequence delivery, has to buffer all the out of order PDUs and reorder them, resulting in a delay spike. Another reason for delay spikes is residual errors in the hybrid-ARQ. To assure reliable delivery, RLC layer retransmissions are scheduled when hybrid-ARQ processes fail to deliver data units. RLC layer ARQ consumes more time than hybrid-ARQ processes, and thus causes delay spikes. Such delay spikes increase the round trip time (RTT) of the mobile network, which leads to long delays in IP packet services [205]. Long packet delays sometimes causes TCP RTOs (Retransmission Timeouts) that further deteriorate network performance [206].

9.1.2 Queuing in Mobile Core Networks

In a mobile core network, data services are supported by the packet switch domain, which consists of an SGSN (Serving GPRS Support Node) and a GGSN (Gateway GPRS Support Node). The SGSN performs mobility management, logical link control, and data unit routing while the GGSN acts as an IP gateway that connects the mobile core network with the WAN (Wide Area Network). Since the mobile core network connects the high speed Internet with the low speed radio network, a queuing delay in the mobile core network is unavoidable. The queuing delay is exacerbated by the large sending buffer in the SGSN. To accommodate the time varying wireless channel and improve channel utilization, the SGSN is usually equipped with large buffers, which result in long queuing delays when the wireless channel rate is low due to bad channel conditions or user mobility [207]. In addition, heavy traffic loads in a cell negatively impact the queuing delay [208]. As a result, the queuing delay in the SGSN dominates mobile core network latency [209]. Excessive queuing delays lead to RTT inflation, retransmission time inflation, SYN timeout, and higher recovery time, thus prolonging packet delivery time.

9.1.3 Network Asymmetry

As compared to the downlink, the uplink of mobile networks has smaller channel rates and is prone to wireless errors [210]. There are several reasons for this asymmetry. Unlike the downlink channel, the uplink channel is non-orthogonal and subject to interference between uplink transmissions. Also, with limited processing ability and transmission power, high order modulation is less useful. In addition, the uplink scheduler and the transmission buffers are located in different places: the former is located in NodeB while the latter is in UE. Such separation requires UE signal buffer status information to the scheduler since the wireless schedulers are very sensitive to the buffer status (application rates) [211]. The additional signaling processes may result in inefficient scheduling. Because of the weaknesses, the uplink performance limits the RTT of the networks [206], and UE with larger uplink traffic experiences longer RTT [212].

9.1.4 Queue Management

Owing to the differences between wireline networks and mobile networks, the queue management scheme designed for wireline networks, for example, RED (Random Early Detection), may face some problems in mobile networks [213]. RED monitors average queue length and drops packets based on statistical probabilities. If the average queue length is smaller than a predefined threshold, all incoming packets are accepted. Otherwise, incoming packets will be dropped based on statistical probabilities which increase as the queue grows. When the queue is full, all incoming packets will be dropped [214]. One problem is that a static minimum threshold[1] cannot effectively maintain the queue size. A static minimum threshold often fails to reflect the capacity of a time varying wireless channel. Moreover, the mobile network has a larger queue draining latency due to its lower bandwidth. As a result, if the mobile network experiences a temporal bandwidth shrink, a queue will build up, and is not easily drained. Another problem relates to the probabilistic discarding. Probabilistic discarding may lead to delayed congestion feedback. In light of the large queue draining delay, delayed feedback can result in over-buffering and excessive queuing delay. In addition, mobile networks keep a per user queue, and thus statistical multiplexing in the queue is lower since the queue is shared

---

[1] If the queue size is larger than the minimum threshold, the incoming packets will be dropped according to a predefined probability.

by at most 2–4 flows. The probability that discarded packets are from the same connection is higher. Therefore, connections that are not granted fair share of the bandwidth will experience longer delay.

9.1.5 First Packet Delay

In mobile networks, a subscriber's first packet tends to experience high jitter due to delay in state transition and delay related to network registration and resource allocation [215]. UEs may be in different RRC (Radio Resource Control) states, which are shown in Figure 9.1. UEs are either in the idle mode or connected mode when they are powered on. When UEs are in the connected mode, they may be in four different states: CELL_DCH (Cell Dedicated Channel), CELL_FACH (Cell Forward Access Channel), CELL_PCH (Cell Page channel), and URA_PCH (UTRAN Registration Area Paging Channel) [216]. The state transitions between different states have a direct impact on the latency. For instance, it usually takes 2–4 seconds to transit from CELL_FACH to CELL_DCH since it requires setting up the DCH (Dedicated Channel) channel [206]. As a result, the first packet tends to experience a longer delay. Another reason of first packet delay is related to network registration and resource allocation. To access the packet switch services, UEs have to undergo a three-phase activation process. The phases are GPRS Attach, PDP Content Activation and RAB Establishment, and Register with IMS [217]. After this process, wireless resources are allocated to the UE, and the UE can then access packet switch services. This activation process introduces additional latency for the first packet.

9.1.6 TCP Flaws

TCP is one of the core protocols of the Internet protocol suites. Many popular applications such as web services, HTTP-based streaming, email, and file transfer rely on TCP to provide reliable point to point communications. Therefore, the performance of TCP has a significant impact on the service delivery latency over wireless networks [218]. In this subsection, we discuss two TCP flaws: TCP slow start and TCP ACK compression, and their impacts on delivery latency.

9.1.6.1 TCP Slow Start
TCP utilizes slow start and congestion avoidance to probe the available bandwidth. At the beginning of one connection, the TCP sender sends data units according to a predefined CWND (Congestion WiNDow), and exponentially increases the CWND at every

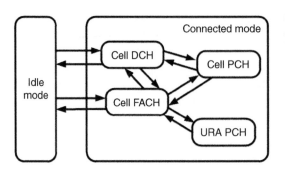

Figure 9.1 The RRC states in UMTS. *Source:* Han 2013 [14]. Reproduced with permission of IEEE.

ACK (acknowledgment) until reaching a predefined threshold. Then, TCP enters the congestion avoidance phase. The slow start of TCP has a significant impact on the user perceived service response time [219]. At the start up phase, the bandwidth is not fully utilized, and it takes several RTTs to ramp the congestion window up to the link BDP (Bandwidth Delay Product) [220]. Moreover, most of the TCP flows generated by popular mobile services, for example web services, are short lived and only experience the slow start phase [221]. Therefore, the user perceived network throughput is limited by the initial CWND rather than the network capacity. Such limitation deteriorates the service latency.

9.1.6.2 TCP ACK Compression

TCP ACK compression is a well known problem in wireless networks [222]. Owing to link asymmetry and in order delivery, several ACKs can be compressed. ACK compression disturbs the synchronization between the TCP sender and the receiver, and worsens network congestion, therefore increasing service response time [205].

9.1.7 Application Misbehavior

Applications optimized for the Internet may misbehave in the wireless environment. One of these misbehaviors is TCP concurrency, which refers to the case where the application opens several TCP connections with the server simultaneously. TCP concurrency is efficient in wireline networks since it reduces the impact of TCP slow start and the delay caused by in-sequence fetching. However, using multiple TCP connections has several flaws over wireless networks [223]. First, the simultaneous TCP connections often cause so-called self-congestion, and delay the new TCP connection setup [219]. Although Huang *et al.* [224] showed that using multiple TCP connections in wireless networks improves web page download time by 30%, the study makes the assumption that the network conditions (the RTT) are unchanged during the experiment. Real world measurement shows that the RTT increases as the number of simultaneous TCP connections increases [212]. RTT inflation has a negative impact on web page download time, and results in degraded network performance. Second, establishing a TCP connection, which requires a three-way handshake, introduces significant latency. Third, the protocol control overhead is high and consequently degrades the performance. Thus, the number of concurrent TCP connections should be optimized according to the wireless network condition; otherwise, it will introduce additional delay rather than improvements.

9.1.8 Mobile Devices

Because of their limited capability, mobile devices themselves are one of the limiting factors for mobile applications. One of the limitations is the low transmission power of the mobile device, which results in a relatively unreliable uplink and TCP misbehavior, for example, ACK compression, thus increasing service response times. Another limitation is the limited computing ability and small buffer size. With its limited computing ability, UE requires a long time to process incoming data, especially multimedia content. Given the small buffer size, the receive buffer will be filled quickly. Thus, the TCP sender will pause sending the data since the receive buffer is not available. This results

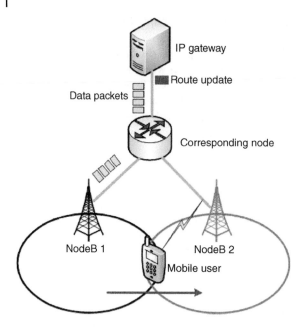

Figure 9.2 Packet loss in the handover process. *Source:* Han 2013 [14]. Reproduced with permission of IEEE.

in a long period of TCP idle during the data transmission, thus significantly impacting the service response time [224].

9.1.9 User Mobility

User mobility, which often triggers the handoff process, impacts network latency by introducing handover delay. As a result, mobile users experience far worse performance than stationary users [225]. Network layer handover latency, which is caused by network registration delay and route update delay, is one of the limitations for IP based mobile networks [226]. In addition, the handover processes can result in high packet loss levels [227]. As shown in Figure 9.2, if data packets arrive at the corresponding point earlier than the route update information, they will be forwarded through the old path to NodeB 1. Since the mobile user is already attached to NodeB 2, data packets sent through NodeB 1 will be lost. These packet losses may result in excessive delays.

9.2 Mobile System Evolution

To reduce network latency and secure subscribers' QoE, mobile communications systems are being enhanced from two perspectives. On the one hand, a new radio access network architecture named EUTRAN (evolved UTMS terrestrial radio access network) has been adopted in 3GPP LTE Advanced. EUTRAN adopts a flat architecture which simplifies signaling processes and reduces network latency. On the other hand, mobile networks and content delivery networks are being integrated to accelerate content delivery in mobile networks.

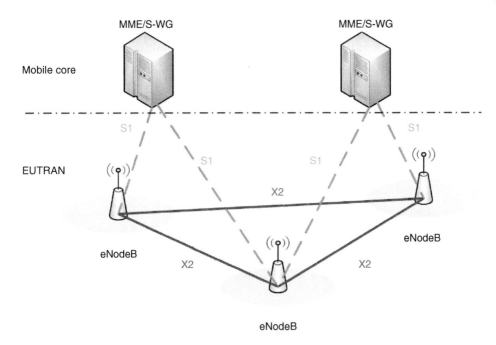

Figure 9.3 Overall system architecture. *Source:* Han 2013 [14]. Reproduced with permission of IEEE.

9.2.1 EUTRAN

As shown in Figure 9.3, EUTRAN applies a flat architecture which allows BSs to communicate with each other through X2 interfaces. EUTRAN eliminates RNC and moves its functions into eNodeB. By simplifying the system architecture, EUTRAN reduces the overall amount of protocol related processing, and thus reduces the latency. In the following, we discuss latency reduction in both the control plane and user plane in EUTRAN.

9.2.1.1 Control Plane Latency Reduction

EUTRAN adopts a simple RRC state machine and limits its states to two: RRC_IDLE and RRC_CONNECTED, as shown in Figure 9.4. In the RRC_IDLE state, there is

Figure 9.4 The RRC states in EPS. *Source:* Han 2013 [14]. Reproduced with permission of IEEE.

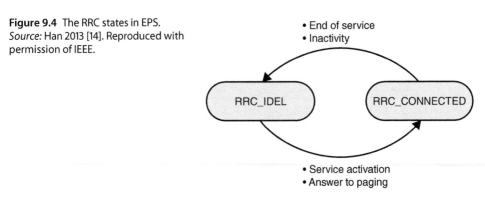

no connection between eNodeB and UE. The state switching from RCC_IDLE to RCC_CONNECTED is triggered by either service activation from UEs or paging messages from BSs. In the RRC_CONNECTED state, there is an active connection between UE and eNodeB, and data or signaling messages can be exchanged over wireless channels. If there is no service activity for a certain duration, UE returns to the RRC_IDLE state. As compared to RRC states in UMTS as shown in Figure 9.1, the simple RRC state machine in EUTRAN helps reduce first packet delay as discussed in Section 9.1. Mohan *et al.* [228] compared control plane latency in UMTS and LTE, and showed that the latency of the mobile originated call setup from the idle state has been improved from 1717 ms in UMTS to 188 ms in LTE. This significant latency reduction improves the responsiveness of mobile networks.

9.2.1.2 User Plane Latency Reduction

As shown in Figure 9.5, the radio access networks of EPS (EUTRAN) have eliminated RNC and relocated its functions to eNodeB. In EUTRAN, compression and ciphering functions are supported by the PDCP (packet data convergence protocol) layer at eNodeB. In addition, both hybrid-ARQ and RLC layer ARQ are accommodated in the

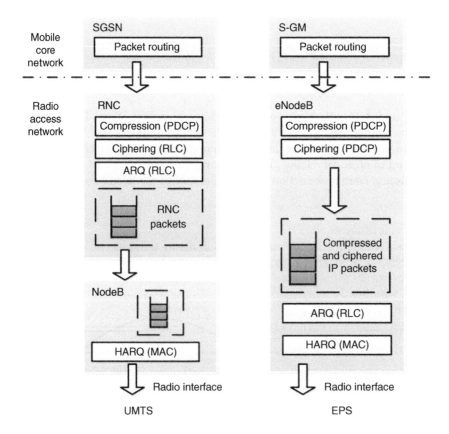

Figure 9.5 UMTS and EPS downlink user plane handling comparison. *Source:* Han 2013 [14]. Reproduced with permission of IEEE.

eNodeB. Therefore, in EUTRAN, only one buffer of header compressed and ciphered IP packets is required for data processing. This system architecture helps reduce latency in the user plane from two aspects. First, this architecture reduces the queuing delay in radio access networks. In UMTS, data processing requires two separated queues at RNC and NodeB, respectively. Hence, a flow control mechanism is required in order to maintain an optimal queue size at NodeB. The flow control mechanism between RNC and NodeB may introduce additional queuing delay. By merging functions of RNC and NodeB into eNodeB, only one buffer is required for data processing in EUTRAN. Therefore, the queuing delay is reduced. Second, this architecture reduces the delay caused by handover processes. As mentioned in Section 9.1, excessive delays may be introduced by handover processes in mobile networks. EUTRAN introduces a new inter-eNodeB interface, X2, to reduce the delay caused by handover processes. The X2 interface enables a flat architecture in EUTRAN that accelerates the network registration procedure during handover. In addition, X2 interfaces allow the data buffer to be forwarded between source and target BSs, thus reducing the probability of packet loss during handover processes [229].

9.2.2 Integrating Mobile Networks and CDN

Mobile networks were originally designed for voice and data communications and are not optimized for content delivery. Therefore, the performance of mobile networks in terms of content delivery cannot be guaranteed. Integrating mobile networks and CDN provides end-to-end solutions for content delivery in mobile networks and enhances network performance. Ericsson and Akamai [230] proposed the mobile cloud accelerator which reserves a portion of network bandwidth for premium content. Therefore, content with higher priority can avoid the delay caused by network congestion and be delivered faster. In addition, Blumofe *et al.* [231] proposed deploying caching proxies into mobile networks to accelerate content delivery.

9.3 Content and Network Optimization

In this section, we present an overview of available content and network optimization techniques, which are classified into content domain techniques, network domain techniques, and cross domain techniques.

9.3.1 Content Domain Techniques

Wireless links usually have constrained bandwidth, which leads to an excessive queuing delay in mobile core networks. Reducing the traffic volume over mobile networks can accelerate wireless transmission, thus alleviating the queuing delay. Content domain techniques aim to reduce the data volume over mobile networks. They include caching, data redundancy elimination, prefetching, and data compression.

9.3.1.1 Caching
Caching stores copies of frequently requested content in subscribers' local caches or in the cache proxy located at the network edge. Instead of being responded to by the original content server, subscribers' requests are responded to by their local cache or

the cache proxy close to them. In this way, caching effectively reduces network traffic and the number of round trips required to download content, thus improving content delivery performance in term of service response time. The performance of the caching scheme is determined by the hit rate of the cache; the higher the hit rate, the better the performance. To maximize the hit rate, Chakravorty *et al.* [223] presented a novel caching scheme that utilizes content hash to eliminate redundant caching. The caching scheme indexes the objects by using content hash, and maps the URLs to the respective hash key. Since the same objects have an identical hash value, they are only cached once even though they may be pointed to by different URLs. In dynamically generated web pages, mapping multiple URLs to the same content is commonplace. A content hash based caching scheme can significantly improve the hit rate of the client cache.

9.3.1.2 Data Redundancy Elimination

Data redundancy elimination techniques have proven to be an efficient way to reduce the traffic load on bandwidth constrained networks [232, 233]. Redundancy elimination algorithms rely on the caches deployed at both end of the link. When there is a packet, the ingress nodes find the common sequences of bytes in the packets in their cache and replace them with a fixed size pointer. Upon receiving the pointer, the egress nodes decode the content from their local caches. However, wireless loss can lead to unsynchronized caches between sender and the receiver, which results in incorrect decoding of the content. To address this problem, the informed marking scheme was proposed to detect packet losses and to prevent redundancy elimination algorithms from using them for encoding [234]. With informed marking, redundancy elimination algorithms can save more than 60% of bandwidth, thus reducing queuing delay in mobile core networks.

9.3.1.3 Prefetching

Prefetching techniques hide network latency from users by predicting the users' next requests, and pre-retrieving the corresponding content. The prefetching system usually consists of two modules: a prediction module that predicts users' next requests and a pre-retrieving module that processes the prefetching. The prediction algorithms can be classified into two major categories: history-based prediction algorithms and content based prediction algorithms [235]. History-based prediction algorithms predict future requests based on the observed pattern of past requests. There are two main techniques for history-based prediction algorithms: Markov process models [236] and data mining techniques [237]. Content based prediction algorithms focus on the content of the web pages that have been requested, and identify the content that is likely to be of particular interests to the users [238]. The recommendations from the prediction module are stored in a hint list, which consists of a set of URLs that are likely to be requested by the user in the near future. The pre-retrieving module fetches the URLs in the hint list when bandwidth is available. Khemmarat *et al.* [239] proposed a video prefetching system which consists of a search result-based prefetching scheme and a recommendation-aware prefetching scheme. The search result-based prefetching scheme utilizes video search results provided by video sharing sites such as YouTube and Youku as hints, while the recommendation-aware prefetching scheme prefetches the video prefix in the recommendation list provided by the video sharing sites.

9.3.1.4 Data Compression

Data compression techniques accelerate content delivery by reducing the data volume generated by applications. For web services, data compression algorithms can be classified into two categories: intra-web page compression and inter-web page compression. Intra-web page compression techniques reduce the data redundancies within the web page. Intra-web page compression can be further classified into lossless compression and lossy compression. The former applies to text or binary based content such as HTML files and Javascript. The latter usually applies to image based content like pictures and video, and it trades image quality for file size.

Inter-web page compression minimizes redundant data transfer between multiple web page requests. Delta encoding is one inter-web page compression technique. It identifies the difference between a newly requested web page and a previous one, and sends only the difference to the web client. Using delta encoding, web clients avoid downloading the whole web page every time it is accessed. Therefore, delta encoding can significantly reduce the amount of traffic over the network, thus accelerating web services [240].

9.3.2 Network Domain Techniques

Network domain techniques, which are designed to optimize mobile networks and communication protocols, can be classified into five categories: handover optimization, queue management techniques, network coding, TCP optimization, and session layer optimization.

9.3.2.1 Handover Optimization Techniques

As shown in network measurement studies, handover delays may cause excessive packet losses and degrade network performance in terms of content delivery. IP soft handover can reduce handover latency and packet loss during the handover processes. It reduces handover latency by providing mobile hosts with information about the new access point and the associated subnet prefix before they switch to the new access point [241]. Thus, subscribers can carry out network registration and address resolution prior to the switch. In addition, as compared to hard handover, it reduces packet loss because it does not disconnect from the previous access point until the route information is updated in the corresponding node. Nurvitadhi *et al.* [227] proposed adaptive semi-soft handover that further improves the latency performance by optimizing tune-in time based on the network conditions.[2]

9.3.2.2 Queue Management Techniques

As discussed in network measurement studies, queue management schemes designed for the Internet face several problems in mobile networks. By addressing these problems, queue management schemes tailored for mobility can enhance network performance. Sagfors *et al.* [213] proposed exploiting knowledge about a time varying link's capacity, and to dynamically set the minimum threshold of RED to reflect network conditions. In addition, they proposed applying deterministic dropping strategies rather than probabilistic dropping strategies. In the deterministic dropping policy, the congestion

---

[2] The tune-in time is the time when the mobile host switches from the old access point to the new one.

is signaled as soon as it is detected by dropping a single packet. The dropping policy incorporates a discard prevention counter algorithm to avoid multiple losses from the same TCP connection.

9.3.2.3 Network Coding Techniques

Network coding has emerged as a promising technique to enhance the robustness and effectiveness of data transmission over lossy wireless networks. In the following, we discuss two potential applications of network coding that can improve mobile network performance.

Improving HARQ Efficiency

HARQ (Hybrid automatic repeat request) is adopted by 4G mobile networks to enhance the reliability of wireless links. The signaling overhead of HARQ may introduce additional latency into mobile networks. Lang *et al.* [242] proposed network coded HARQ (NC-HARQ) which integrates network coding into HARQ to increase its efficiency in terms of throughput, thus reducing the signaling overhead. NC-HARQ applies to scenarios where there are consecutive erroneous packets. For example, if there are two consecutive erroneous packets, instead of retransmitting them individually, NC-HARQ codes the packets using XOR into one retransmission packet. In this case, NC-HARQ only requires three transmissions to handle two consecutive erroneous packets while the original HARQ mechanism requires four transmissions. Therefore, NC-HARQ enhances HARQ throughput.

Enhancing TCP performance

Most popular Internet applications, for example, web services and video streaming, are based on TCP to achieve reliable data transfer. However, TCP does not perform well in mobile networks because of unreliable wireless links. Therefore, enhancing TCP performance in mobile networks is crucial for accelerating mobile content delivery. Network coding is one of the promising technologies that may enhance network performance in lossy networks. Therefore, integrating network coding into TCP may enhance TCP performance in lossy networks. Sundararajan *et al.* [243] proposed TCP-Network Coding (TCP-NC) which takes advantage of network coding to mask losses from the congestion control algorithm and achieves effective congestion control over lossy wireless networks. As shown in Figure 9.6, TCP-NC implements a network coding layer between the Internet protocol layer and the TCP layer. The network coding layer accepts packets from the TCP layer and buffers them into an encoding buffer until they are acknowledged by the receiver. The sender sends a linear combination of the packets in the encoding buffer. Upon receiving the linear combination, the network coding layer at the receiver side finds out which packets have been decoded based on the received packets. The decoded packets define an ordering of the degree of freedom [243] (i.e., a linear combination that reveals one unit of new information) which is consistent with the packet sequence number. The receiver acknowledges the degree of freedom to the sender, which allows the TCP congestion window at the sender side to advance. From the TCP sender's view, it appears that the transmitted packets wait in a fictitious queue until all degrees of freedom have been acknowledged. Therefore, TCP-NC translates the lossiness of the wireless link into an additional queuing delay. Thus, TCP-NC adopts TCP-Vegas, which is a delay-triggered congestion control mechanism [244], at its TCP

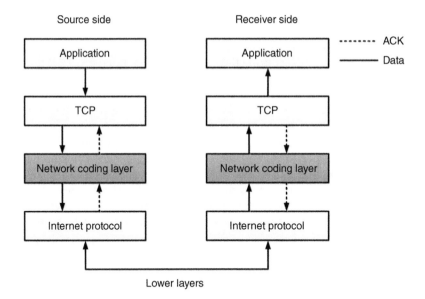

Figure 9.6 New network coding layer in the protocol stack. *Source:* Han 2013 [14]. Reproduced with permission of IEEE.

layer. Kim *et al.* [245] evaluated the performance of TCP-NC in lossy wireless networks, and showed that TCP-NC is able to increase window size fast and maintain a large window size despite wireless losses.

9.3.2.4 TCP Optimization

Beside network coding, other solutions have been proposed to optimize TCP performance in mobile networks. These solutions can be classified into two categories. The first category is to design transmission control mechanisms to maximize the utilization of the available bandwidth. The second category is to design ACK mechanisms to provide accurate feedback on network conditions to the TCP sender.

Transmission Control Mechanisms

To speed up TCP connections, several TCP variants have been introduced. Zhang and Qiu [246] proposed TCP/SPAND, which significantly reduces latency for short transfers. TCP/SPAND utilizes shared network performance information to estimate the fair share of network resource. Then, based on both this estimation and the transfer data size, the TCP sender will determine an optimal CWND size that minimizes the number of round trips required for the transfer. However, the optimized CWND size can be potentially large. To limit the maximum burstiness of the outgoing traffic caused by the large CWND, TCP/SPAND applies a leaky-bucket based pacing scheme to smoothly send out the packets in the initial CWND.

Chakravorty and Pratt [221] designed TCP CWND Clamping to eliminate the latency caused by TCP slow start. This TCP variant fixes the initial CWND to an estimation of the BDP. After sending out the packets in the initial CWND, the TCP CWND Clamping goes into a self-clocking state in which it clocks out one segment each time it receives an ACK for an equivalent amount of data from the receiver. This design reduces the queuing delay since it limits the CWND to BDP, and enables quick recovery from loss by limiting the data over the link. Goff *et al.* [247] proposed "Freeze-TCP" to prevent the TCP sender from dropping the CWND during the handoff process. In Freeze-TCP, the mobile user monitors the signal strength, detects the handoff, and predicts a temporary disconnection. Then, the mobile user signals to the TCP sender to freeze the TCP flow to prevent the reduction of CWND due to excessive packet losses during the handoff process.

Rodriguez and Fridman [248] proposed TCP connection sharing which takes advantage of the concurrent TCP connections to estimate the RTT and congestion window size. The enhanced TCP stack caches the RTTs and congestion window sizes of current or recently expired TCP connections, according to which TCP estimates the initial CWND of new connections to the same mobile host. In this way, the starting parameters can be effectively estimated, thus reducing the latency incurred by the TCP slow start phase.

TCP-Mobile Edge

We designed a new TCP algorithm, TCP-ME (Mobile Edge) [249], to accelerate content delivery in mobile networks. Considering the QoS mechanisms of mobile networks, TCP-ME is designed to differentiate the packet loss caused by wireless errors, the traffic conditioning mechanism of mobile core networks, and Internet congestion, and then to react to the packet loss accordingly. To detect wireless errors, we proposed to mark the ACK (Acknowledge) packets in the uplink direction at the BS, and the marking threshold is a function of the instantaneous downlink queue and the number of consecutive HARQ retransmissions. Inspired by TCP-New Jersey [250], the ECN mechanism is modified with deterministic marking to detect Internet congestion. The packet loss caused by traffic conditioners of mobile networks is detected if the incoming DUPACK is not marked. TCP-ME adapts the inter-packet intervals when the packet loss is caused by wireless errors or admission control mechanism. If the packet loss is due to Internet congestion, TCP-ME applies TCP-New Reno's congestion window adaptation algorithm [251]. Simulation results show that TCP-ME can improve web service response time by about 80% in mobile networks.

To detect packet loss caused by wireless errors, we implement an ACK marking algorithm on BSs. The proposed algorithm utilizes the CE bit in the IP header of the uplink ACKs, which is not used in current TCP/IP protocols [252], to notify the wireless loss. In the proposed scheme, we define the CE bit in the IP header of the uplink ACKs as a wireless loss alarm (WLA) bit. The WLA bit is set when the equivalent length of the downlink queue of the BS is greater than a predefined threshold. A similar idea appears in [253], in which the authors proposed to mark the outgoing ACKs when the router queue length exceeded a threshold, to reduce the time required to notify the sender about the congestion. However, our proposed ACK marking algorithm, instead of detecting wireless congestion, aims to feed back wireless errors to the server through marking the WLA bit. The equivalent downlink queue length is calculated as

$$L_E = \begin{cases} \dfrac{L_I}{1 - (n/\lambda)^3}, & \text{for } n < \lambda; \\ L_{th}, & \text{for } n \geq \lambda, \end{cases} \qquad (9.1)$$

where L_E is the equivalent queue length, L_I is the instantaneous downlink queue length when the uplink ACK arrives at the BS, L_{th} is the ACK marking threshold, n is the number of consecutive HARQ retransmissions, and λ is the maximum allowed HARQ retransmission. The intuition behind the equivalent queue length calculation is that the more consecutive HARQ retransmissions there are, the higher the probability of wireless errors, and thus the larger the equivalent queue length. In addition, BSs usually keep individual queues for the attached users, so the equivalent queue length indicates wireless conditions of individual subscribers. Because the pure ACK packets should not be marked as ECN capable [252], the WLAs bit are only marked at BSs. Therefore, the marked ACK packets can effectively reflect wireless errors. When the TCP sender receives ACKs with the WLA bit marked, the TCP sender knows that the TCP receiver is experiencing bad wireless channels, and that the packet loss indicated by the DUPACKs is more likely to be being caused by wireless errors.

To detect Internet congestion, we inherit the congestion warning (CW) mechanism of TCP-Jersey [254], in which the router marks all the packets when the average queue length exceeds a threshold, and the TCP sender that receives the marked packets will determine whether to adjust its transmission strategy. The non-probabilistic packet marking helps the TCP sender detect packet loss due to network congestion. When the TCP sender receives DUPACK with the ECE bit marked in the TCP header, the TCP sender knows for sure that the network is in a congested state and the packet loss indicated by the DUPACKs is more likely to be caused by congestion.

In mobile networks, the maximum bit rate of traffic admitted into the network is limited by the traffic conditioner on the gateway of the mobile networks. The token bucket algorithm is recommended as the traffic conditioner by 3GPP [255]. Let r be the token rate and b be the bucket size. Data are said to be conformed if the amount of data submitted during any time period T does not exceed $b + rT$. The bucket size is recommended to equal the maximum SDU (Service Data Unit) size. Let m be the SDU size; if the packet inter-arrival time is less than $\frac{m}{r}$, packet data will be dropped.[3] To avoid packet loss caused by the traffic conditioner, the TCP sender must estimate its packet inter-arrival time on the gateway, and adjust the packet sending interval accordingly. ACK-triggered TCP algorithms do not consider the effect of traffic conditioners, and thus they do not adapt the inter-packet interval according to an estimation of packet inter-arrival times. However, estimating packet inter-arrival time is not trivial for two reasons. First, the TCP sender does not know the token rate of the traffic conditioner at the gateway. Second, the crossing traffic on the intermediate routers may unpredictably change the packet intervals. As shown in Figure 9.7(a), a small inter-packet interval, T_1, at the TCP sender is prolonged by crossing traffic, and results in a large inter-packet arrival interval, T_1', at the gateway. Figure 9.7(b) shows an opposite scenario. Even though the TCP sender keeps a large inter-packet interval, the intermediate routers may shorten the interval, which results in packet loss on the gateway.

Therefore, a large inter-packet interval cannot guarantee that the packets are conformed by the traffic conditioner on the gateway. TCP-ME adjust its packet sending interval according to the indication of packet loss caused by the traffic conditioner. As

---

[3] Here, we assume the nonconformant packets will be dropped rather than being marked as preferential dropping.

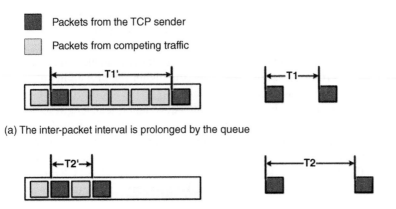

■ Packets from the TCP sender

□ Packets from competing traffic

(a) The inter-packet interval is prolonged by the queue

(b) The inter-packet interval is shortened by the queue

Figure 9.7 The dynamic of the inter-packet interval. *Source:* Han 2012 [249]. Reproduced with permission of IEEE.

was presented in the previous subsection, TCP-ME recognizes wireless errors by checking whether the WLA bit of the DUPACK is marked, and detects Internet congestion by checking whether the CW is marked. If neither the WLA bit nor the CW bit is marked, TCP-ME renders the cause of the packet loss as the inter-packet arrival interval being smaller than the tokens' arrival interval. In this case, TCP-ME sets the maximum sending rate, R_{max}, to the current sending rate, R_n, which is equal to $1/T_n$, where T_n is the current packet sending interval, and reduces the sending rate by a factor of α. To make sure the sending rate is only reduced once in one round trip, TCP-ME sets the recovery sequence number, S_{rec}, to the maximum unacked sequence number, S_{unack}^{max}, when the sending rate is reduced. When a TCP sender receives DUPACKs that indicate a packet loss due to the traffic conditioner, the TCP sender checks the sequence number of the lost packet first. Only when the sequence number is larger than S_{rec} does the TCP sender reduce the sending rate. The sending rate adaptation algorithm is illustrated in Table 9.1.

Here, ε is the parameter that controls the transition between additive rate increasing and proper rate searching. C is the additive rate increasing step. β, which is less than 1, is the factor for proper rate searching. A larger β indicates a more aggressive rate increase. Since $SRI()$ reacts to packet loss caused by both wireless errors and the traffic conditioner, TCP-ME differentiates the sources of the packet loss. If the packet loss is caused by the traffic conditioner, $SRD()$ sets R_n as R_{max} since R_n is more likely reflecting the upper bound of the allowed sending rate. Otherwise, if the packet loss is caused by wireless errors, the TCP sender does not update R_{max} since the packet loss does not indicate the upper limit of the token rate on traffic conditioners. However, the TCP sender reduces the sending rate by a factor of α for two reasons. First, if the WLA is marked, this indicates that the TCP receiver is experiencing a bad wireless condition. Therefore, reducing the packet sending rate can avoid excessive packet loss from wireless errors. Second, TCP-ME reduces the sending rate rather than increasing the congestion window since reducing the sending rate can also increase the probability that the packets are admitted by the traffic conditioner on the gateway mobile network gateway.

Table 9.1 Sending rate adaptation algorithm.

Sending Rate Increasing

SRI()

{

    if $(R_n >= R_{\max} - \varepsilon)$

        $R_{n+1} = R_n + C;$

    else

        $R_{n+1} = R_n + \beta \times (R_{\max} - R_n);$

    end if

}

Sending Rate Decreasing

SRD()

{

    $S_{\text{rec}} = S_{\text{unack}}^{\max}$

    if $(!IP.WLA)$

        $R_{\max} = R_n;$

        $R_n = \alpha \times R_n;$

    else

        $R_n = \alpha \times R_n;$

    end if

}

The key advantage of TCP-ME is to differentiate the causes of packet loss and react to them accordingly. If the packet loss is caused by network congestion, TCP-New Reno's congestion control mechanism is applied to adjust the congestion window and to retransmit the lost packet. If the packet loss is caused by wireless errors or the traffic conditioner, TCP-ME adjusts its transmitting rate as is presented in the previous subsection and retransmits the lost packet immediately without modifying the congestion window. The flowchart of the TCP-ME sender's response to normal ACK and DUPACK are illustrated in Figure 9.8 and Figure 9.9, respectively.

We design simulations to show its advantages over the other TCP flavors in terms of service response time in mobile networks. The network topology for the simulation is shown in Figure 9.10. We use FTP as the cross traffic in wired networks and the mobile network. The mobile FTP client is served by the mobile FTP server while the other FTP clients are served by the (normal) FTP server. In the simulation, the web page contains a 1 MB HTML file, 10 media images whose sizes are uniformly distributed between 500 and 2000 bytes, and 50 small images whose sizes are uniformly distributed between 10 and 400 bytes. In the mobile networks, we model the wireless channel using free space as the pathloss model and ITU pedestrian A as the multipath channel model. The parameters of the rate adaptive algorithm are $\varepsilon = 50 \, bits \, sec^{-1}$, $C = 2000 \, bits \, sec^{-1}$, $\beta = 0.15$, and $\alpha = 0.9$.

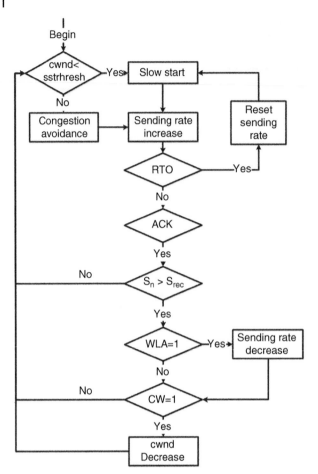

Figure 9.8 TCP sender's response to ACK. *Source:* Han 2012 [249]. Reproduced with permission of IEEE.

Figure 9.11 shows the average web page response times of different TCP flavors. The bars are the page response times, and the lines with plus markers are the standard deviation of the page response time. In the simulation scenarios, the average page response time of TCP-ME is about 30 seconds while the average page response time of the other simulated TCP flavors including TCP-Westwood, TCP-New Reno, TCP-SACK, and TCP-Reno are more than 150 seconds. As compared to the other simulated TCP flavors, TCP-ME reduces the page response time by about 80%. In addition, the standard deviation of TCP-ME is small, which indicates a relative steady performance in term of downloading the object.

Figure 9.12 shows the response time of the inline objects from the HTML file. TCP-ME has the shortest object response time when compared to the other TCP flavors. The average object response time of TCP-New Reno is the largest among the simulated TCP flavors. However, the average page response time of TCP-New Reno is less than that of TCP-Westwood, and is almost the same as that of TCP-SACK and TCP-Reno. This is because TCP-New Reno is more aggressive during the fast recovery period, which may result in more packet loss due to the effect of the traffic conditioners. Therefore, some of the packets of the TCP-New Reno flows suffer long delays, thus leading to a larger

Figure 9.9 TCP sender's response to DUPACK. *Source:* Han 2012 [249]. Reproduced with permission of IEEE.

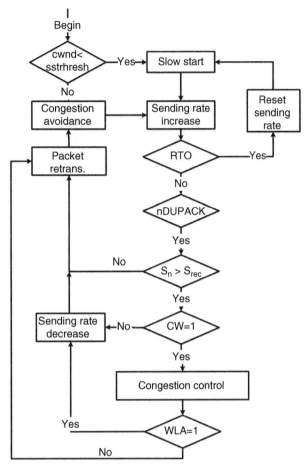

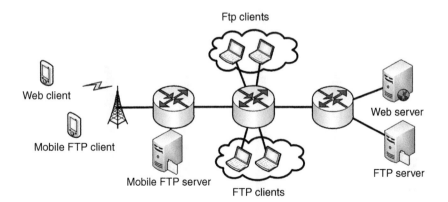

Figure 9.10 The network topology for the simulation. *Source:* Han 2012 [249]. Reproduced with permission of IEEE.

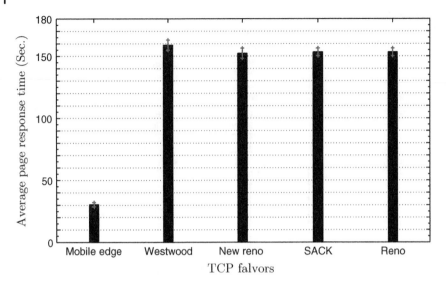

Figure 9.11 Web page response time. *Source:* Han 2012 [249]. Reproduced with permission of IEEE.

average object response time. Because the average object response time is much smaller than the average page response time, several individual large object response times do not have a significant impact on the average page response time.

Figure 9.13 shows the amount of traffic dropped at the mobile gateway which implements the traffic conditioners. Owing to the rate adaptive algorithm, TCP-ME can effectively estimate the token rate of the traffic conditioners, and thus the average traffic drop of TCP-ME flows is smaller than the other TCP flavor. TCP-ME improves

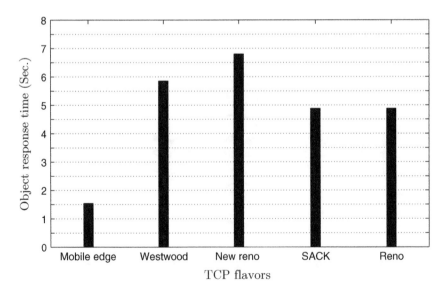

Figure 9.12 Response time for inlined object. *Source:* Han 2012 [249]. Reproduced with permission of IEEE.

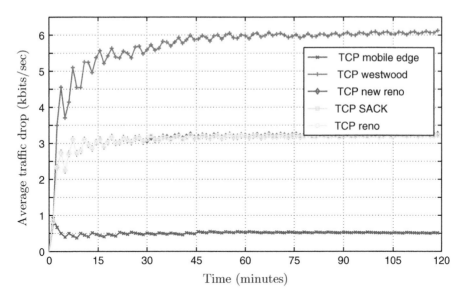

Figure 9.13 Average traffic drop at the gateway. *Source:* Han 2012 [249]. Reproduced with permission of IEEE.

the performance of the service response time by avoiding excessive packet loss. This also indicates that TCP-ME does not try to steal bandwidth from users with other TCP flavors. The average packet drops of TCP-New Reno, TCP-Reno, and TCP-SACK are almost the same, and they overlap.

Figure 9.14 shows the average number of retransmissions on the web server. TCP-ME has the least retransmissions. That is to say, given the same size of web page, the server

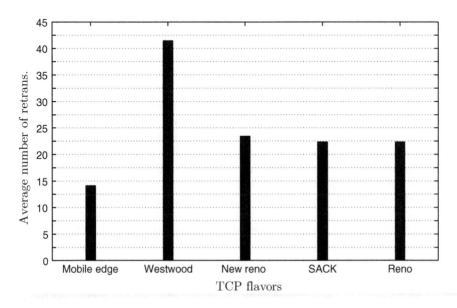

Figure 9.14 Number of retransmitted packets on web server. *Source:* Han 2012 [249]. Reproduced with permission of IEEE.

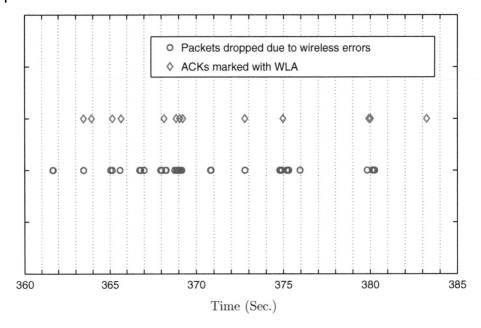

Figure 9.15 Test of ACK marking mechanism at BS. *Source:* Han 2012 [249]. Reproduced with permission of IEEE.

using TCP-ME sends out the least amount of traffic. Therefore, the improvement gained from using TCP-ME does not sacrifice the performance of the other users utilizing different TCP flavors.

Figure 9.15 shows the effectiveness of our ACK marking mechanism. The diamonds indicate the time when the ACKs are marked while the circles show when packets are lost due to wireless errors. The packet loss statistics are collected at the client side while the marked ACKs are recorded at the server side. The wireless errors are not fed back to the TCP sender if there are no uplink ACKs when the packet loss happens. Therefore, there are fewer marked ACKs than packets losses caused by wireless errors. The ACK marking mechanism cannot perfectly feedback the wireless errors. However, in general, the packet losses caused by wireless loss in mobile networks are not significant since the network applies HARQ and ARQ retransmission to enhance the reliability of the wireless transmissions. In addition, we keep the marking mechanism aggressive by choosing a small marking threshold. An aggressive marking strategy helps in token rate estimation and adaptation since it increases the probabilities that the TCP sender detects packet loss caused by traffic conditioners.

Figure 9.16 shows the performance of the TCP-ME's sending rate adaptive algorithm. The token rate of the traffic conditioner at the gateway is 384 kbps, which is a recommended value for best-effort services in WiMAX networks. The initial sending rate of TCP-ME is 200 kbps. The median value of the sending rate is about 392 kbps, which indicates that the sending rate of TCP-ME is larger than the token rate of the traffic conditioner most of the time. This is because crossing traffic may enlarge the packet intervals as discussed in the previous section. Therefore, although the web server has a larger sending rate, the packet arrival rates seen by the gateway are about the same as the token rates of the traffic conditioners.

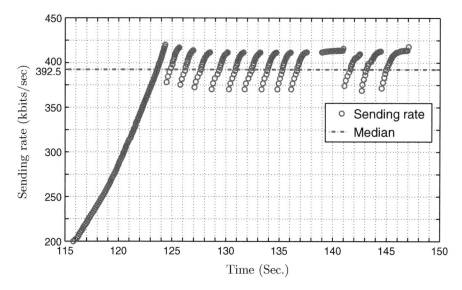

Figure 9.16 The sending rate adaptation of TCP-ME. *Source:* Han 2012 [249]. Reproduced with permission of IEEE.

ACK Mechanisms

As discussed in the measure studies, in mobile networks, hybrid-ARQ and RLC retransmission can cause increased delay and rate variability, which, together with the requirement of in order delivery, translate to ACK compression. Bursty ACK arrivals leads to the release of a burst of packets from the TCP sender, which may result in multiple packet losses, thus degrading the service response time. Several methods have been proposed to address the ACK compression problem, such as ACK congestion control [256], ACK filtering [257], and ACK prioritization [258]. Regarding mobile networks, Chan and Ramjee [259] introduced an ACK regulator to alleviate the impact of ACK compression. The regulation algorithm assures that the TCP sender is operated mainly in the congestion avoidance phase by limiting the maximum number of buffer overflow losses to one. In the algorithm, the ACK regulator sends an arriving ACK back toward the TCP sender only when the available buffer space is enough for at least one data packet. Therefore, the ACK regulator allows at most one buffer overflow packet loss. The ACK regulator can significantly improve TCP throughput while slightly increasing the RTT. As a result, it reduces the overall service response time.

The methods based on ACK management may, however, reduce the performance of the TCP sender due to delayed ACKs. To minimize the impact of ACK regulation on the TCP sender, Ming-Chit *et al.* [260] proposed ACE (Acknowledgment based on CWND Estimation), which varies the number of acknowledged packets per ACK according to the sender's CWND. The larger the CWND, the more acknowledged packets per ACK, and vice versa. Therefore, the number of ACKs sent on the uplink is reduced without much impact on the TCP sender.

9.3.2.5 Session Layer Techniques

The goals of session layer optimization are three-fold: to reduce the number of DNS lookups, to optimize the number of concurrent TCP connections, and to minimize TCP idle time.

Reducing DNS lookups

Web based Internet applications usually access content from different servers. To download the content, DNS protocol is frequently used to resolve server names. Considering the large RTT of wireless networks, frequent DNS queries introduce significant delays to web services. Therefore, reducing the number of DNS lookups can accelerate web based applications. One method to achieve this is to configure the web browser to explicitly point to a cache proxy. In this way, the mobile user only needs to look up the DNS of the cache proxy, and only opens connections with the cache proxy. The cache proxy will perform all its DNS queries over wireline networks, thus reducing the service response latency. Another method is to rewrite the URLs contained in the HTML [261]. In this method, the rewriting proxy intercepts the HTML file from the web server, and adds the IP address of the proxy server as a prefix to the URLs. As a result, the web clients recognize the proxy server as the only web content server, and thus frequent DNS queries are avoided.

Reducing concurrent TCP links

Concurrent TCP connections may degrade web performance in wireless networks. Session layer techniques include URL rewriting, DNS rewriting, and content bundling. As discussed in the previous paragraph, following URL rewriting, web clients only open TCP connections with the proxy server, thus minimizing the number of TCP connections. In DNS rewriting, the DNS rewriting proxy intercepts the DNS servers' responses and attaches the IP address of the cache proxy at the top of the returned IP lists [261]. As a result, the cache proxy is the first choice for the web client. The IP addresses of the original servers work as backups in case the cache proxy is unavailable due to mobility. Therefore, the number of TCP connections can be reduced in the same way as explained for URL rewriting. Content bundling groups all the embedded objects in a single file. As a result, web clients do not have to establish multiple TCP connections to download the embedded objects from different servers.

Reducing TCP idle time

With their limited processing abilities, mobile devices may require a relatively long time to parse received data and then send the next request. As a result, the TCP sender may stay in an idle state waiting for incoming requests. Waiting in the TCP idle state delays the response time of web based applications. To address this problem, Gomez *et al.* [262] introduced the "parse and push proxy", which parses the HTML for the mobile users and pushes the enlisted objects toward them prior to the requests. On the client side, mobile users always check their local cache before requesting content; if the content has been pushed into their local cache by the proxy, the mobile users retrieve the content from the cache; otherwise, they send the requests to the server. This method reduces both the TCP idle time and the number of round trips required to download web pages.

9.3.3 Cross Domain Techniques

Cross domain techniques optimize the interactions between applications/content and the underlying networks. These techniques can be classified into two categories: content adaptation techniques and protocol adaptation techniques.

9.3.3.1 Content Adaptation Techniques

Content adaptation techniques accelerate mobile web services by tailoring the original content to fit mobile networks and mobile devices. The benefits of content adaptation are two fold: for the network, content adaptations help to reduce the data volume of web pages, thus alleviating the long latency caused by low bandwidth; for the mobile device, by fitting the content to mobile devices' capabilities such as screen size, memory size, and processing speed, content adaptations save the time spent on rendering the complex web pages on mobile devices [263]. Content adaptation techniques can be classified into three categories: proxy based content adaptation, application aware adaptation, and page layout adaptation [264].

- **Proxy based adaptation:** This approach does not require any modifications to either the content server or the web browsers. All web communications are redirected to the proxy, where the content is adapted according to the characteristics of mobile devices and the network conditions. The proxy decides the appropriate adaptations to mobile devices [265].
- **Application aware adaptation:** In application aware adaptation, applications and the operating system cooperate to adapt the content for subscribers. The operating systems carry out centralized content adaptations at the system level such as monitoring resource levels and arbitrating application concurrency, while individual applications take care of application specific adaptation such as selecting the fidelity of the content [266].
- **Web page layout adaptation:** In web page layout adaptation, the layouts of pages are reconstructed to fit users' preferences. Web page layout adaptations can be implemented either in the proxy [267] or in the mobile devices [268]. If the adaptations are implemented in the proxy, the proxy will re-author the web content according to predefined rules and policies. If the adaptations are installed on mobile devices, they provide subscribers a summary of the web content, from which they can choose to view extended views of the web page.

9.3.3.2 Protocol Adaptation Techniques

Interactions between application behavior and the underlying protocols have a significant impact on network performance, and if not properly considered, may completely negate the improvement achieved by optimizing the underlying protocols [269]. Therefore, application awareness is required to accelerate delivery in wireless networks. Application aware acceleration techniques can recognize application types and optimize the underlying protocol according to the characteristics of different applications. A^3 [269] is an application aware acceleration software that can recognize FTP, CIFS, SMTP and HTTP protocols based on the session messages, and optimize them accordingly. A^3 integrates five techniques: transaction prediction and prefetching to predict users' next requests and pre-retrieve them, redundant and aggressive retransmission to protect session control messages from loss, prioritized fetching to fetch important data first with respect to application performance, infinite buffering to prevent a network connection termination because of the small receiving buffer of mobile devices, and application aware encoding (using application specific information to better encode or compress the data). Upon recognizing the application, these techniques are specified according to the application's characteristics, thus achieving better performance than application-blind acceleration.

9.4 Mobile Data Offloading

Mobile data services usually experience longer latency than wireline networks because mobile networks have lower bandwidth and less reliable connections than wireline networks. Mobile data offloading enables subscribers to utilize high rate networks, such as WiFi, to retrieve content, thus accelerating content delivery. There are two mechanisms that can be applied to offload mobile data traffic: direct data offloading and network aggregation.

9.4.1 Direct Data Offloading

Existing mobile data traffic offloading mechanisms can be classified into two categories. The first category of mechanisms is offloading mobile traffic through opportunistic communications. In this method, instead of downloading content from infrastructure, mobile users can share their content with each other via peer-to-peer collaborations. The other explores the use of metro scale WiFi networks to offload mobile data traffic.

9.4.1.1 Offloading Through Opportunistic Communications

This type of traffic offloading mechanism relies on mobile users to disseminate the content. Given a set of mobile users who request some content from the servers, the offloading mechanism selects a subset of users and delivers the content to them through cellular networks. This subset of users further disseminates the content through opportunistic ad hoc communications to users outside the subset. The key designing issue is to select the subset of target users in order to maximize traffic offloading. Han *et al.* [23] proposed a mechanism to select the subset of users based on either users' activities or mobilities. However, their mechanism does not guarantee the performance of content delivery in terms of delay. To meet delay requirements, Whitbeck *et al.* [270] proposed pushing the content to the users through cellular networks when ad hoc communication fails to deliver the content within some target time. Li *et al.* [271] analyzed the mobile data offloading problem mathematically. They formulated the offloading problem as a submodular function maximization problem, and proposed several algorithms to achieve the optimal solution. To encourage mobile users to participate in traffic offloading, Zhou *et al.* [25] proposed an incentive framework that motivates users to leverage their delay tolerance for cellular data offloading. Mashhadi *et al.* [24] proposed a proactive caching mechanism for mobile users in order to offload mobile traffic. When the local storage does not have the requested content, the proactive caching mechanism will set a target delay for this request, and explores opportunities to retrieve data from neighboring mobile nodes. The proactive cache mechanism requests data from the cellular network when the target delay is exceeded.

9.4.1.2 Offloading Through Metro Scale WiFi

The other type of mechanism is to exploit metro scale WiFi networks to offload cellular data traffic. Lee *et al.* [21] showed from their measurement results that a user is in WiFi coverage for 70% of the time on average, and if users can tolerate a two hour delay in data transfer, the network can offload about 70% cellular traffic to WiFi networks. Deshpande *et al.* [16] compared the performance of mobile and metro scale WiFi for vehicular network access and showed that even though suffering frequent disconnections, WiFi delivers high throughput when connected. Gass *et al.* [272] pointed out that the full potential of WiFi access points in term of the transmission rate has not yet been reached and is limited to the rate of the backhaul connection to which the access point is connected. To eliminate backhaul bottlenecks, Gass *et al.* [273] proposed In-Motion proxy and Data Rendezvous protocols to enable download of a large amount of data during a short period using opportunistic WiFi connections. The In-Motion proxy approach prefetches the requested data from the original server to its local cache, and streams the data to the users when they are connected to WiFi access points. In-Motion proxy enables large data transfers to be completed with several short connection durations. Data Rendezvous protocol eliminates the bottlenecks between the In-Motion proxy and WiFi access points through access point prediction and selection. Balasubramanian *et al.* [22] proposed offloading delay tolerant traffic, such as email and file transfer, to WiFi networks. When WiFi networks are not available or experiencing blackouts, data traffic is quickly switched back to mobile networks to avoid violating the applications' tolerance threshold.

We designed a content pushing system which pushes content to mobile users through opportunistic WiFi connections [15]. The system responds to a user's pending requests or predicted future requests, codes the requested content by using Fountain codes, predicts the user's route, and prelocates the coded content to the WiFi access points along the user's route. When the user connects to these WiFi access points, the requested content is delivered to the user via the WiFi connections.

9.4.2 Network Aggregation

Network aggregation techniques enable users to utilize multiple radio interfaces to download content and thus reduce content delivery delay. In this subsection, we describe recent network aggregation methods: pTCP [274], MAR [275], and super-aggregation [276].

9.4.2.1 pTCP

pTCP is a parallel TCP structure that aggregates bandwidth of different networks. As shown in Figure 9.17, the pTCP structure contains pTCP, which acts as a central engine that interacts with applications, IP networks, and TCP-v. pTCP creates one TCP-v as a virtual TCP pipe for each radio interface. TCP-v relies on the traditional TCP protocol to achieve reliable transmission. pTCP maintains the central send and receive buffers across all pipelines and schedules the transmission over all the TCP-v pipelines according to their link conditions.

9.4.2.2 MAR

MAR is a mobile access router that provides Internet access to users on a commuter bus through aggregating the bandwidth of multiple Wireless Wide Area Networks (WWANs). Mobile users use short range wireless interfaces such as WiFi and Bluetooth to connect to MAR. MAR routes users' requests through the aggregated radio links. The

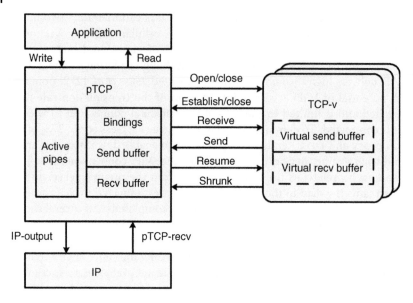

Figure 9.17 pTCP structure. *Source:* Han 2013 [14]. Reproduced with permission of IEEE.

advantage of MAR is that it does not require any modification to either users' devices or content servers.

9.4.2.3 Super-aggregation

Super-aggregation aims to intelligently aggregate multiple radio interfaces to achieve better network performance. Instead of simply aggregating radio interfaces, super-aggregation assigns different radio interfaces to different tasks based on their link characteristics. It contains three principles: selective offloading, proxying, and mirroring. With respect to selective offloading, super-aggregation offloads the ACK messages of WiFi networks to mobile networks in order to alleviate ACK congestion in the uplink of the WiFi networks. It utilizes the mobile link as a monitoring proxy that reports blackout events to the WiFi networks. Upon receiving such a report, the WiFi link freezes the current TCP status to avoid CWND reduction. Regarding mirroring, super-aggregation utilizes mobile links as redundant links to fetch lost packets.

9.5 Web Content Delivery Acceleration System

The procedure for provisioning wireless web services is shown in Figure 9.18. Before web clients can fetch data from the server, they have to conduct DNS query and TCP handshake processes, which costs at least two round trips. Given the large RTT of mobile networks, such processes introduce significant delays to the service response time. In addition, the clients are prone to opening multiple TCP connections simultaneously, thus resulting in RTT inflation as discussed in the previous section. To accelerate web content delivery, we have to reduce the number of round trips required to download

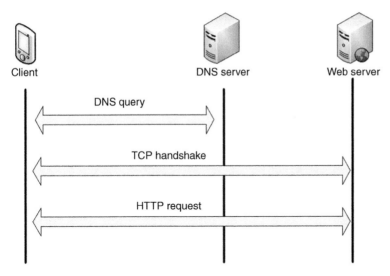

Figure 9.18 Web connection procedure. *Source:* Han 2013 [14]. Reproduced with permission of IEEE.

web pages, and to reduce RTTs. In this section, we overview existing web content delivery acceleration systems.

9.5.1 Web Acceleration System

Web based applications are among the most popular mobile applications. Therefore, accelerating web content delivery is essential in mobile networks. Web acceleration systems that integrate several optimization techniques at different layers in the OSI stack are effective tools for web delivery acceleration. In this subsection, we discuss and compare several existing web acceleration systems for mobile networks.

9.5.1.1 GPRSWeb

GPRSWeb [223] is a proxy based web optimization system. As shown in Figure 9.19, GPRSWeb applies a dual proxy architecture consisting of a link-aware middleware in mobile devices and a server proxy located on the wired–wireless border. At the client side, the connection manager takes users' requests from the web browser and passes them to the connection module. This module checks whether the requested content is cached by the cache manager. If there is no matched content, the connection module invokes a server stub to issue a request to the server proxy, and a response handler to process the reply. The response handler also interacts with the cache manager to update the cache state. In the server proxy, client stubs receive clients' requests and pass them to the server control module. This module checks the server cache for responses to clients' requests. If the objects are not cached in the server cache, the server controller invokes HTTP stubs to download the objects from the remote Internet server. Then, these objects are sent to the client stub through the response sender. The client stubs coordinate data compression and other optimizations before the response is finally sent back to the client proxy. GPRSWeb has four main mechanisms to improve web performance over GPRS networks. First, GPRSWeb applies a UDP based transport layer

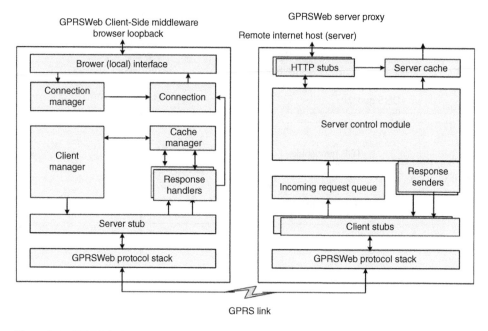

GPRSWeb Client-Side middleware browser loopback

GPRSWeb server proxy

Remote internet host (server)

Figure 9.19 GPRSWeb. *Source:* Han 2013 [14]. Reproduced with permission of IEEE.

protocol to accelerate web content delivery. The GPRSWeb transport protocol runs between the server proxy the client side middleware. By incorporating the UDP based protocol, GPRSWeb mitigates the problems of TCP, such as slow startup and ACK compression, and achieves better bandwidth utilization. By relying on RLC layer retransmission, the transport protocol realizes reliable message transfer. Second, extended caching is implemented to reduce the number of round trips required to download web pages. Extended caching is an SHA (Secure Hash Algorithm) based caching protocol that can effectively increase the hit rates of client-side caching. Third, GPRSWeb applies data compression techniques to reduce the wireless network traffic volume. It utilizes gzip to compress the raw data while using delta encoding algorithms to update the cached content. Fourth, the parse and push mechanism is implemented to reduce the idle period of the wireless link during a web page download. The proxy server parses the HTML files and pushes the relevant content into clients' local caches before they are requested.

9.5.1.2 WebAccel

WebAccel [277] is a client side web service enhancement system that integrates three optimization techniques: prioritized fetching, object reordering, and connection management. The prioritized fetching mechanism fetches objects according to their priority levels. In WebAccel, on-screen objects are granted higher priority than off-screen ones. As illustrated in Figure 9.20, the on-screen objects—objects1, 2, and 3—are fetched first. In this way, WebAccel addresses the screen contention problem and thus improves the user perceived response time. The object reordering algorithm improves

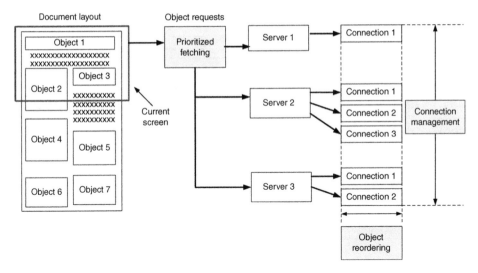

Figure 9.20 WebAccel. *Source:* Han 2013 [14]. Reproduced with permission of IEEE.

bandwidth utilization through inter-connection reordering and intra-connection reordering. Inter-connection reordering reschedules the object requests across the connections to balance the load distribution over concurrent TCP connections. Intra-connection reordering reorders the sequence of object requests according to the TCP status. For example, WebAccel sends small objects in the slow start phase of TCP and sends larger objects in the congestion avoidance phase. The connection management mechanism maintains an optimal number of concurrent TCP connections to improve bandwidth utilization as well as to mitigate the service delay caused by self-congestion. Figure 9.20 shows the connection management mechanism maintaining three TCP connections.

9.5.1.3 Nett-Gain

The Nett-Gain platform is a commercial mobile network acceleration system which provides optimization at both the transport layer and application layer [278]. The system protocol stack is shown in Figure 9.21. It supports both a dual proxy model and a clientless model that does not require middleware at the client side. At the application layer, Nett-Gain enhances the transmission rate controller by estimating the network availability using the packet delay and delay derivatives as indicators. Such indicators can reflect the network availability in a timely and quantitative manner, and thus enable an accurate insight into network congestion. The dual proxy model adopts WBST (Wireless Boosted Session Transport) [279], which is a UDP/IP based transport protocol. WBST does not require a three-way handshake and eliminates TCP slow start. For the clientless mode, TCP+ is applied to maintain the semantics and form of the standard TCP with the improved rate controller. At the application layer, the Nett-Gain platform optimizes the HTTP protocol by incorporating GetAll for the dual proxy mode and HTTP+ for the clientless mode. However, details of the techniques used are not disclosed. In addition,

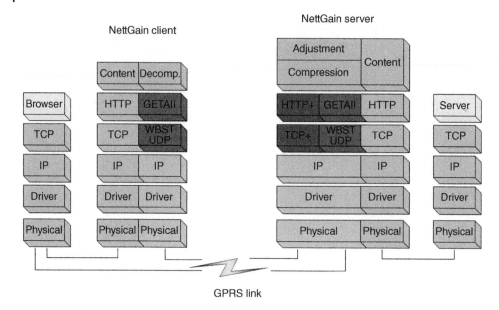

Figure 9.21 Nett-Gain platform. *Source:* Han 2013 [14]. Reproduced with permission of IEEE.

Nett-Gain applies various compression techniques, such as image reduction, animation reduction, header and request compression, and so on.

9.5.1.4 Macara

The Bytemobile Macara platform [262] is another commercial web content delivery accelerator, which accelerates mobile networks by reducing the number of round trips for data transfer in transactions using TCP and HTTP. The HTTP protocol is replaced by the Macara Dynamic Interleaving technology, which minimizes the latency of web page downloading. The platform imposes the use of multiple TCP connections and merges multiple data requests and responses. Extended caching and content format adaptation is also applied.

9.5.1.5 NPS

NPS (Non-interfering Prefetching System) [280] is a prefetching system for web services that can avoid interference between prefetch and demand requests at the server as well as in the networks. NPS only uses spare resources to prefetch web content. To avoid interference with demand requests at the server level, NPS monitors and restricts the prefetch load on the server. To avoid interference at the network level, it applies TCP-Nice [281] which is a congestion control protocol designed for low priority network traffic. NPS gives lower priority to prefetching requests than demand requests to avoid delay incurred by the prefetching requests in rendering of demand pages. Aggressive prefetching may result in cache pollution, and thus reduces cache efficiency. To address this problem, NPS applies two heuristics to regulate prefetching. First, it limits the ratio of prefetched bytes to demand bytes sent to a client. Second, it sets the *Expires HTTP* header to a value in the relatively near future to encourage clients to evict the prefectched content earlier.

9.5.1.6 pTHINC

pTHINC [282] is a PDA thin client system that leverages the more powerful web server to render the web pages, and then send simple screen updates to the PDA for display. It consists of a client reviewer and a pTHINC server. The client reviewer takes input commands from users and sends them to the pTHINC server. The server processes the command, virtualizes and resizes the display, and then sends the screen updates back to the clients. pTHINC provides the user a persistent web session that allows the user to reconnect to the same session after a disconnection. This function benefits mobile users who experience intermittent wireless connections. pTHINC can achieve up to eight times better performance than the native browser in terms of web service latency.

9.5.1.7 Chameleon

Chameleon [283] is a URICA (Usage-awaRe Interactive Content Adaptation) prototype that adapts image-rich web pages for browsing over bandwidth limited wireless links. It trades image fidelity for image loading latency. URICA consists of three components: the client application, the adaptation proxy, and the content server. The adaptation proxy adapts the images according to predictions from the prediction engine, which uses statistics-based policies and a personalized adaptation schedule. The client application allows users to send feedback to the adaptation proxy if they are not satisfied with the image quality. On receiving the feedback, the adaptation server enhances the image fidelity accordingly and saves the user's preferences for future reference. Chameleon can save up to 65% latency and up to 80% bandwidth consumption for mobile web browsing.

9.5.1.8 Silo

Silo [284] is designed to minimize the number of HTTP requests needed to render a web page. Silo leverages Javascript and DOM (Document Object Model) storage to minimize the number of HTTP requests. Silo applies DOM storage to allow a web page to maintain a key/value database on the client. As illustrated in Figure 9.23, when a web client

Figure 9.22 The standard HTTP protocol. *Source:* Han 2013 [14]. Reproduced with permission of IEEE.

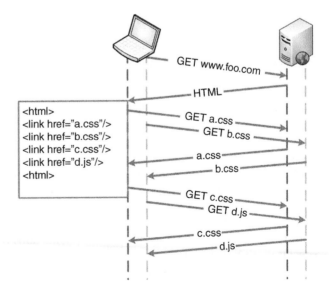

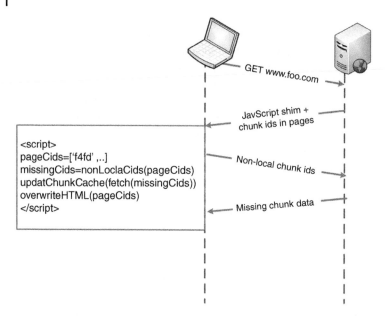

```
<script>
pageCids=['f4fd' ,..]
missingCids=nonLoclaCids(pageCids)
updatChunkCache(fetch(missingCids))
overwriteHTML(pageCids)
</script>
```

Figure 9.23 The Silo protocol. *Source:* Han 2013 [14]. Reproduced with permission of IEEE.

requests a silo enabled web page, the server responds with a small Javascript codelet that checks the available chunks on the client side and sends back the information to the server. Then, the server sends a list of the chunks contained in the web page as well as the missing chunks to the client. As compared to the standard HTTP protocol shown in Figure 9.22, Silo reduces the number of round trips required to fetch the objects contained in the web page by minimizing the HTTP requests, and avoids TCP slow start for unnecessary HTTP connections.

The optimization techniques that are adopted in web acceleration systems are summarized in Table 9.2.

9.6 Multimedia Content Delivery Acceleration

Media streaming protocols can be classified into two categories: push-based protocols and pull-based protocols [285]. Push-based protocols are UDP based, for example, RTSP, in which the media server continually streams packets to the client until the session is torn down. Pull-based protocols are TCP based, for example, HTTP, in which the client requests content actively from the media server. HTTP streaming, which relies on TCP, is preferred for Internet video applications due to its lower cost and easier implementation [286]. Popular video sharing websites, for example YouTube, YouKu, and MySpace, almost exclusively apply HTTP streaming. Leading solution providers, such as Microsoft [286], Apple [287], and Adobe [288], promote video delivery applications based on HTTP. The standardization organizations IETF and 3GPP also promote specifications based on HTTP-based streaming [289, 290]. In this section, we give an overview of techniques to accelerate HTTP-based multimedia content delivery.

Table 9.2 Optimization techniques adopted in various web acceleration systems

Systems	Proxy Module	Acceleration Techniques
GPRSWeb [223]	Dual proxies	UDP based transport protocol, SHA-based caching, parse and push, data compression, delta encoding.
WebAccel [277]	Client side proxy	Object reordering, TCP optimization, prioritized fetching, inter-connection load balancing.
Nett-Gain [278]	Dual or single proxies	TCP and HTTP optimization, image reduction, animation reduction, request compression.
Macara [262]	Dual proxies	TCP and HTTP optimization, caching, content adaptation.
NPS [280]	Single proxy	TCP optimization, interference-aware prefetching.
pTHINC [282]	Dual proxies	Thin client protocol, content adaptation.
Chameleon [283]	Dual proxies	Image quality feedback protocol, content adaptation.
Silo [284]	Client side proxy	Silo application layer protocol, DOM storage caching.

9.6.1 Adaptive Streaming

Two approaches have been proposed for HTTP-based video streaming: progressive download and adaptive streaming. The progressive download approach simply transfers the entire video file as soon as possible. The client requests the video content with certain bit rates according to the client's available bandwidth. The server responds with the requested video content. When the client's playback buffer is filled, the client starts playing the media. At the same time, it continues downloading the media in the background with the same bit rates until it finishes the download or the user tears down the connection. The client can play the media smoothly as long as the playback rate is not greater than the download rate. However, to ensure good performance, streaming over TCP requires network bandwidth that is twice that of the video rate [291]. Such bandwidth requirements are unable to be guaranteed, especially in mobile networks. As a result, if the available bandwidth shrinks, the client may suffer a long latency or frequent interruption during playback. Addressing the problem with progressive download, adaptive streaming allows either the client or server to adjust media bit rates according to the available bandwidth, and can alleviate the performance deterioration caused by bandwidth variations. Adaptive streaming can be achieved through stream switching mechanisms, in which the video content is encoded into different bit rates and quality levels, and is partitioned into fragments of a few seconds in length. The outgoing video segments are switched among alternate encodings of a stream based on either the server's estimations or the client's requests.

Adaptive streaming is an essential video content delivery technique in mobile networks, where users usually experience unstable wireless channels with bursty packet losses that result in varying available bandwidth. To accommodate adverse channel conditions, adaptive streaming is applied to switch to lower bit rate streaming, thus reducing congestion and sustaining the video streaming [292]. Therefore, video adaptive mechanisms are the key component of HTTP-based adaptive streaming. They are expected to map the channel conditions to the bit rates of the video files in a timely and accurate manner. If this is not the case, the large reaction delay of the rate-based adaptive mechanism can either degrade user perceived quality or introduce a delay spike and freeze the streaming. Adaptive mechanisms can be implemented either at the client side or at the server side. We discuss both these mechanisms.

9.6.1.1 Client Side Mechanisms

Client side adaptive algorithms are usually implemented in the video player. The server informs the clients about the available audio and video bit rates and resolutions through the manifest file. The player requests the media content with appropriate bitrate according to the estimated available bandwidth. We discuss the adaptive algorithms of two commercialized video players: Microsoft Silverlight and Netflix Player.

The Microsoft Silverlight smooth streaming player starts by requesting the lowest bitrates, and gradually increases the requested bitrates up to the highest available. The smooth streaming player uses throughput averaged over several fragments to estimate the available bandwidth rather than using the latest fragment throughput. Such a rate adaptation mechanism introduces a reaction delay. As a result, Microsoft Silverlight tends to neglect short bandwidth spikes and reacts only to relatively large spikes [293].

Compared to the Microsoft Silverlight smooth streaming player, the Netflix player is more aggressive [293]. It estimates negative bandwidth spikes by using a smoothed version of the underlying per-fragment throughput while estimating positive bandwidth spikes based on instantaneous measurements. Therefore, the Netflix player attempts to deliver the highest possible encoding rate even when the bandwidth is not sufficient, and it neglects negative bandwidth spikes. To compensate for the lack of available bandwidth, Netflix uses a large playback buffer size, which results in a long start up latency.

Although both the Microsoft Silverlight smooth streaming player and the Netflix player fail to adjust their bitrates to the available bandwidth in a timely manner, they perform very well in wired networks. Because wired networks usually have high bandwidth, the playback buffer can be filled quickly after short lived negative bandwidth spikes. However, wireless networks, which have only limited bandwidth, may require a relatively long time to fill the playback buffer. As a result, streaming can be frozen while the playback buffer drains away.

One reason for the large reaction delay is that the video players tend to estimate the available bandwidth based on a smoothed version of the underlying throughput rather than the latest throughput. These estimators work well in wired networks, but they may cause problems in wireless networks as discussed above. In addition to available bandwidth estimators, the protocol latency delays the application layer actions in adapting to the available bandwidth. To improve bandwidth utilization, the servers usually maintain a large send buffer. The data in the send buffer may wait for several RTTs until they can be sent out, and thus the data format may not reflect current network conditions. Tuning the send buffer can significantly reduce the TCP protocol latency [294], and therefore may reduce the reaction delay.

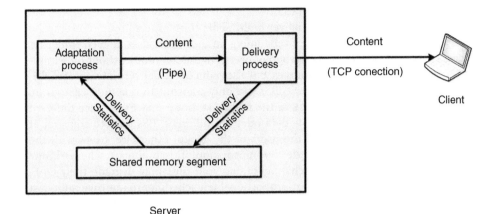

Figure 9.24 Application layer bandwidth estimation. *Source:* Han 2013 [14]. Reproduced with permission of IEEE.

9.6.1.2 Server Side Algorithms

There are three major categories of server side adaptive algorithms: application layer bandwidth estimation based algorithms, TCP stack-based bandwidth estimation based algorithms, and priority streaming algorithms [295].

- **Application layer bandwidth estimation**: This approach estimates the available bandwidth based on the time taken to transmit a specific media content item in a blocking TCP socket [296]. As shown in Figure 9.24, the delivery process sends the media content to a blocking TCP socket and writes the delivery rate statistics to shared memory segments periodically. The application process reads the statistics from the shared memory segments, and configures the media transcoder accordingly.
- **TCP based bandwidth estimation**: This method utilizes TCP statistics such as current CWND, RTT and ACKs to estimate the available bandwidth [254]. Argyriou [297] proposed a simple model that estimates the available bandwidth and expected content deterioration based on TCP statistics, and then encodes the media content accordingly.
- **Priority Streaming**: Priority streaming [298] is designed to improve the quality of video in a period of time, which is guarded by the so called deadline. The video stream is split into segments. The video syntax elements such as slices, frames, and layers are reordered according to their priorities. Each segment has a sending deadline, after which the server discards the current segment and switches to the next segment. The video quality depends on how many video syntax elements are received for the video segment before its sending deadline.

9.6.2 Other Methods

Besides adaptive streaming, many other techniques can be applied to enhance multimedia delivery over mobile networks. In this subsection, we discuss several popular techniques that can effectively improve delivery latency.

9.6.2.1 Multicast Scheduling Optimization

Multicast is an efficient method to deliver multimedia content to a group of users who request the same content. Instead of scheduling an individual user at each time slot, multicast enables BSs to schedule a group of users. Therefore, multicast not only minimizes the utilization of network resources, but also reduces the delivery latency of multimedia content. However, in multicast, BSs can only transmit to one multicast group at one data rate. Because subscribers in the multicast group may experience different channel fading, the users' achievable data rates are different. BSs have to transmit at the lowest achievable data rate of the users in the group; otherwise, users with the largest channel fading cannot decode the information. This multicast scheme limits the throughput of users with low channel fading, and thus may degrade their QoE. Won *et al.* [299] proposed a multicast proportional fair scheduler to improve multicast throughput. The idea of the multicast scheduler is to let users with poor channel quality receive a base level of service and to enhance the service quality of users with good channel quality. The multicast proportional fair scheduler improves multicast throughput and accelerates content distribution in mobile networks.

9.6.2.2 Multiple HTTP/TCP Connections

In this approach, the client opens multiple HTTP/TCP connections with the server. One of the advantages of this method is that multiple connections can avoid the delay caused by TCP slow start. The other advantage is that multiple connections can alleviate the throughput reduction effect caused by packet loss since the probabilities that all concurrent TCP connections experience packet loss are small. Therefore, the use of multiple HTTP/TCP connections can alleviate streaming performance deterioration caused by packet loss [300, 301] and insufficient bandwidth [302]. However, in wireless networks, concurrent TCP connections may result in self-congestion, and thus introduce additional latency.

9.6.2.3 Priority-Based Protection

In this method, media content items are given different protection in terms of FEC and retransmission opportunities according to their importance. In [303], frame priorities are associated with RLC layer protection in mobile networks. Important frames are given higher priorities, and thus gain more protection in terms of more retransmission opportunities. In [304], different layers of the layered video content are granted different error protection by limiting the number of retransmission attempts according to their priorities.

9.6.2.4 Multiple Path Aggregation

As discussed in the previous section, network aggregation can increase the bandwidth and reliability of wireless networks. Such advantages benefit media delivery in wireless networks. Miu *et al.* [305] proposed using multiple channels simultaneously, or to switch among them based on the channel conditions, in order to reduce media transmission latency over WLAN. In [306, 307], the authors proposed exploiting the diversity of multiple WWANs and to aggregate lower capacity wide area data channels together to create a single high bandwidth channel for multimedia applications. Kaspar *et al.* [308] used HTTP range requests to download video segments over multiple wireless links.

9.6.2.5 Cooperative Delivery

In wireless networks, users may experience bad channel quality due to wireless fading and the distance between the users and BSs. This can cause packet losses and introduces delay. Relying on nearby wireless peers to relay the packets can be one effective method against erroneous channel conditions. Li *et al.* [309] proposed a multi-source streaming system that leverages wireless peers by having them form a joint sender group with the server. Lu *et al.* [310] proposed an opportunistic retransmission protocol by using "overhearing" nodes as relays to retransmit failed packets on behalf of the source.

9.7 Summary

In this chapter, we have presented an overview of content delivery acceleration solutions in mobile networks. By studying live network measurements, we have identified the major obstacles that delay content delivery over mobile networks. Then, we have classified the solutions for accelerating content delivery into three categories: mobile system evolution, content and network optimization, and mobile data offloading, and have discussed content delivery acceleration solutions in each category. Considering web browsing and video streaming as two of the most important services in future mobile networks, we have provided an overview of web content delivery acceleration systems, and discussed HTTP-based adaptive streaming techniques. Therefore, this chapter is a useful reference for further investigation.

In recent years, tremendous efforts have been made to accelerate service delivery in mobile networks from both academia and industry. However, mobile network latency is still too large to satisfy users' expectations. There are still many open challenges to be answered. In terms of web service delivery, to the best of our knowledge, most existing web acceleration systems do not integrate network layer acceleration techniques. Furthermore, the interactions of different content delivery acceleration approaches are seldom studied. In addition, the question of how to implement a prefetching mechanism with a reasonable pricing policy is still under investigation.

9.8 Questions

9.1 From the networking perspective, what are the major obstacles for timely delivering content in mobile networks?

9.2 What are the network domain techniques for expediting content delivery in mobile networks?

9.3 What are the content domain techniques for expediting content delivery in mobile networks?

References

1 Hasan, Z., Boostanimehr, H., and Bhargava, V. (2011) Green cellular networks: A survey, some research issues and challenges. *IEEE Communications Surveys and Tutorials*, **13** (4), 524–540.

2 Gürandr, G. and Alagöz, F. (2011) Green wireless communications via cognitive dimension: an overview. *IEEE Network*, **25** (2), 50–56.

3 Auer, G., Giannini, V., Desset, C., Godor, I., Skillermark, P., Olsson, M., Imran, M., Sabella, D., Gonzalez, M., Blume, O., and Fehske, A. (2011) How much energy is needed to run a wireless network? *IEEE Wireless Communications*, **18** (5), 40–49.

4 Correia, L., Zeller, D., Blume, O., Ferling, D., Jading, Y., Auer, G., and Van Der Perre, L. (2010) Challenges and enabling technologies for energy aware mobile radio networks. *IEEE Communications Magazine*, **48** (11), 66–72.

5 Han, T. and Ansari, N. (2013) On greening cellular networks via multicell cooperation. *IEEE Wireless Communications Magazine*, **20** (1), 82–89.

6 Etoh, M., Ohya, T., and Nakayama, Y. (2008) Energy consumption issues on mobile network systems, in *International Symposium on Applications and the Internet*, Jul. 28, 2008, Turku, Finland.

7 Piro, G., Miozzo, M., Forte, G., Baldo, N., Grieco, L., Boggia, G., and Dini, P. (2013) HetNets powered by renewable energy sources: Sustainable next-generation cellular networks. *IEEE Internet Computing*, **17** (1), 32–39.

8 Marsan, M., Chiaraviglio, L., Ciullo, D., and Meo, M. (2009) Optimal energy savings in cellular access networks, in *IEEE International Conference on Communications (Workshops)*, Jun. 14, 2009, Dresden, Germany.

9 Niu, Z. (2011) Tango: traffic-aware network planning and green operation. *IEEE Wireless Communications*, **18** (5), 25–29.

10 Zhou, J., Li, M., Liu, L., She, X., and Chen, L. (2010) Energy source aware target cell selection and coverage optimization for power saving in cellular networks, in *IEEE/ACM International Conference on Green Computing and Communications*, Dec. 18, 2010, Hangzhou, China.

11 Han, T. and Ansari, N. (2012) ICE: Intelligent cell breathing to optimize the utilization of green energy. *IEEE Communications Letters*, **16** (6), 866–869.

12 Han, T. and Ansari, N. (2012) Optimizing cell size for energy saving in cellular networks with hybrid energy supplies, in *IEEE Global Telecommunications Conference*, Dec. 3, 2012, Anaheim, CA, USA.

13 Heliot, F., Imran, M., and Tafazolli, R. (2011) Energy efficiency analysis of idealized coordinated multi-point communication system, in *IEEE Vehicular Technology Conference*, May 15, 2011, Budapest, Hungary.

14 Han, T., Ansari, N., Wu, M., and Yu, H. (2013) On accelerating content delivery in mobile networks. *IEEE Communications Surveys Tutorials*, **15** (3), 1314–1333.

15 Han, T. and Ansari, N. (2012) Opportunistic content pushing via WiFi hotspots, in *the 3rd IEEE International Conference on Network Infrastructure and Digital Content*, Sep. 21, 2012, Beijing, China.

16 Deshpande, P., Hou, X., and Das, S.R. (2010) Performance comparison of 3G and metro-scale WiFi for vehicular network access, in *the 10th Annual Conference on Internet Measurement*, Nov. 1, 2010, Melbourne, Australia.

17 Han, T. and Ansari, N. (2014) Powering mobile networks with green energy. *IEEE Wireless Communications*, **21** (1), 90–96.

18 Han, T. and Ansari, N. (Aug 8, 2013) On optimizing green energy utilization for cellular networks with hybrid energy supplies. *IEEE Transactions on Wireless Communications*, **12** (8), 3872–3882.

19 Kim, H., de Veciana, G., Yang, X., and Venkatachalam, M. (2012) Distributed α-optimal user association and cell load balancing in wireless networks. *IEEE/ACM Transactions on Networking*, **20** (1), 177–190.

20 Jo, H.S., Sang, Y.J., Xia, P., and Andrews, J. (2012) Heterogeneous cellular networks with flexible cell association: A comprehensive downlink sinr analysis. *IEEE Transactions on Wireless Communications*, **11** (10), 3484–3495.

21 Lee, K., Rhee, I., Lee, J., Yi, Y., and Chong, S. (2010) Mobile data offloading: how much can WiFi deliver? *SIGCOMM Computing and Communications Review*, **41** (4), 425–426.

22 Balasubramanian, A., Mahajan, R., and Venkataramani, A. (2010) Augmenting mobile 3G using WiFi, in *the 8th International Conference on Mobile Systems, Applications, and Services*, Jun. 15, 2010, San Francisco, CA, USA.

23 Han, B., Hui, P., Kumar, V.A., Marathe, M.V., Pei, G., and Srinivasan, A. (2010) Cellular traffic offloading through opportunistic communications: a case study, in *Proceedings of the 5th ACM Workshop on Challenged Networks*, Sep. 24, 2010, Chicago, Illinois, USA.

24 Mashhadi, A. and Hui, P. (Oct, 2010), Proactive caching for hybrid urban mobile networks. URL http://www.cs.ucl.ac.uk/research/researchnotes/documents/RN_10_05_000.pdf.

25 Zhuo, X., Gao, W., Cao, G., and Dai, Y. (2011) Win-coupon: An incentive framework for 3G traffic offloading, in *the 19th IEEE International Conference on Network Protocols*, Oct. 17, 2011, Vancouver, BC, Canada.

26 Corroy, S., Falconetti, L., and Mathar, R. (2012) Dynamic cell association for downlink sum rate maximization in multi-cell heterogeneous networks, in *IEEE International Conference on Communications*, June. 10, 2012, Ottawa, Canada.

27 Fooladivanda, D. and Rosenberg, C. (2013) Joint resource allocation and user association for heterogeneous wireless cellular networks. *IEEE Transactions on Wireless Communications*, **12** (1), 248–257.

28 Madan, R., Borran, J., Sampath, A., Bhushan, N., Khandekar, A., and Ji, T. (2010) Cell association and interference coordination in heterogeneous lte-a cellular networks. *IEEE Journal on Selected Areas in Communications*, **28** (9), 1479–1489.

29 Bletsas, A., Khisti, A., Reed, D., and Lippman, A. (2006) A simple cooperative diversity method based on network path selection. *IEEE Journal on Selected Areas in Communications*, **24** (3), 659–672.

30 Zhao, Y., Adve, R., and Lim, T.J. (2006) Improving amplify-and-forward relay networks: optimal power allocation versus selection. *IEEE Transactions on Wireless Communications*, **6** (8), 3114–3123.

31 Wang, B., Han, Z., and Liu, K.J.R. (2007) Distributed relay selection and power control for multiuser cooperative communication networks using buyer/seller game, in *IEEE International Conference on Computer Communications*, May 6, 2007, Barcelona, Spain, pp. 544–552.

32 Sharma, S., Shi, Y., Hou, Y., and Kompella, S. (2011) An optimal algorithm for relay node assignment in cooperative Ad Hoc networks. *IEEE/ACM Transactions on Networking*, **19** (3), 879–892.

33 Yang, D., Fang, X., and Xue, G. (2012) HERA: An optimal relay assignment scheme for cooperative networks. *IEEE Journal on Selected Areas in Communications*, **30** (2), 245–253.

34 Sustainable energy use in mobile communications. URL http://www.howgreenisit. co.uk/files/sustainable_energy_WP_revA_070619.pdf, (accessed on Apr. 22, 2015), (2007).

35 Orange Green Strategy for AMEA Zone, 2nd ITU Green Standards Week, (Sep. 17, 2012).

36 Andrews, J., Singh, S., Ye, Q., Lin, X., and Dhillon, H. (2014) An overview of load balancing in HetNets: old myths and open problems. *IEEE Wireless Communications*, **21** (2), 18–25.

37 Haykin, S. (2005) Cognitive radio: brain-empowered wireless communications. *IEEE Journal on Selected Areas in Communications*, **23** (2), 201–220, doi:10.1109/JSAC.2004.839380.

38 Evolved universal terrestrial radio access (E-UTRA) and evolved universal terrestrial radio access network (E-UTRAN), (Oct. 3, 2011).

39 Ericsson mobile cloud accelerator. URL http://www.ericsson.com/ourportfolio/ telecom-operators/mobile-cloud-accelerator?nav=marketcategory002, (2011).

40 E-Plus, Nokia Siemens Networks build Germany's first off-grid base station. URL http://www.nokiasiemensnetworks.com/news-events/press-room/press-releases/e-plus-nokia-siemens-networks-build-germany-s-first-off-grid-base-station.

41 Irmer, R., Droste, H., Marsch, P., Grieger, M., Fettweis, G., Brueck, S., Mayer, H.P., Thiele, L., and Jungnickel, V. (2011) Coordinated multipoint: Concepts, performance, and field trial results. *IEEE Communications Magazine*, **49** (2), 102–111.

42 Sawahashi, M., Kishiyama, Y., Morimoto, A., Nishikawa, D., and Tanno, M. (2010) Coordinated multipoint transmission/reception techniques for LTE-advanced [Coordinated and Distributed MIMO]. *IEEE Wireless Communications*, **17** (3), 26–34.

43 Jayaweera, S., Bkassiny, M., and Avery, K. (2011) Asymmetric cooperative communications based spectrum leasing via auctions in cognitive radio networks. *IEEE Transactions on Wireless Communications*, **10** (8), 2716–2724.

44 Fan, L., Lei, X., Fan, P., and Hu, R. (2012) Outage probability analysis and power allocation for two-way relay networks with user selection and outdated channel state information. *IEEE Communications Letters*, **16** (5), 638–641.

45 Laneman, J.N. and Wornell, G.W. (2003) Distributed space-time-coded protocols for exploiting cooperative diversity in wireless networks. *IEEE Transactions on Information Theory*, **49** (10), 2415–2425.

46 Zou, Y., Yao, Y.D., and Zheng, B. (2012) Opportunistic distributed space-time coding for decode-and-forward cooperation systems. *IEEE Transactions on Signal Processing*, **60** (4), 1766–1781.

47 Herlich, M. and Karl, H. (2011) Reducing power consumption of mobile access networks with cooperation, in *the 2nd International Conference on Energy-Efficient Computing and Networking*, May 31, 2011, New York City, NY.

48 Goiri, I., Le, K., Nguyen, T.D., Guitart, J., Torres, J., and Bianchini, R. (2012) Greenhadoop: leveraging green energy in data-processing frameworks, in *the 7th ACM European Conference on Computer Systems*, Apr. 10, 2012, Bern, Switzerland, pp. 57–70.

49 Ho, C.K. and Zhang, R. (2012) Optimal energy allocation for wireless communications with energy harvesting constraints. *IEEE Transactions on Signal Processing*, **60** (9), 4808–4818.

50 Maghraby, H.A.M., Shwehdi, M., and Al-Bassam, G. (2002) Probabilistic assessment of photovoltaic (PV) generation systems. *IEEE Transactions on Power Systems*, **17** (1), 205–208.

51 Yang, J. and Ulukus, S. (2012) Optimal packet scheduling in an energy harvesting communication system. *IEEE Transactions on Communications*, **60** (1), 220–230.

52 Han, T. and Ansari, N. (2013) Energy agile packet scheduling to leverage green energy for next generation cellular networks, in *IEEE International Conference on Communications*, Jun. 9, 2013, Budapest, Hungary.

53 Farbod, A. and Todd, T.D. (2007) Resource allocation and outage control for solar-powered WLAN mesh networks. *IEEE Transactions on Mobile Computing*, **6** (8), 960–970.

54 Kwak, J., Son, K., Yi, Y., and Chong, S. (2012) Greening effect of spatio-temporal power sharing policies in cellular networks with energy constraints. *IEEE Transactions on Wireless Communications*, **11** (12), 4405–4415.

55 Han, T. and Ansari, N. (2013) Green-energy aware and latency aware user associations in heterogeneous cellular networks, in *IEEE Global Telecommunications Conference*, Dec. 9, 2013, Atlanta, GA, USA.

56 Yucek, T. and Arslan, H. (2009) A survey of spectrum sensing algorithms for cognitive radio applications. *IEEE Communications Surveys Tutorials*, **11** (1), 116–130.

57 Akyildiz, I., Lee, W.Y., Vuran, M.C., and Mohanty, S. (2008) A survey on spectrum management in cognitive radio networks. *IEEE Communications Magazine*, **46** (4), 40–48.

58 Akyildiz, I.F., Lee, W.Y., Vuran, M.C., and Mohanty, S. (2006) Next generation/dynamic spectrum access/cognitive radio wireless networks: A survey. *Computer Networks*, **50** (13), 2127–2159.

59 Gur, G. and Alagoz, F. (2011) Green wireless communications via cognitive dimension: an overview. *IEEE Network*, **25** (2), 50–56.

60 Domenico, A.D., Strinati, E.C., and Capone, A. (2014) Enabling green cellular networks: A survey and outlook. *Computer Communications*, **37**, 5–24.

61 Deng, R., He, S., Chen, J., Jia, J., Zhuang, W., and Sun, Y. (2012) Energy-efficient spectrum sensing by optimal periodic scheduling in cognitive radio networks. *IET Communications*, **6** (6), 676–684.

62 Zhao, Q. and Sadler, B. (2007) A survey of dynamic spectrum access. *IEEE Signal Processing Magazine*, **24** (3), 79–89.

63 Yucek, T. and Arslan, H. (2009) A survey of spectrum sensing algorithms for cognitive radio applications. *IEEE Communications Surveys Tutorials*, **11** (1), 116–130.

64 Gan, X., Xu, M., and Li, H. (2012) Energy efficient sequential sensing in multi-user cognitive ad hoc networks: A consideration of an adc device. *Journal of Communications and Networks*, **14** (2), 188–194.

65 Axell, E., Leus, G., and Larsson, E. (2010) Overview of spectrum sensing for cognitive radio, in *the 2nd International Workshop on Cognitive Information Processing*, Jun. 14, 2010, Naregno, Elba Island, Italy, pp. 322–327.

66 Wang, L., Wang, J., Ding, G., Song, F., and Wu, Q. (2011) A survey of cluster-based cooperative spectrum sensing in cognitive radio networks, in *Cross Strait Quad-Regional Radio Science and Wireless Technology Conference*, Jul. 26, 2011, Harbin, Heilongjiang, China.

67 Wei, J. and Zhang, X. (2010) Energy-efficient distributed spectrum sensing for wireless cognitive radio networks, in *IEEE Conference on Computer Communications Workshops*, Mar. 15, 2010, San Diego, CA, USA.

68 Wang, B., Liu, K., and Clancy, T. (2010) Evolutionary cooperative spectrum sensing game: how to collaborate? *IEEE Transactions on Communications*, **58** (3), 890–900.

69 Lai, J., Dutkiewicz, E., Liu, R.P., and Vesilo, R. (2012) Comparison of cooperative spectrum sensing strategies in distributed cognitive radio networks, in *IEEE Global Communications Conference*, Dec. 3, 2012, Anaheim, CA, USA.

70 Chien, W.B., Yang, C.K., and Huang, Y.H. (2011) Energy-saving cooperative spectrum sensing processor for cognitive radio system. *IEEE Transactions on Circuits and Systems I: Regular Papers*, **58** (4), 711–723.

71 Sun, X. and Tsang, D. (2013) Energy-efficient cooperative sensing scheduling for multi-band cognitive radio networks. *IEEE Transactions on Wireless Communications*, **12** (10), 4943–4955.

72 Jiang, H., Lai, L., Fan, R., and Poor, H. (2009) Optimal selection of channel sensing order in cognitive radio. *IEEE Transactions on Wireless Communications*, **8** (1), 297–307.

73 Chen, Y., Zhao, Q., and Swami, A. (2009) Distributed spectrum sensing and access in cognitive radio networks with energy constraint. *IEEE Transactions on Signal Processing*, **57** (2), 783–797.

74 Lee, W.Y. and Akyildiz, I. (2012) Spectrum-aware mobility management in cognitive radio cellular networks. *IEEE Transactions on Mobile Computing*, **11** (4), 529–542.

75 Wang, L.C. and Wang, C.W. (2008) Spectrum handoff for cognitive radio networks: Reactive-sensing or proactive-sensing?, in *IEEE International Performance, Computing and Communications Conference*, Dec. 7, 2008, Austin, Texas, USA.

76 Wang, C.W. and Wang, L.C. (2012) Analysis of reactive spectrum handoff in cognitive radio networks. *IEEE Journal on Selected Areas in Communications*, **30** (10), 2016–2028.

77 Li, Q., Feng, Z., Li, W., and Gulliver, T. (2013) Joint temporal and spatial spectrum sharing in cognitive radio networks: A region-based approach with cooperative spectrum sensing, in *IEEE Wireless Communications and Networking Conference*, Apr. 7, 2013, Shanghai, China.

78 Huang, J., Berry, R.A., and Honig, M.L. (2006) Auction-based spectrum sharing. *Mobile Networks and Applications*, **11** (3), 405–418.

79 Stotas, S. and Nallanathan, A. (2011) Enhancing the capacity of spectrum sharing cognitive radio networks. *IEEE Transactions on Vehicular Technology*, **60** (8), 3768–3779.

80 Zhang, Q., Jia, J., and Zhang, J. (2009) Cooperative relay to improve diversity in cognitive radio networks. *IEEE Communications Magazine*, **47** (2), 111–117.

81 Etkin, R., Parekh, A., and Tse, D. (2007) Spectrum sharing for unlicensed bands. *IEEE Journal on Selected Areas in Communications*, **25** (3), 517–528.

82 Zhang, R., Liang, Y.C., and Cui, S. (2010) Dynamic resource allocation in cognitive radio networks. *IEEE Signal Processing Magazine*, **27** (3), 102–114.

83 Niyato, D. and Hossain, E. (2008) Spectrum trading in cognitive radio networks: A market-equilibrium-based approach. *IEEE Wireless Communications*, **15** (6), 71–80.

84 Illanko, K., Naeem, M., Anpalagan, A., and Androutsos, D. (2012) Low complexity energy efficient power allocation for green cognitive radio with rate constraints, in *IEEE Global Communications Conference*, Dec. 3, 2012, Anaheim, CA, USA.

85 Buzzi, S. and Saturnino, D. (2011) A game-theoretic approach to energy-efficient power control and receiver design in cognitive cdma wireless networks. *IEEE Journal of Selected Topics in Signal Processing*, **5** (1), 137–150.

86 Han, T. and Ansari, N. (2013) Auction-based energy-spectrum trading in green cognitive cellular networks, in *IEEE International Conference on Communications*, Jun. 9, 2013, Budapest, Hungary.

87 Hasan, Z., Bansal, G., Hossain, E., and Bhargava, V. (2009) Energy-efficient power allocation in ofdm-based cognitive radio systems: A risk-return model. *IEEE Transactions on Wireless Communications*, **8** (12), 6078–6088.

88 Gao, S., Qian, L., and Vaman, D. (2009) Distributed energy efficient spectrum access in cognitive radio wireless ad hoc networks. *IEEE Transactions on Wireless Communications*, **8** (10), 5202–5213.

89 Wang, J., Ghosh, M., and Challapali, K. (2011) Emerging cognitive radio applications: A survey. *IEEE Communications Magazine*, **49** (3), 74–81.

90 Ge, M. and Wang, S. (2013) Energy-efficient power allocation for cooperative relaying cognitive radio networks, in *IEEE Wireless Communications and Networking Conference*, Apr. 7, 2013, Shanghai, China, pp. 691–696.

91 Li, P., Guo, S., Cheng, Z., and Vasilakos, A. (2013) Joint relay assignment and channel allocation for energy-efficient cooperative communications, in *IEEE Wireless Communications and Networking Conference*, Apr. 7, 2013, Shanghai, China.

92 Huang, S., Chen, H., Zhang, Y., and Chen, H.H. (2013) Sensing-energy tradeoff in cognitive radio networks with relays. *IEEE Systems Journal*, **7** (1), 68–76.

93 Shaat, M. and Bader, F. (2012) Asymptotically optimal subcarrier matching and power allocation for cognitive relays with power and interference constraints, in *IEEE Wireless Communications and Networking Conference*, Apr. 1, 2012, Paris, France, pp. 663–668.

94 Zhao, G., Yang, C., Li, G., Li, D., and Soong, A. (2011) Power and channel allocation for cooperative relay in cognitive radio networks. *IEEE Journal of Selected Topics in Signal Processing*, **5** (1), 151–159.

95 Chen, D., Ji, H., and Li, X. (2011) Optimal distributed relay selection in underlay cognitive radio networks: An energy-efficient design approach, in *IEEE Wireless Communications and Networking Conference*, Mar. 28, 2011, Quintana-Roo, Mexico.

96 Luo, C., Min, G., Yu, F., Chen, M., Yang, L., and Leung, V. (2013) Energy-efficient distributed relay and power control in cognitive radio cooperative communications. *IEEE Journal on Selected Areas in Communications*, **31** (11), 2442–2452.

97 Chen, D., Ji, H., and Leung, V. (2011) Energy-efficient cross-layer enhancement of multimedia transmissions over cognitive radio relay networks, in *IEEE Wireless Communications and Networking Conference*, Mar. 28, 2011, Quintana-Roo, Mexico, pp. 856–861.

98 Cao, B., Mark, J., Zhang, Q., Lu, R., Lin, X., and Shen, X. (2013) On optimal communication strategies for cooperative cognitive radio networking, in *IEEE International Conference on Computer Communications*, Apr. 14, 2013, Turin, Italy.

99 Fudenberg, D. and Tirole, J. (1991) *Game Theory*, MIT Press, Cambridge, MA.

100 Huang, X., Han, T., and Ansari, N. (2015) On green energy powered cognitive radio networks. *IEEE Communications Surveys & Tutorials*, **17** (2), 827–842.

101 Andrews, J., Claussen, H., Dohler, M., Rangan, S., and Reed, M. (2012) Femtocells: Past, present, and future. *IEEE Journal on Selected Areas in Communications*, **30** (3), 497–508.

102 Cheung, W.C., Quek, T., and Kountouris, M. (2012) Throughput optimization, spectrum allocation, and access control in two-tier femtocell networks. *IEEE Journal on Selected Areas in Communications*, **30** (3), 561–574.

103 ElSawy, H. and Hossain, E. (2014) Two-tier hetnets with cognitive femtocells: Downlink performance modeling and analysis in a multichannel environment. *IEEE Transactions on Mobile Computing*, **13** (3), 649–663.

104 Wildemeersch, M., Quek, T., Slump, C., and Rabbachin, A. (2013) Cognitive small cell networks: Energy efficiency and trade-offs. *IEEE Transactions on Communications*, **61** (9), 4016–4029.

105 Yang, C., Li, J., Sheng, M., and Liu, Q. (2012) Green heterogeneous networks: a cognitive radio idea. *IET Communications*, **6** (13), 1952–1959, doi:10.1049/iet-com.2011.0801.

106 Xie, R., Yu, F., and Ji, H. (2012) Energy-efficient spectrum sharing and power allocation in cognitive radio femtocell networks, in *IEEE International Conference on Computer Communications*, Mar. 25, 2012, Orlando, FL, USA.

107 Le, T., Mayaram, K., and Fiez, T. (2008) Efficient far-field radio frequency energy harvesting for passively powered sensor networks. *IEEE Journal of Solid-State Circuits*, **43** (5), 1287–1302.

108 Kansal, A., Hsu, J., Zahedi, S., and Srivastava, M.B. (2007) Power management in energy harvesting sensor networks. *ACM Transactions on Embedded Computing Systems*, **6** (4), 32.

109 Han, T. and Ansari, N. (2013) On optimizing green energy utilization for cellular networks with hybrid energy supplies. *IEEE Transactions on Wireless Communications*, **12** (8), 3872–3882.

110 Zhang, R. and Ho, C.K. (2013) Mimo broadcasting for simultaneous wireless information and power transfer. *IEEE Transactions on Wireless Communications*, **12** (5), 1989–2001.

111 Gunduz, D., Stamatiou, K., Michelusi, N., and Zorzi, M. (2014) Designing intelligent energy harvesting communication systems. *IEEE Communications Magazine*, **52** (1), 210–216.

112 Luo, S., Zhang, R., and Lim, T.J. (2013) Optimal save-then-transmit protocol for energy harvesting wireless transmitters. *IEEE Transactions on Wireless Communications*, **12** (3), 1196–1207.

113 Sudevalayam, S. and Kulkarni, P. (2011) Energy harvesting sensor nodes: Survey and implications. *IEEE Communications Surveys Tutorials*, **13** (3), 443–461.

114 Sharma, V., Mukherji, U., Joseph, V., and Gupta, S. (2010) Optimal energy management policies for energy harvesting sensor nodes. *IEEE Transactions on Wireless Communications*, **9** (4), 1326–1336.

115 Ozel, O., Tutuncuoglu, K., Yang, J., Ulukus, S., and Yener, A. (2011) Transmission with energy harvesting nodes in fading wireless channels: Optimal policies. *IEEE Journal on Selected Areas in Communications*, **29** (8), 1732–1743.

116 Huang, C., Zhang, R., and Cui, S. (2013) Throughput maximization for the gaussian relay channel with energy harvesting constraints. *IEEE Journal on Selected Areas in Communications*, **31** (8), 1469–1479.

117 Liu, L., Zhang, R., and Chua, K.C. (2013) Wireless information transfer with opportunistic energy harvesting. *IEEE Transactions on Wireless Communications*, **12** (1), 288–300.

118 Nasir, A., Zhou, X., Durrani, S., and Kennedy, R. (2013) Relaying protocols for wireless energy harvesting and information processing. *IEEE Transactions on Wireless Communications*, **12** (7), 3622–3636.

119 Blasco, P., Gunduz, D., and Dohler, M. (2013) A learning theoretic approach to energy harvesting communication system optimization. *IEEE Transactions on Wireless Communications*, **12** (4), 1872–1882.

120 Yin, S., Zhang, E., Yin, L., and Li, S. (2013) Optimal saving-sensing-transmitting structure in self-powered cognitive radio systems with wireless energy harvesting, in *IEEE International Conference on Communications*, Jun. 9, 2013, Budapest, Hungary.

121 Park, S., Kim, H., and Hong, D. (2013) Cognitive radio networks with energy harvesting. *IEEE Transactions on Wireless Communications*, **12** (3), 1386–1397.

122 Deng, R., Chen, J., Yuen, C., Cheng, P., and Sun, Y. (2012) Energy-efficient cooperative spectrum sensing by optimal scheduling in sensor-aided cognitive radio networks. *IEEE Transactions on Vehicular Technology*, **61** (2), 716–725.

123 Park, S., Lee, S., Kim, B., Hong, D., and Lee, J. (2011) Energy-efficient opportunistic spectrum access in cognitive radio networks with energy harvesting, in *the 4th International Conference on Cognitive Radio and Advanced Spectrum Management*, Oct. 26, 2011, Barcelona, Spain.

124 Park, S., Heo, J., Kim, B., Chung, W., Wang, H., and Hong, D. (2012) Optimal mode selection for cognitive radio sensor networks with rf energy harvesting, in *IEEE 23rd International Symposium on Personal Indoor and Mobile Radio Communications*, Sep. 9, 2012, Sydney, Australia.

125 Chung, W., Park, S., Lim, S., and Hong, D. (2013) Optimal transmit power control for energy-harvesting cognitive radio system, in *IEEE 78th Vehicular Technology Conference*, Sep. 2, 2013, Las Vegas, Nevada, USA.

126 Gao, X., Xu, W., Li, S., and Lin, J. (2013) An online energy allocation strategy for energy harvesting cognitive radio systems, in *International Conference on Wireless Communications Signal Processing*, Oct. 24, 2013, Hangzhou, China.

127 Yu, R., Zhang, Y., Gjessing, S., Yuen, C., Xie, S., and Guizani, M. (2011) Cognitive radio based hierarchical communications infrastructure for smart grid. *IEEE Network*, **25** (5), 6–14.

128 Qiu, R., Hu, Z., Chen, Z., Guo, N., Ranganathan, R., Hou, S., and Zheng, G. (2011) Cognitive radio network for the smart grid: Experimental system architecture, control algorithms, security, and microgrid testbed. *IEEE Transactions on Smart Grid*, **2** (4), 724–740.

129 Bu, S., Yu, F., and Qian, Y. (2013) Energy-efficient cognitive heterogeneous networks powered by the smart grid, in *IEEE International Conference on Computer Communications*, Apr. 14, 2013, Turin, Italy.

130 Ansari, N. and Han, T. (2016) FreeNet: Spectrum and energy harvesting wireless networks. *IEEE Network*, **30** (1), 66–71.

131 Lu, X., Wang, P., Niyato, D., and Hossain, E. (2014) Dynamic spectrum access in cognitive radio networks with RF energy harvesting. *IEEE Wireless Communications*, **21** (3), 102–110.

132 Siomina, I., Varbrand, P., and Yuan, D. (2006) Automated optimization of service coverage and base station antenna configuration in UMTS networks. *IEEE Wireless Communications*, **13** (6), 16–25.

133 Doppler, K., Rinne, M., Wijting, C., Ribeiro, C., and Hugl, K. (2009) Device-to-device communication as an underlay to LTE-advanced networks. *IEEE Communications Magazine*, **47** (12), 42–49.

134 Windfi: Renewable energy wireless basestation. URL http://www.wirelesswhitespace.org/projects/wind-fi-renewable-energy-basestation.aspx.

135 Ansari, N., Zhang, C., Rojas-Cessa, R., Sakarindr, P., and Hou, E. (2008) Networking for critical conditions. *IEEE Wireless Communications*, **15** (2), 73–81.

136 Badawy, G., Sayegh, A., and Todd, T. (2010) Energy provisioning in solar-powered wireless mesh networks. *IEEE Transactions on Vehicular Technology*, **59** (8), 3859–3871.

137 Rajan, D., Sabharwal, A., and Aazhang, B. (2004) Delay-bounded packet scheduling of bursty traffic over wireless channels. *IEEE Transactions on Information Theory*, **50** (1), 125–144.

138 Comcast xfinity hotspots map. URL http://hotspots.wifi.comcast.com/, (accessed on Apr. 22, 2015).

139 Han, T. and Ansari, N. (2014) Enabling mobile traffic offloading via energy spectrum trading. *IEEE Transactions on Wireless Communications*, **13** (6), 3317–3328.

140 Arnold, O., Richter, F., Fettweis, G., and Blume, O. (2010) Power consumption modeling of different base station types in heterogeneous cellular networks, in *IEEE Future Network and Mobile Summit*, Jun. 16, 2010, Florence, Italy.

141 Rappaport, T.S. (2002) *Wireless Communications: Principles and Practice*, Prentice Hall.

142 Garey, M.R. and Johnson, D.S. (1979) *Computer and Intractability: A Guide to the Thoery of NP-Completeness*, W.H. Freeman and Company: New York, NY, USA.

143 Boyd, S. and Vandenberghe, L. (2004) *Convex Optimization*, Cambridge University Press: Cambridge, UK.

144 Evolution of land mobile radio (including personal) ccommunications: COST 231. URL http://www.cost.eu/COST_Actions/ict/Actions/231, (accessed on Apr. 22, 2015).

145 Cisco Aironet 1550 Series Outdoor Access Point. URL http://www.cisco.com/en/US/prod/collateral/wireless/ps5679/ps11451/data_sheet_c78-641373.pdf.

146 Cai, L., Poor, H., Liu, Y., Luan, T., Shen, X., and Mark, J. (2011) Dimensioning network deployment and resource management in green mesh networks. *IEEE Wireless Communications*, **18** (5), 58–65.

147 Peng, C., Lee, S.B., Lu, S., Luo, H., and Li, H. (2011) Traffic-driven power saving in operational 3G cellular networks, in *the 17th International Conference on Mobile Computing and Networking*, Sep. 19, 2011, Las Vegas, Nevada, USA.

148 System advisor model (SAM). URL https://sam.nrel.gov/, (accessed on Apr. 22, 2015).

149 PVWatts. URL http://www.nrel.gov/rredc/pvwatts/, (accessed on Apr. 22, 2015).

150 Lorincz, J., Garma, T., and Petrovic, G. (2012) Measurements and modelling of base station power consumption under real traffic loads. *Sensors*, **12** (4), 4281–4310.

151 Marsan, M., Bucalo, G., Di Caro, A., Meo, M., and Zhang, Y. (2013) Towards zero grid electricity networking: Powering BSs with renewable energy sources, in *IEEE International Conference on Communications Workshops*, Jun. 9, 2013, Budapest, Hungary.

152 Kleinrock, L. (1976) *Queueing Systems: Computer applications*, John Wiley & Sons, Inc.: Hoboken, NJ, USA.

153 Capozzi, F., Piro, G., Grieco, L., Boggia, G., and Camarda, P. (2013) Downlink packet scheduling in LTE cellular networks: Key design issues and a survey. *IEEE Communications Surveys Tutorials*, **15** (2), 678–700.

154 Bertsekas, D.P. (2009) *Convex Optimization Theory*, Athena Scientific.

155 Han, T. and Ansari, N. (2016) Provisioning green energy for base stations in heterogeneous networks. *IEEE Transactions on Vehicular Technology*, **65** (7), 5439–5448.

156 UCSD Solar Resource Web Application. URL http://solar.ucsd.edu/datasharing/.

157 Han, T. and Ansari, N. (2016) A traffic load balancing framework for software-defined radio access networks powered by hybrid energy sources. *IEEE/ACM Transactions on Networking*, **24** (2), 1038–1051.

158 Han, T. and Ansari, N. (2017) Network utility aware traffic loading balancing in Backhaul-constrained cache-enabled small cell networks with hybrid power supplies. *IEEE Transactions on Mobile Computing*, DOI: 10.1109/TMC.2017.2652464.

159 Wang, L. and Kuo, G.S. (2013) Mathematical modeling for network selection in heterogeneous wireless networks: A tutorial. *IEEE Communications Surveys Tutorials*, **15** (1), 271–292.

160 LTE; general packet radio service (GPRS) enhancements for evolved universal terrestrial radio access network (E-UTRAN) access (3GPP TS 23.401 version 11.9.0 release 11). URL http://www.etsi.org/deliver/etsi_ts/123400_123499/123401/11.09.00_60/ts_123401v110900p.pdf, (accessed on Apr. 22, 2015).

161 Ye, Q., Rong, B., Chen, Y., Al-Shalash, M., Caramanis, C., and Andrews, J. (2013) User association for load balancing in heterogeneous cellular networks. *IEEE Transactions on Wireless Communications*, **12** (6), 2706–2716.

162 Aryafar, E., Keshavarz-Haddad, A., Wang, M., and Mung, C. (2013) RAT selection games in HetNets, in *IEEE International Conference on Computer Communications*, Apr. 14, 2013, Turin, Italy.

163 Damnjanovic, A., Montojo, J., Wei, Y., Ji, T., Luo, T., Vajapeyam, M., Yoo, T., Song, O., and Malladi, D. (2011) A survey on 3GPP heterogeneous networks. *IEEE Wireless Communications*, **18** (3), 10–21.

164 Singh, S., Dhillon, H., and Andrews, J. (2013) Offloading in heterogeneous networks: Modeling, analysis, and design insights. *IEEE Transactions on Wireless Communications*, **12** (5), 2484–2497.

165 Pantisano, F., Bennis, M., Saad, W., and Debbah, M. (2014) Cache-aware user association in backhaul-constrained small cell networks, in *the 12th International Symposium on Modeling and Optimization in Mobile, Ad Hoc, and Wireless Networks*, May 12, 2014, Hammamet, Tunisia.

166 Nakamura, T., Nagata, S., Benjebbour, A., Kishiyama, Y., Hai, T., Xiaodong, S., Ning, Y., and Nan, L. (2013) Trends in small cell enhancements in lte advanced. *IEEE Communications Magazine*, **51** (2), 98–105.

167 Wang, X., Chen, M., Taleb, T., Ksentini, A., and Leung, V. (2014) Cache in the air: exploiting content caching and delivery techniques for 5G systems. *IEEE Communications Magazine*, **52** (2), 131–139.

168 Poularakis, K., Iosifidis, G., and Tassiulas, L. (2014) Approximation algorithms for mobile data caching in small cell networks. *IEEE Transactions on Communications*, **62** (10), 3665–3677.

169 Shanmugam, K., Golrezaei, N., Dimakis, A., Molisch, A., and Caire, G. (2013) Femtocaching: Wireless content delivery through distributed caching helpers. *IEEE Transactions on Information Theory*, **59** (12), 8402–8413.

170 Monserrat, J., Droste, H., Bulakci, O., Eichinger, J., Queseth, O., Stamatelatos, M., Tullberg, H., Venkatkumar, V., Zimmermann, G., Dotsch, U., and Osseiran, A. (2014) Rethinking the mobile and wireless network architecture: The METIS research into 5G, in *European Conference on Networks and Communications*, Jun. 23, 2014, Bologna, Italy.

171 Niu, Z., Wu, Y., Gong, J., and Yang, Z. (2010) Cell zooming for cost-efficient green cellular networks. *IEEE Communications Magazine*, **48** (11), 74–79.

172 Lopez-Perez, D., Guvenc, I., De la Roche, G., Kountouris, M., Quek, T., and Zhang, J. (2011) Enhanced intercell interference coordination challenges in heterogeneous networks. *IEEE Wireless Communications*, **18** (3), 22–30.

173 Son, K., Chong, S., and Veciana, G. (2009) Dynamic association for load balancing and interference avoidance in multi-cell networks. *IEEE Transactions on Wireless Communications*, **8** (7), 3566–3576.

174 Raj, M., Kant, K., and Das, S. (2013) Energy adaptive mechanism for P2P file sharing protocols, in *Euro-Par 2012: Parallel Processing Workshops*, *Lecture Notes in Computer Science*, vol. 7640 (eds I. Caragiannis, M. Alexander, R. Badia, M. Cannataro, A. Costan, M. Danelutto, F. Desprez, B. Krammer, J. Sahuquillo, S. Scott, and J. Weidendorfer), Springer Berlin Heidelberg, pp. 89–99.

175 Imon, S., Khan, A., Di Francesco, M., and Das, S. (2015) Energy-efficient randomized switching for maximizing lifetime in tree-based wireless sensor networks. *IEEE/ACM Transactions on Networking*, **23** (5), 1401–1415.

176 HIT photovoltaic module. URL http://panasonic.net/ecosolutions/solar/hit/, (accessed on Apr. 22, 2015).

177 Gomaa, H., Messier, G., Williamson, C., and Davies, R. (2013) Estimating instantaneous cache hit ratio using markov chain analysis. *IEEE/ACM Transactions on Networking*, **21** (5), 1472–1483.

178 Breslau, L., Cao, P., Fan, L., Phillips, G., and Shenker, S. (1999) Web caching and Zipf-like distributions: evidence and implications, in *the Eighteenth Annual Joint Conference of the IEEE Computer and Communications Societies*, Mar 25,1999. New York City, New York, USA.

179 Rodriguez, P., Spanner, C., and Biersack, E.W. (2001) Analysis of web caching architectures: Hierarchical and distributed caching. *IEEE/ACM Transactions on Networking*, **9** (4), 404–418.

180 Jelenkovic, P. and Radovanovic, A. (2003) Asymptotic insensitivity of least-recently-used caching to statistical dependency, in *the Twenty-Second Annual Joint Conference of the IEEE Computer and Communications.*, Mar. 30, 2003, San Francisco, CA, USA.

181 Zhang, Y., Ansari, N., Wu, M., and Yu, H. (2012) On wide area network optimization. *IEEE Communications Surveys Tutorials*, **14** (4), 1090–1113.

182 Riordan, C. and Hulstron, R. (1990) What is an air mass 1.5 spectrum? [solar cell performance calculations], in *IEEE Photovoltaic Specialists Conference*, May 25, 1990, Kissimmee, FL, USA.

183 Han, T. and Ansari, N. (2014) Smart grid enabled mobile networks: Jointly optimizing bs operation and power distribution, in *IEEE International Conference on Communications*, Jun. 10, 2014, Sydney, Australia.

184 Vytelingum, P., Ramchurn, S.D., Voice, T.D., Rogers, A., and Jennings, N.R. (2010) Trading agents for the smart electricity grid, in *the 9th International Conference on Autonomous Agents and Multiagent Systems*, May 09, 2010, Toronto, Canada.

185 Bu, S., Yu, F., Cai, Y., and Liu, X. (2012) When the smart grid meets energy-efficient communications: Green wireless cellular networks powered by the smart grid. *IEEE Transactions on Wireless Communications*, **11** (8), 3014–3024.

186 Lavaei, J. and Low, S. (2012) Zero duality gap in optimal power flow problem. *IEEE Transactions on Power Systems*, **27** (1), 92–107.

187 Munkres, J.R. (1993) *Elements of Algebraic Topology*, Perseus Books Pub., New York.

188 Grant, M. and Boyd, S. (2012), CVX: Matlab software for disciplined convex programming, version 2.0 beta, http://cvxr.com/cvx.

189 Han, T. and Ansari, N. (2013) Heuristic relay assignments for green relay assisted device to device communications, in *IEEE Global Communications Conference*, Dec. 9, 2013, Atlanta, GA, USA.

190 Nosratinia, A., Hunter, T., and Hedayat, A. (2004) Cooperative communication in wireless networks. *IEEE Communications Magazine*, **42** (10), 74–80.

191 Golrezaei, N., Shanmugam, K., Dimakis, A., Molisch, A., and Caire, G. (2012) Wireless video content delivery through coded distributed caching, in *the 2012 IEEE International Conference on Communications*, Jun. 10, 2012, Ottawa, Canada.

192 Lei, L., Zhong, Z., Lin, C., and Shen, X. (2012) Operator controlled device-to-device communications in lte-advanced networks. *IEEE Wireless Communications*, **19** (3), 96–104.

193 Sidiropoulos, N., Davidson, T., and Luo, Z.Q. (2006) Transmit beamforming for physical-layer multicasting. *IEEE Transactions on Signal Processing*, **54** (6), 2239–2251, doi:10.1109/TSP.2006.872578.

194 Lozano, A. (2007) Long-term transmit beamforming for wireless multicasting, in *IEEE International Conference on Acoustics, Speech and Signal Processing*, Apr. 16, 2007, Hawaii, United States.

195 Liu, K., Sadek, A.K., Su, W., and Kwasinski, A. (2009) *Cooperative Communications and Networking*, Cambridge University Press: New York, NY.

196 Han, T. and Ansari, N. (2011) Energy efficient wireless multicasting. *IEEE Communications Letters*, **15** (6), 620–622.

197 Arnold, O., Richter, F., Fettweis, G., and Blume, O. (2010) Power consumption modeling of different base station types in heterogeneous cellular networks, in *Proc. IEEE Future Network and Mobile Summit 2010*, Jun. 2010, Florence, Italy, pp. 1–8.

198 Liu, K.J.R., Sadek, A.K., Su, W., and Kwasinski, A. (2009) *Cooperative Communications and Networking*, Cambridge University Press: Cambridge, UK.

199 Han, T. and Ansari, N. (2014) Offloading mobile traffic via green content broker. *IEEE Internet of Things Journal*, **1** (2), 161–170.

200 Ambrosy, A., Blume, O., Klessig, H., and Wajda, W. (2011) Energy saving potential of integrated hardware and resource management solutions for wireless base stations, in *IEEE International Symposium on Personal Indoor and Mobile Radio Communications*, Sep. 11, 2011, Toronto, Canada.

201 Burkard, R., Dell'Amico, M., and Martello, S. (2009) *Assignment Problems*, Society for Industrial and Applied Mathematics: Philadelphia, PA, USA.

202 Toyoda, Y. (1975) A simplified algorithm for obtaining approximate solutions to zero-one programming problems. *Management Science*, **21** (12), 1417–1427.

203 Dahlman, E., Parkvall, S., Sköld, J., and Beming, P. (2008) *3G Evolution: HSPA and LTE for Mobile Broadband*, Academic Press, Elsevier.

204 Cano-Garcia, J., Gonzalez-Parada, E., and Casilari, E. (2006) Experimental analysis and characterization of packet delay in UMTS networks, in *Next Generation Teletraffic and Wired / Wireless Advanced Networking*, vol. 4003, Springer Berlin Heidelberg, pp. 396–407.

205 Mutter, A., Necker, M., and Lück, S. (2004) IP-packet service time distributions in UMTS radio access networks, in *the 10th Open European Summer School and IFIP WG 6.3 Workshop*, Jun. 14, 2004, Tampere, Finland.

206 Prokkola, J., Perälä, P.H.J., Hanski, M., and Piri, E. (2009) 3G/HSPA performance in live networks from the end user perspective, in *IEEE International Conference on Communications*, Jun. 14, 2009, Dresden, Germany.

207 Liu, X., Sridharan, A., Machiraju, S., Seshadri, M., and Zang, H. (2008) Experiences in a 3G network: interplay between the wireless channel and applications, in *the 14th ACM International Conference on Mobile Computing and Networking*, Sep. 14, 2008, San Francisco, CA, USA.

208 Tan, W.L., Lam, F., and Lau, W.C. (2008) An empirical study on the capacity and performance of 3G networks. *IEEE Transactions on Mobile Computing*, 7 (6), 737–750.

209 Romirer-Maierhofer, P., Ricciato, F., and Coluccia, A. (2008) Explorative analysis of one-way delays in a mobile 3G network, in *the 16th IEEE Workshop on Local and Metropolitan Area Networks*, Sep. 3, 2008, Cluj-Napoca, Romania.

210 Lee, Y. (2006) Measured TCP Performance in CDMA 1x EV-DO Network, in *Proceedings of the 7th International Conference on Passive and Active Measurement, PAM '06*, Mar. 2006, Adelaide, Australia.

211 Mattar, K., Sridharan, A., Zang, H., Matta, I., and Bestavros, A. (2007) TCP over CDMA2000 networks: A cross-layer measurement study, in *Passive and Active Network Measurement*, vol. 4427, Springer Berlin / Heidelberg, pp. 94–104.

212 Kilpi, J. and Lassila, P. (2006) Micro- and macroscopic analysis of RTT variability in GPRS and UMTS networks, in *Networking 2006: Networking Technologies, Services, and Protocols; Performance of Computer and Communication Networks; Mobile and Wireless Communications Systems*, vol. 3976, Springer Berlin / Heidelberg, pp. 1176–1181.

213 Sagfors, M., Ludwig, R., Meyer, M., and Peisa, J. (2003) Queue management for TCP traffic over 3G links, in *IEEE Wireless Communications and Networking Conference*, Mar. 16, 2003, New Orleans, LA, USA.

214 Floyd, S. and Jacobson, V. (1993) Random early detection gateways for congestion avoidance. *IEEE/ACM Transactions on Networking*, **1**, 397–413.

215 Pentikousis, K., Palola, M., Jurvansuu, M., and Perala, P. (2005) Active goodput measurements from a public 3G/UMTS network. *IEEE Communications Letters*, **9** (9), 802–804.

216 Holma, H. and Toskala, A. (2000) *WCDMA for UMTS: Radio Access for Third Generation Mobile Communications*, John Wiley & Sons.

217 Chen, J.C. and Zhang, T. (2004) *IP-Based Next-Generation Wireless Networks: System, Architecture, and Protocols*, John Wiley & Sons.

218 Nishiyama, H., Ansari, N., and Kato, N. (2010) Wireless loss-tolerant congestion control protocol based on dynamic AIMD theory. *Wireless Communications*, **17** (2), 7–14.

219 Benko, P., Malicsko, G., and Veres, A. (2004) A large-scale, passive analysis of end-to-end TCP performance over GPRS, in *the 23rd Annual Joint Conference of the IEEE Computer and Communications Societies*, Mar. 7, 2004, Hong Kong, China.

220 Chakravorty, R. and Pratt, I. (2002) Performance issues with general packet radio service. *Journal of Communications and Networks (JCN)*, **4**, 266–281.

221 Chakravorty, R. and Pratt, I. (2002) WWW performance over GPRS, in *Mobile and Wireless Communications Network, 2002. 4th International Workshop on*, Sep. 2002, Stockholm, Sweden.

222 Tian, Y., Xu, K., and Ansari, N. (2005) TCP in wireless environments: problems and solutions. *IEEE Communications Magazine*, **43** (3), S27–S32.

223 Chakravorty, R., Clark, A., and Pratt, I. (2005) Optimizing web delivery over wireless links: design, implementation, and experiences. *IEEE Journal on Selected Areas in Communications*, **23** (2), 402–416.

224 Huang, J., Xu, Q., Tiwana, B., Mao, Z.M., Zhang, M., and Bahl, P. (2010) Anatomizing application performance differences on smartphones, in *the 8th International Conference on Mobile Systems, Applications, and Services*, Jun. 15, 2010, San Francisco, CA, USA.

225 Jang, K., Han, M., Cho, S., Ryu, H.K., Lee, J., Lee, Y., and Moon, S.B. (2009) 3G and 3.5G wireless network performance measured from moving cars and high-speed trains, in *the 1st ACM workshop on Mobile Internet through Cellular Networks*, Sep. 20, 2009, Beijing, China.

226 Lau, C.K. (2005) Improving mobile IP handover latency on end-to-end TCP in UMTS/WCDMA networks, in *ACM Conference on Emerging Network Experiment and Technology*, Oct. 24, 2005, Toulouse, France.

227 Nurvitadhi, E., Lee, B., Yu, C., and Kim, M. (2007) Adaptive semi-soft handoff for cellular IP networks. *Int. J. Wire. Mob. Comput.*, **2**, 109–119.

228 Mohan, S., Kapoor, R., and Mohanty, B. (2011), Latency in HSPA data networks. URL http://www.qualcomm.com/documents/files/latency-in-hspa-data-networks.pdf.

229 Lescuyer, P. and Lucidarme, T. (2008) *Evolved Packet System (EPS): The LTE and SAE Evolution of 3G UMTS*, John Wiley & Sons Ltd: Chichester, UK.

230 Ericsson (2012), Mobile cloud accelerator named best broadband cloud solution. URL https://www.ericsson.com/news/121030-award-mobile-cloud-accelerator_ 244159017_c.

231 Blumofe, R.D., Kanitkar, V., and Walther, D.S. (2012), Extending a content delivery network (cdn)into a mobile or wireline network.

232 Zhang, Y. and Ansari, N. (2014) On protocol-independent data redundancy elimination. *IEEE Communications Surveys Tutorials*, **16** (1), 455–472.

233 Anand, A., Muthukrishnan, C., Akella, A., and Ramjee, R. (2009) Redundancy in network traffic: findings and implications, in *the 11th International Joint Conference on Measurement and Modeling of Computer Systems*, Jun. 15, 2009, Seattle, WA, USA.

234 Lumezanu, C., Guo, K., Spring, N., and Bhattacharjee, B. (2010) The effect of packet loss on redundancy elimination in cellular wireless networks, in *the 10th Annual Conference on Internet Measurement*.

235 Yao, J.T., Domènech, J., Pont-Sanjuán, A., Sahuquillo, J., and Gil, J.A. (2010) Evaluation, analysis and adaptation of web prefetching techniques in current web, in *Web-based Support Systems*, Springer London, pp. 239–271.

236 Zhu, J., Hong, J., and Hughes, J.G. (2002) Using Markov models for web site link prediction, in *the 13th ACM Conference on Hypertext and Hypermedia*, Jun. 11, 2002, College Park, MD, USA.

237 Huang, Y.F. and Hsu, J.M. (2008) Mining web logs to improve hit ratios of prefetching and caching. *Know.-Based Syst.*, **21**, 62–69.

238 Davison, B.D. (2002) Predicting web actions from HTML content, in *the 13th ACM Conference on Hypertext and Hypermedia*, Jun. 11, 2002, College Park, MD, USA.

239 Khemmarat, S., Zhou, R., Gao, L., and Zink, M. (2011) Watching user generated videos with prefetching, in *the 2nd Annual ACM Conference on Multimedia Systems*, Feb. 22, 2011, San Jose, CA, USA.

240 Housel, B., Samaras, G., and Lindquist, D. (1998) WebExpress: A client/intercept based system for optimizing web browsing in a wireless environment. *Mobile Networks and Applications*, **3**, 419–431.

241 Koodli, R. (2005), Fast handovers for mobile IPv6. URL https://www.ietf.org/rfc/ rfc4068.txt.

242 Lang, Y., Wübben, D., Dekorsy, A., Braun, V., and Doetsch, U. (2012) Improved HARQ based on network coding and its application in LTE, in *IEEE Wireless Communications and Networking Conference*, Apr. 1, 2012, Paris, France.

243 Sundararajan, J., Shah, D., Médard, M., Jakubczak, S., Mitzenmacher, M., and Barros, J. (2011) Network coding meets TCP: Theory and implementation. *Proceedings of the IEEE*, **99** (3), 490–512.

244 Brakmo, L.S. and Peterson, L.L. (1995) TCP Vegas: end to end congestion avoidance on a global internet. *IEEE Journal on Selected Areas in Communications*, **13** (8), 1465–1480.

245 Kim, M., Médard, M., and Barros, J. (2011) Modeling network coded TCP throughput A simple model and its validation, in *the 5th International ICST Conference on*

Performance Evaluation Methodologies and Tools Communications, May 16, 2011, ENS, Cachan. France.

246 Zhang, Y. and Qiu, L. (2000) Speeding up short data transfers: Theory, architectural support, and simulation results, *Tech. Rep.*, Cornell University, Ithaca, NY, USA.

247 Goff, T., Moronski, J., Phatak, D., and Gupta, V. (2000) Freeze-TCP: a true end-to-end TCP enhancement mechanism for mobile environments, in *the 19th Annual Joint Conference of the IEEE Computer and Communications Societies*, Mar. 26, 2000, Tel-Aviv, Israel.

248 Rodriguez, P. and Fridman, V. (2004) Performance of PEPs in cellular wireless networks, in *Web Content Caching and Distribution*, Springer Netherlands, pp. 19–38.

249 Han, T., Ansari, N., Wu, M., and Yu, H. (2012) TCP-Mobile Edge: Accelerating content delivery in mobile networks, in *IEEE International Conference on Communications*, Jun. 10, 2012, Ottawa, Canada.

250 Xu, K., Tian, Y., and Ansari, N. (2005) Improving TCP performance in integrated wireless communications networks. *Computer Networks*, **47** (2), 219–237.

251 Jacobson, V. (1988) Congestion avoidance and control. *SIGCOMM Computer Communication Review*, **18** (4), 314–329.

252 Ramakrishnan, K., Networks, T., Floyd, S., and Black, D. (2001), The Addition of Explicit Congestion Notification (ECN) to IP. URL http://www.ietf.org/rfc/rfc3168.txt.

253 Li, Z. and Zhang, G. (Nov. 14, 2006) A novel differentiated marking strategy based on explicit congestion notification for wireless tcp, in *IEEE Region 10 Conference*, Hong Kong, China.

254 Xu, K., Tian, Y., and Ansari, N. (2004) TCP-Jersey for wireless IP communications. *IEEE Journal on Selected Areas in Communications*, **22** (4), 747–756.

255 3GPP TS 23.107 (version 10.0.0 release 10): Quality of service (qos) concept and architecture. Technical Specification, (2011).

256 Floyd, S., Arcia, A., Ros, D., and Iyengar, J. (2009), Adding acknowledgement congestion control to TCP. URL https://tools.ietf.org/html/draft-floyd-tcpm-ackcc-06.

257 Balakrishnan, H., Katz, R.H., and Padmanbhan, V.N. (1999) The effects of asymmetry on TCP performance. *Mobile Networks and Applications*, **4**, 219–241.

258 Kalampoukas, L., Varma, A., and Ramakrishnan, K.K. (1998) Improving TCP throughput over two-way asymmetric links: analysis and solutions. *SIGMETRICS Performance Evaluation Review*, **26**, 78–89.

259 Chan, M.C. and Ramjee, R. (2002) TCP/IP performance over 3G wireless links with rate and delay variation, in *the 8th annual international conference on Mobile computing and networking*, Sep. 23, 2002, Atlanta, GA, USA.

260 Ming-Chit, I.T., Jinsong, D., and Wang, W. (2000) Improving TCP performance over asymmetric networks. *SIGCOMM Computing and Communications Review*, **30**, 45–54.

261 Rodriguez, P., Mukherjee, S., and Ramgarajan, S. (2004) Session level techniques for improving web browsing performance on wireless links, in *the 13th International Conference on World Wide Web*, May 17, 2004, New York, NY, USA.

262 Gomez, C., Catalan, M., Viamonte, D., Paradells, J., and Calveras, A. (2008) Web browsing optimization over 2.5G and 3G: end-to-end mechanisms vs. usage of performance enhancing proxies. *Wireless Communication and Mobile Computing*, **8**, 213–230.

263 Chen, Y., Xie, X., Ma, W.Y., and Zhang, H.J. (2005) Adapting web pages for small-screen devices. *IEEE Internet Computing*, **9** (1), 50–56.

264 Mohomed, I. (2009) *Interactive Content Adaptation*, PhD. thesis, University of Toronto.

265 Knutsson, B., Lu, H., Mogul, J., and Hopkins, B. (2003) Architecture and performance of server-directed transcoding. *ACM Transactions on Internet Technology*, **3**, 392–424.

266 Noble, B. and Satyanarayanan, M. (1999) Experience with adaptive mobile applications in odyssey. *Mobile Networks and Applications*, **4**, 245–254.

267 Moshchuk, A., Gribble, S.D., and Levy, H.M. (2008) Flashproxy: transparently enabling rich web content via remote execution, in *the 6th International Conference on Mobile Systems, Applications, and Services*, Jun. 17, 2008, Breckenridge, CO, USA.

268 Buyukkokten, O., Garcia-Molina, H., and Paepcke, A. (2001) Accordion summarization for end-game browsing on PDAs and cellular phones, in *SIGCHI Conference on Human Factors in Computing Systems*, Mar. 31, 2001, Seattle, WA, United States.

269 Zhuang, Z., Chang, T.Y., Sivakumar, R., and Velayutham, A. (2009) Application-aware acceleration for wireless data networks: Design elements and prototype implementation. *IEEE Transactions on Mobile Computing*, **8**, 1280–1295.

270 Whitbeck, J., Amorim, M., Lopez, Y., Leguay, J., and Conan, V. (2011) Relieving the wireless infrastructure: When opportunistic networks meet guaranteed delays, in *the 12th IEEE International Symposium on a World of Wireless, Mobile and Multimedia Networks*, Jun. 22, 2011, Lucca, Italy.

271 Li, Y., Su, G., Hui, P., Jin, D., Su, L., and Zeng, L. (2011) Multiple mobile data offloading through delay tolerant networks, in *the 6th ACM Workshop on Challenged Networks*.

272 Gass, R. and Diot, C. (2010) An experimental performance comparison of 3G and Wi-Fi, in *the 11th International Conference on Passive and Active Measurement*, Apr. 7, 2010, Zurich, Switzerland.

273 Gass, R. and Diot, C. (2010) Eliminating backhaul bottlenecks for opportunistically encountered Wi-Fi hotspots, in *the 71st IEEE Vehicular Technology Conference*, May 16, 2010, Taipei, Taiwan.

274 Hsieh, H.Y. and Sivakumar, R. (2002) A transport layer approach for achieving aggregate bandwidths on multi-homed mobile hosts, in *the 8th Annual International Conference on Mobile Computing and Networking*, Sep. 23, 2002, Atlanta, Georgia, USA.

275 Rodriguez, P., Chakravorty, R., Chesterfield, J., Pratt, I., and Banerjee, S. (2004) MAR: a commuter router infrastructure for the mobile internet, in *the 2nd International Conference on Mobile Systems, Applications, and Services*, Jun. 6, 2004, Boston, MA, USA.

276 Tsao, C.L. and Sivakumar, R. (2009) On effectively exploiting multiple wireless interfaces in mobile hosts, in *ACM Conference on Emerging Network Experiment and Technology*, Dec. 1, 2009, Rome, Italy.

277 Chang, T.Y., Zhuang, Z., Velayutham, A., and Sivakumar, R. (2008) WebAccel: Accelerating web access for low-bandwidth hosts. *Computer Networks*, **52**, 2129–2147.

278 Grozdanovic, J. and Lukovic, D. (2007) Implementation of Nett-Gain in GPRS system in MTS 064. *Telekom Srbija, MTS 064*.

279 Flash Netwoks (2005) Nettgain technology-white paper. *white paper*.

280 Kokku, R., Yalagandula, P., Venkataramani, A., and Dahlin, M. (2003) NPS: A non-interfering deployable web prefetching system, in *the 4th USENIX Symposium on Internet Technologies and Systems*, Mar. 26, 2003, Seattle, WA, USA.

281 Venkataramani, A., Kokku, R., and Dahlin, M. (2002) TCP Nice: a mechanism for background transfers. *SIGOPS Operating Systems Review*, **36**, 329–343.

282 Kim, J., Baratto, R.A., and Nieh, J. (2006) pTHINC: a thin-client architecture for mobile wireless web, in *the 15th International Conference on World Wide Web*, May 23, 2006, Edinburgh, Scotland.

283 Mohomed, I., Cai, J.C., and de Lara, E. (2006) URICA: Usage-awaRe Interactive Content Adaptation for mobile devices. *SIGOPS Operating Systems Review*, **40**, 345–358.

284 Mickens, J. (2010) Silo: exploiting JavaScript and DOM storage for faster page loads, in *USENIX Conference on Web Application Development*, Jun. 22, 2010, Boston, MA, USA.

285 Begen, A., Akgul, T., and Baugher, M. (2011) Watching video over the web, part i: Streaming protocols. *IEEE Internet Computing*, **15** (2), 54–63.

286 Zambelli, A. (2009), IIS smooth streaming technical overview. White paper.

287 HTTP live streaming overview - networking, Internet & web. White paper, (2010).

288 HTTP dynamic streaming on the Adobe flash platform. White paper, (2010).

289 Pantos, R. and May, W. (2010), HTTP live streaming (draft-pantos-http-live-streaming-05). URL http://tools.ietf.org/html/draft-pantos-http-live-streaming-05.

290 Transparent end-to-end packet-switched streaming service (PSS). *ETSI TS 126 234 V9.5.0 (2011–01); Protocols and codecs*, (2011).

291 Wang, B., Kurose, J., Shenoy, P., and Towsley, D. (2008) Multimedia streaming via TCP: An analytic performance study. *ACM Transactions on Multimedia Computing, Communications and Applications*, **4**, 16:1–16:22.

292 Zhu, X. and Girod, B. (2007) Video streaming over wireless networks, in *European Signal Processing Conference*, Sep. 3, 2007, Poznan, Poland.

293 Akhshabi, S., Begen, A.C., and Dovrolis, C. (2011) An experimental evaluation of rate-adaptation algorithms in adaptive streaming over HTTP, in *Proceedings of the 2nd Annual ACM Conference on Multimedia Systems, MMSys '11*, Feb. 23, 2011, San Jose, CA, USA.

294 Goel, A., Krasic, C., and Walpole, J. (2008) Low-latency adaptive streaming over TCP. *ACM Transactions on Multimedia Computing, Communications and Applications*, **4**, 20:1–20:20.

295 Kuschnig, R., Kofler, I., and Hellwagner, H. (2010) An evaluation of TCP-based rate-control algorithms for adaptive internet streaming of H.264/SVC, in *the 1st Annual ACM Conference on Multimedia Systems*, Feb. 22, 2010, Phoenix, Arizona, USA.

296 Prangl, M., Kofler, I., and Hellwagner, H. (2008) Towards QoS improvements of TCP-based media delivery, in *the 4th International Conference on Networking and Services*, Mar. 16, 2008, Gosier, Guadeloupe.

297 Argyriou, A. (2007) Real-time and rate-distortion optimized video streaming with TCP. *Image Communincations*, **22**, 374–388.

298 Krasic, C., Walpole, J., and Feng, W.c. (2003) Quality-adaptive media streaming by priority drop, in *the 13th International Workshop on Network and Operating Systems Support for Digital Audio and Video*, Jun. 1, 2003, Monterey, CA, USA.

299 Won, H., Cai, H., Eun, D.Y., Guo, K., Netravali, A., Rhee, I., and Sabnani, K. (2009) Multicast scheduling in cellular data networks. *IEEE Transactions on Wireless Communications*, **8** (9), 4540–4549.

300 Kuschnig, R., Kofler, I., and Hellwagner, H. (2011) Evaluation of HTTP-based request-response streams for internet video streaming, in *the 2nd Annual ACM Conference on Multimedia Systems*, Feb. 22, 2011, San Jose, CA, USA.

301 Kuschnig, R., Kofler, I., and Hellwagner, H. (2010) Improving internet video streamilng performance by parallel TCP-based request-response streams, in *the 7th IEEE Conference on Consumer Communications and Cetworking Conference*, Jan. 9, 2010, Las Vegas, Nevada, USA.

302 Nguyen, T. and Cheung, S.C.S. (2005) Multimedia streaming using multiple TCP connections, in *the 24th IEEE International Performance, Computing, and Communications Conference*, Apr. 7, 2005, Phoenix, AZ, USA.

303 Chakravorty, R., Banerjee, S., and Ganguly, S. (2006) MobiStream: Error-resilient video streaming in wireless WANs using virtual channels, in *the 25th IEEE International Conference on Computer Communications*, Apr. 23, 2006, Barcelona, Spain.

304 Li, Q. and van der Schaar, M. (2004) Providing adaptive QoS to layered video over wireless local area networks through real-time retry limit adaptation. *IEEE Transactions on Multimedia*, **6** (2), 278–290.

305 Miu, A., Apostolopoulos, J.G., Tan, W.t., and Trott, M. (2003) Low-latency wireless video over 802.11 networks using path diversity, in *IEEE International Conference on Multimedia and Expo*, Jul. 6, 2003, Baltimore, MD, USA.

306 Chesterfield, J., Chakravorty, R., Crowcroft, J., Rodriguez, P., and Banerjee, S. (2004) Experiences with multimedia streaming over 2.5G and 3G networks, in *the First International Workshop on Broadband Wireless Multimedia: Algorithms, Architectures and Applications*, Oct. 29, 2004, San Jose, CA, USA.

307 Chesterfield, J., Chakravorty, R., Pratt, I., Banerjee, S., and Rodriguez, P. (2005) Exploiting diversity to enhance multimedia streaming over cellular links, in *the 24th Annual Joint Conference of the IEEE Computer and Communications Societies*, Mar. 13, 2005, Miami, FL, USA.

308 Kaspar, D., Evensen, K., Engelstad, P., Hansen, A.F., Halvorsen, P., and Griwodz, C. (2010) Enhancing video-on-demand playout over multiple heterogeneous access networks, in *the 7th IEEE Conference on Consumer Communications and Cetworking Conference*, Jan. 9, 2010, Las Vegas, Nevada, USA.

309 Li, D., Chuah, C.N., Cheung, G., and Yoo, S.J.B. (2005) MUVIS: multi-source video streaming service over wlans. *Journal of Communication and Networks (JCN)*, **7**, 144–156.

310 Lu, M.H., Steenkiste, P., and Chen, T. (2010) Robust wireless video streaming using hybrid spatial/temporal retransmission. *IEEE Journal on Selected Areas in Communications*, **28**, 476–487.

Index

Green Mobile Networks: A Networking Perspective, First Edition. Nirwan Ansari and Tao Han.
© 2017 John Wiley & Sons Ltd. Published 2017 by John Wiley & Sons Ltd.

List of Abbreviations

3GPP	3rd generation partnership project
ACE	Acknowledgment based on CWND estimation
AF; DF	Amplify-and-forward; decode-and-forward
AMC	Adaptive modulation and coding
ARQ; HARQ	Automatic repeat-request; hybrid ARQ
BA	Bandwidth allocation
BDP	Bandwidth delay product
BES	Base station energy sharing
BESS	Binary energy system sizing
BM	Battery minimization
BPO	Base station operation and power distribution optimization
BS	Base station
CAPEX; OPEX	Capital expenditure; operational expenditure
CBR	Constant bit rate
CELL_DCH	Cell dedicated channel
CELL_FACH	Cell forward access channel
CELL_PCH	Cell page channel
CF	Cluster formation
CH	Cluster head
CoMP	Coordinated multi-point
COS	Content owner selection
CR	Cognitive radio
CRE	Cell range expansion
CSS	Cooperative sensing scheduling
CW	Congestion warning
CWND	Congestion window
D2D	Device-to-device
DAC; ADC	Digital-to-analog converter; analog-to-digital converter
DC; AC	Direct current; alternating current
DCH	Dedicated channel
DRA	Dynamic resource allocation
DRB	Data rate bias
DSA	Dynamic spectrum access

EC-constraint	Energy causality constraint
EDR	Energy depleting ratio
EDS	Energy dependent set
EE	Energy efficiency
EH	Energy harvesting
ELLA	Energy loss and latency aware
EA	Energy Allocation
EPS	Evolved packet system
ESG	Energy saving greedy
ESM	Energy savings maximization
EST	Energy spectrum trading
EUTRAN (UTRAN)	Evolved UTMS terrestrial radio access network
FC	Fusion center
GALA, vGALA	Green energy aware and latency aware; virtual GALA
GAP	Green energy aware problem
GCB	Green content broker
GEO	Green energy optimization
GEP	Green energy provisioning
GESS	Green energy system sizing
GPRS	General packet radio service
GRA	Green relay assignment
GUA	General user association
HetNets	Heterogeneous networks
HTO	Heuristic traffic offloading
ICE	Intelligent cell breathing
ICT	Information and communications technology
IoT	Internet of things
ISP	Internet service provider
IT	Interference temperature
KKT	Karush–Kuhn–Tucker
LAP	Latency aware problem
LEM	Largest EDR minimization
LM	Latency minimization
LOEP	Loss of energy probability
LOLP	Loss of load probability
LTE	Long term evolution
MAC	Media access control
MBS	Macro base station
MDP	Markov decision process
MEA	Multi-stage energy allocation
MEB	Multi-BS energy balancing
MIMO	Multi-input-multi-output
MINLP	Mixed integer non-linear programming
MNO	Mobile network operator
MRC	Maximal ratio combining
MWBM	Maximum weight bipartite matching

NC	No cache
NLOS	Non-line-of-sight
NPS	Non-interfering prefetching system
NUA	Network utility aware
O-DSTC	Opportunistic distributed space-time coding
OFDM; OFDMA	Orthogonal frequency-division multiplexing; orthogonal frequency-division multiple access
P2P; P2MP	Point to point; point to multi-point
PBR	Power-bandwidth ratio
PBS	Primary base station
PCA	Provisioning cost aware
PCM; SPCM; HPCM; WPCM	Power consumption minimization; simplified PCM; heuristic PCM; weighted PCM
PDCP	Packet data convergence protocol
POMDP	Partially observable Markov decision process
PS	Processor sharing
QB	QoS bound
QoE	Quality of experience
QoS	Quality of service
RA	Resource allocation
RAN; RANC	Radio access network; RAN controller
RAT	Radio access technology
RED	Random early detection
RF	Radio frequency
RLC	Radio link control
RN	Relay node
RR	Round robin
RRC	Radio resource control
RSSI	Received-signal-strength-indication
RTO	Retransmission timeouts
RTT	Round trip time
SAM	System advisor model
SBS	Secondary base stations
SCBS	Small cell base station
SCN	Small cell network
SCS	Serving content selection
SD	Source-destination
SDR	Secondary data rate
SDU	Service data unit
SE	Spectrum, efficiency
SGSN; GGSN	Serving GPRS support node; gateway GPRS support node
SHA	Secure hash algorithm
SI	Side information
SINR	Signal interference noise ratio
SN	Source node
SNR	Signal to noise ratio

SoftRAN	Software-defined radio access network
SSF	Strongest signal first
TDMA; FDMA	Time division multiple access; frequency division multiple access
TOM	Traffic offloading maximization
UA	User association
UE	User equipment
UMTS	Universal mobile telecommunications system
URA_PCH	UTRAN registration area paging channel
URICA	Usage-aware interactive content adaptation
WBST	Wireless boosted session transport
WEM	Weighted energy minimization
WLA	Wireless loss alarm
WUA; AWUA	Weighted user–BS association; approximate WUA
X2	Inter-eNode B interfaces